# Litigating Climate Change in the Global South

# Litigating Climate Change in the Global South

JOLENE LIN
*Associate Professor, Faculty of Law, National University of Singapore, Singapore*

JACQUELINE PEEL
*Director, Melbourne Climate Futures and Professor of Law, Melbourne Law School, The University of Melbourne, Australia*

OXFORD
UNIVERSITY PRESS

OXFORD
UNIVERSITY PRESS

Great Clarendon Street, Oxford, OX2 6DP,
United Kingdom

Oxford University Press is a department of the University of Oxford.
It furthers the University's objective of excellence in research, scholarship,
and education by publishing worldwide. Oxford is a registered trade mark of
Oxford University Press in the UK and in certain other countries



Published in the United States of America by Oxford University Press
198 Madison Avenue, New York, NY 10016, United States of America

British Library Cataloguing in Publication Data
Data available

Library of Congress Control Number: 2023951617

ISBN 978-0-19-284389-0

DOI: 10.1093/9780191926525.001.0001

Printed and bound by
CPI Group (UK) Ltd, Croydon, CR0 4YY

MIX
Paper | Supporting
responsible forestry
FSC® C013604
FSC
www.fsc.org

# Foreword

In an era where the consequences of climate change echo globally, the imperative to address its impacts with urgency, empathy, and inclusivity becomes increasingly evident. The book before you, *Litigating Climate Change in the Global South*, emerges as a beacon guiding us through the intricate landscape of climate change litigation within a vital, yet often underrepresented, context.

The development of a Global South narrative is critical, especially in the case of climate change litigation. This narrative is essential to bring forward the region's experience to ensure that the requirements and measures necessary within the Global South are addressed, and that the voices and stories of an area grappling with formidable challenges are heard.

Climate change knows no boundaries, and its impact is felt across societies, economies, and ecosystems. Yet, as this book compellingly illustrates, the response to this global crisis must be as diverse as the communities it affects. This work underscores the significance of understanding how courts in the Global South have responded to climate change litigation, illuminating both the progress made and the hurdles yet to be overcome.

In Pakistan's constitutional courts, the Supreme Court and High Court, a human rights-based approach has encouraged the enactment of specific climate change legislation and commissions to focus on climate change rights and climate justice to bring out the spirit of the Global South narrative. This narrative seeks to transform challenges into opportunities and vulnerabilities into strengths.

Climate change affects everyone: our environment, health, and the economy. This book emerges as a crucial companion in the ongoing pursuit of climate-based governance. It unpacks the complexities of climate litigation, its impact, and the difference it has made, including the tools that can be used and how to involve stakeholders. The book also examines the framework of governance, constitutional rights, and the law. It starts the conversation and relies on Global South scholarship, a more relatable regional perspective.

The book brings out a wealth of expertise in climate litigation scholarship, enriched further by the authors' direct involvement with litigants and environmental advocates offering an insightful glimpse into the ever-evolving landscape of ecological jurisprudence.

In the spirit of intellectual exploration, this book introduces a framework that tracks continents and perspectives. Its purpose goes beyond documentation; it aspires to spark conversations, catalyse inquiries, and encourage further research to refine our understanding of climate litigation's role in shaping a sustainable future.

As the number of climate-related cases rises across Global South nations, the relevance of this book grows exponentially as it also contributes to a collective effort to address climate change's global footprint.

Mrs Justice Ayesha A Malik
Supreme Court of Pakistan
21 August 2023

# Preface and Acknowledgements

Litigation concerning efforts to deal with the problem of climate change, often referred to as simply 'climate litigation', has been on a rise worldwide in the past decade. According to the Grantham Research Institute on Climate Change and the Environment which publishes an annual survey of climate litigation trends globally, climate cases now number more than two thousand as of 2023. This number has doubled since 2015, with approximately one-quarter of all cases being filed between 2020 and 2022.

This turn to courts in the climate change context is driven by a sense of urgency to take action. Global anthropogenic greenhouse gas emissions from fossil use in 2022 were the highest ever recorded. This is the case despite everything that has been known about their effects for decades. In the years since the United Nations Framework Convention on Climate Change was signed in 1992, humanity has released more carbon into the atmosphere than during the rest of human history combined. As we look forward, the future scenario is worrying. According to the United Nations, even assuming that all Member States meet their current pledged emissions reductions under the Paris Agreement, the Earth is still on track to experience 2.1 to 2.9 degrees Celsius of warming by the end of the century. This is a level of temperature increase that would entail catastrophic consequences, including potential triggering of tipping point scenarios that could lead to even further warming through massive and uncontrollable natural releases of greenhouse gases. Yet, the political branch of government in most states, as well as the international political processes, have not only failed to stop this growth in greenhouse gas emissions but continue to have policies that facilitate fossil fuel use and extraction. Transnational corporations continue to prioritize windfall profits at the expense of destroying the ecological foundations of the global economic and social order. It is against this background that we see litigants calling upon courts around the world to make declarations of climate responsibility and to hold corporations and governments accountable for their actions pertaining to climate change.

The majority of climate litigation has occurred and continues to occur in the wealthy developed countries, particularly the United States of America, Australia, the United Kingdom, and the European Union Member States. However, climate litigation is rapidly emerging in the developing countries of the so-called 'Global South' and relatively little is understood about these developments. This book seeks to shed light on the emergence of climate litigation in the Global South, with a focus on answering key questions such as where climate litigation is flourishing, who are the actors involved, and what impact litigation is having on climate change policy.

This book draws upon our first substantial collaboration as co-authors, 'Transnational Climate Litigation: The Contribution of the Global South', published in June 2019 in the *American Journal of International Law*. In this initial attempt to shine a light on the Global South's contribution to transnational climate litigation, we examined a total of 34 cases. As a result of a tremendous growth in climate litigation (particularly in the Global South) in the period between 2019 and mid-2023, improvements in our data collection, and the support of increasingly dense collaborative networks that have given rise to what can be described as a global climate litigation community, this book examines a total of 128 cases. This number does not include the growing number of cases with truly transnational dimensions (eg cases filed in Europe by multi-actor coalitions challenging fossil fuel projects in Africa) and cases in international tribunals. We also drew upon and developed our other co-authored publications: 'The Farmer or the Hero Litigator? Modes of Climate Litigation in the Global South' (in *Litigating the Climate Emergency: How Human Rights, Courts and Legal Mobilization Can Bolster Climate Action*, edited by César Rodríguez-Garavito, Cambridge University Press, 2022) and 'Climate Change Adaptation Litigation: A View from Southeast Asia' (in *Climate Change Litigation in the Asia Pacific*, edited by Jolene Lin and Douglas A Kysar, Cambridge University Press, 2020).

In undertaking the research for this project, we benefited greatly from funding provided under a Singapore Ministry of Education Academic Research Fund Tier One Grant: 'Climate Change Litigation and the Global South' (Project Number: T1/C2-2/2020). This grant was for an initial 36 months (2020–23). We were granted a 12-month extension due to Jolene going on maternity leave in 2021 and the Covid-19 pandemic that threw academic research and teaching into upheaval in many places. Lockdowns and stringent Covid travel restrictions in both Australia and Singapore where we are located meant that we were unable to conduct interviews and research in our case study jurisdictions as planned. We are very grateful to our interviewees in Brazil, India, Indonesia, Thailand, and the United Kingdom (to name just a few of the jurisdictions included) who generously undertook online interviews with us for the purposes of this project. In accordance with the ethics procedures of our home institutions of the National University of Singapore and the University of Melbourne, interviewees are not named in this book to protect the confidentiality of their responses. We wish to acknowledge their generosity in the giving of their time, expertise, and experience. We appreciate the many insights they offered into climate litigation in the Global South and the influence it has had or may have in the future.

The grant funding also enabled us to organize a workshop at the National University of Singapore on 22 and 23 May 2023, where we presented draft chapters and received valuable feedback on our work. Colleagues at the workshop also provided wonderful insights on climate litigation in their home jurisdictions, further enriching our understanding of the phenomenon. We would like to thank Gitanjali N Gill, Arpitha Kodiveri, Louis Kotzé, Grizelda Mayo-Anda, Sroyon Mukherjee, César Rodríguez-Garavito, Joana Setzer, Linda Yanti Sulistiawati, Maria Antonia Tigre, Agung Wardana, Zhao Yue, and Jing Jing Zhang for their valuable contributions at this workshop. The Singapore Ministry of Education funding also allowed Jolene to employ research assistants to whom she is indebted for their research efforts and assistance with footnoting. These wonderful individuals are Phoon Yi Hao, Rachel Wam, Robin Louis Buxton-Leow, Julia Liaw, and Chieng Hui Jie. At Melbourne, Jackie would like to acknowledge the fantastic research work and assistance with footnoting provided by Thea Shields, who also helped compile the detailed case study material on case impact discussed in the final chapter of the book.

Other colleagues who provided helpful feedback for the book and for earlier formative publications whom we would particularly like to acknowledge include Rebekkah Markey-Towler, Doug Kysar, Alice Palmer, Hari Osofsky, Alex Wang, Lisa Benjamin, and Lisa Vanhala. We would also like to register our heartfelt appreciation to Justice Ayesha Malik of the Supreme Court of Pakistan for taking time out of her tremendously demanding schedule to pen the Foreword for this book. Jolene would also like to express heartfelt gratitude to Joanne Scott for her guidance, steadfast support, and encouragement to embrace intellectual challenges.

We would like to thank the editorial team at Oxford University Press for their support throughout this project. We appreciate their careful management and understanding during the exigencies of Covid and valuable assistance with copyediting and production of proofs.

Last, but not least, we would like to thank our respective families for their love and support during the writing of this book. Jolene would like to thank her husband Thomas Wind, son Lucas, and daughter Saskia. Jackie is grateful to her husband Michael Findlay, daughter Aly, and son Will. We dedicate this book to our children and our children's children and hope for a better climate future for them and the generations to come around the world.

# Contents

# 1
# Litigating Climate Change in the Global South

## 1. INTRODUCTION

A farmer suing his government for violating his constitutional rights through failing to implement the country's declared climate policy frameworks. A coalition of environmental non-governmental organizations (NGOs) working with local communities to oppose the authorization of a large, new coal-fired power plant. Political parties bringing suit to the highest court of the land to take issue with the non-operationalization of the government's financial mechanism for funding climate mitigation and adaptation actions. Young people opposing large-scale deforestation which threatens their climate future and that of future generations. These are just some of the examples of a new phenomenon of climate cases emerging in developing countries of the so-called 'Global South'. This body of Global South cases now numbers more than 120 cases filed, decided, or pending. And it is a fast-growing field: more than three-quarters of the Global South climate cases have been filed in the past five years.

Climate litigation in developed countries of the 'Global North' is a longer standing and well-studied phenomenon. Cases in some jurisdictions, like the United States and Australia, date back to the early 1990s,[1] and there are several books considering this case law,[2] alongside a vast number of scholarly articles and other papers.[3] In 2022, the Intergovernmental Panel on Climate Change (IPCC), in

[1] eg *City of Los Angeles v National Highway Traffic Safety Administration* 912 F2d 478 (DC Cir 1990); *Foundation on Economic Trends v Watkins* 731 F Supp 530, 530 (DDC 1990); *Greenpeace Australia v Redbank Power Company* (1994) 86 LGERA 143; *Gray v Minister for Planning* [2006] NSWLEC 720.

[2] See eg Ivano Alogna, Christine Bakker, and Jean-Pierre Gauci, 'Climate Change Litigation: Global Perspectives—An Introduction' in Ivano Alogna, Christine Bakker, and Jean-Pierre Gauci (eds), *Climate Change Litigation: Global Perspectives* (Brill Nijhoff 2021); Joana Setzer, Nicola Silbert, and Lisa Vanhala, 'The Effectiveness of Climate Change Litigation' in F Sindico and others (eds), *Edward Elgar Research Handbook on Climate Change Litigation* (Edward Elgar forthcoming); Cinnamon Piñon Carlarne, 'The Essential Role of Climate Litigation and the Courts in Averting Climate Crisis' in Benoît Mayer and Alexander Zahar (eds), *Debating Climate Law* (CUP 2021).

[3] Jacqueline Peel and Hari Osofsky, 'Climate Change Litigation' (2020) 16(1) Annual Review of Law and Social Science 21, 32; Brian J Preston, 'Climate Change Litigation' (2011) 5(1) Carbon and Climate Law Review 3.

*Litigating Climate Change in the Global South*. Jolene Lin and Jacqueline Peel, Oxford University Press.
 DOI: 10.1093/9780191926525.003.0001

the Working Group III Mitigation contribution to its Sixth Assessment Report, assessed the significance of climate litigation for the first time.[4] Nonetheless, despite this scholarly and scientific interest in climate litigation, relatively few studies focus on climate case law emerging in the Global South, its distinctive characteristics, and the contribution it is making, and may make, to global climate governance including implementation of the international climate treaty, the Paris Agreement.[5]

The purpose of this book is to shed light on the emerging and accelerating climate litigation in Global South countries and it is the first monograph-length work to provide a comprehensive assessment of this jurisprudence. These climate cases, like those in the Global North, are mainly taking place in domestic courts, although more recently a number of cases have been initiated before regional and international courts and tribunals, including several involving Global South actors.[6] Our guiding hypothesis in undertaking this project was that, without capturing and analysing the richness of climate cases emerging in the Global South, we will only ever have a partial understanding of what climate litigation is, how it comes about, and the role that it plays in addressing the global problem of climate change. As the number of climate cases across Global South countries has exploded in recent years, that task has only assumed more importance, although this complicates the work required to produce a comprehensive account and assessment.

As authors, we bring to this book a long history of climate litigation scholarship and of working directly with litigants, environmental groups, and other supporters of these cases.[7] We are scholars based in countries of the Southern hemisphere but are not Global South scholars. With this book, our aim is to report on and offer initial, systematic analysis of what we see as an interesting and important climate legal phenomenon. In this exercise, we have benefited enormously from the scholarship and work of Global South authors and have drawn on our extensive networks and contacts with practitioners and lawyers working in Global South countries to identify cases and understand their significance. Already, since an initial article on Global South climate litigation that we published in the *American Journal of International Law* in 2019,[8] scholarship by Global South authors on climate litigation in the Global South has expanded considerably,[9] including the seminal article

[4] Intergovernmental Panel on Climate Change, *Climate Change 2022: Mitigation of Climate Change* (Contribution of Working Group III to the Sixth Assessment Report of the Intergovernmental Panel on Climate Change, April 2022) (hereafter WGIII Report).

[5] Paris Agreement (opened for signature 22 April 2016, entered into force 4 November 2016) 3156 UNTS 54113.

[6] See discussion in Chapter 3, Section 4 of this book.

[7] Jacqueline Peel and Jolene Lin, 'Transnational Climate Litigation: The Contribution of the Global South' (2019) 113(4) American Journal of International Law 679; Peel and Osofsky (n 3); Jolene Lin and Douglas A Kysar (eds), *Climate Change Litigation in the Asia Pacific* (CUP 2020).

[8] Peel and Lin (n 7).

[9] Global Network for Human Rights and the Environment project compiling Global South climate litigation database of scholarship at <https://gnhre.org/?page_id=14314> accessed 7 May 2023.

published, also in 2019, by Joana Setzer and Lisa Benjamin.[10] We hope that this book encourages further attention to climate cases in the Global South and a continued flourishing of Global South scholarship on this topic.

In this introductory chapter, we outline the approach and structure of the book. Broadly speaking, the book seeks to answer the following questions:

- *Why* a study of Global South climate litigation is significant, especially in the context of broader international law and governance efforts to address climate change;
- *Where* such climate cases are occurring, including how we are understanding the scope of the 'Global South';
- *What* is encompassed by climate change litigation, which touches on the issue of how such cases are defined and categorized;
- *How* Global South climate litigation is occurring, including trends in the case law and indications of what socio-political, legal, and institutional conditions tend to promote the emergence of such cases in a country;
- *Who* are the actors involved in litigating climate change in the Global South and what roles they play; and
- *What then* are the emerging impacts of this case law in making a meaningful contribution to addressing climate change, and what might be the areas in which we see further development of this jurisprudence in the future.

The remainder of this chapter explains how we have scoped and undertaken our study of Global South climate litigation.

Section 2 outlines the international law and policy context in which climate litigation is occurring and how this case law might contribute to global climate governance. The major international climate instrument is the Paris Agreement of 2015, which sets global climate goals but itself contains no enforcement mechanism. This positions climate litigation, largely undertaken in domestic courts, as a potential mechanism for ensuring the accountability of governments and potentially other actors, such as corporations, for playing their part in contributing to the global climate response, as well as addressing the local consequences of climate change.

Section 3 addresses the question of how we are understanding the 'Global South' and why this terminology has been used. We look to both law and development scholarship, as well as commonly used categorizations of countries under international climate treaties to define the boundaries of the regions and countries we

[10] Joana Setzer and Lisa Benjamin 'Climate Litigation in the Global South: Constraints and Innovations' (2020) 9(1) Transnational Environmental Law 77. Joana Setzer is a Brazilian scholar based at LSE in London. Lisa Benjamin is a scholar from the Bahamas now based at Lewis and Clark Law School in the United States.

are including within the analysis. Within this grouping of Global South countries, not all have seen the emergence of climate cases. 'Global South climate litigation' is thus a shorthand term used in the book to describe climate cases decided, filed, or pending in particular countries located in the Global South.

Section 4 explains further our understanding of climate change litigation, which is elaborated in Chapter 2. Given the breadth and wide-ranging effects of the problem of climate change just about any environmental or natural resource litigation could be 'climate' litigation and there are very large number of cases that now mention the words 'climate change'. As this section articulates, however, we take a narrower view that focuses on cases which place climate issues at their centre ('core cases') or include climate issues as one amongst many issues ('periphery cases') but exclude cases that simply mention climate change in passing ('incidental cases').

Section 5 details the methodology applied in the book in the identification of cases and analysis of emerging trends, dealt with in depth in Chapters 3 and 4. It outlines how, applying the understandings of the Global South and climate litigation elaborated in the earlier sections, we derive a 'Global South docket' of 128 cases, as well as a set of 'transnational cases' which involve Global South plaintiffs but are being litigated in Global North courts or in regional and international tribunals.

Section 6 outlines the remaining structure of the book and Section 7 concludes.

## 2. WHY CLIMATE LITIGATION MATTERS FOR CLIMATE GOVERNANCE

Climate litigation sits within a broader system of laws and governance for addressing climate change that operates at multiple levels of governance from the local and national, to the regional, transnational, and international.[11] At the international level, climate governance is concentrated in the international climate regime, which has its foundations in the 1992 United Nations Framework Convention on Climate Change (UNFCCC) and the later 2015 Paris Agreement.[12]

The Paris Agreement plays an important 'direction-setting' role in international climate governance through its provisions articulating goals for maximum temperature rise over pre-industrial levels (1.5–2°C) and placing obligations on all states parties to strive to reduce greenhouse gas emissions to 'net zero' in the second half of the century 'on the basis of equity, and in the context of sustainable development and efforts to eradicate poverty'.[13] However, the achievement of those

[11] Peel and Lin (n 7).

[12] United Nations Framework Convention on Climate Change (adopted 9 May 1992, entered into force 21 March 1994), S Treaty Doc No 102-38, 1771 UNTS 107; Paris Agreement (n 5). In addition, the Kyoto Protocol to the United Nations Framework Convention on Climate Change (adopted 11 December 1997, entered into force 16 February 2005) 2303 UNTS 162, although it was effectively superseded by the Paris Agreement after the end of the 2nd commitment period in 2020.

[13] Paris Agreement (n 5) arts 2 and 4.1—provisions discussed further in Chapter 2.

goals is entrusted to states parties' 'Nationally Determined Contributions' (NDCs), which allows significant scope for a 'self-differentiated' contribution to the global climate response, set in light of national priorities and circumstances.[14] States parties have obligations to submit their NDCs and to update them on a regular basis,[15] but the Paris Agreement does not impose any substantive obligations on states parties to meet the commitments of their NDCs. In addition, while subsequent NDCs should reflect progression on what went before and the party's highest possible ambition,[16] there is no specific direction on the level of ambition required and the share of action to be undertaken by each state party in order to keep to the Agreement's temperature goals.

This structure of the international climate governance system can be seen as creating an accountability gap which climate litigation may seek to address.[17] In its assessment of climate litigation, highlighted earlier, the IPCC noted:

> Climate-related litigation, for example by governments, private sector, civil society and individuals, is growing—with a large number of cases in some developed countries, and with a much smaller number in some developing countries—and in some cases, has influenced the outcome and ambition of climate governance.[18]

How then does climate litigation 'influence' climate governance?

On one view, climate litigation is a local and localized phenomenon such that its governance consequences would be felt purely within the particular legal system and broader legal culture within which the case takes place.[19] These consequences might relate to the adequacy of legal and policy measures adopted by a country to achieve its NDC submitted under the Paris Agreement or may concern the overall ambition of that NDC and its compliance with international climate commitments. Joana Setzer, Catherine Higham, and Emily Bradeen label such cases 'government framework litigation', noting that a majority of these cases to date have been filed in Global North countries.[20] More often in Global South countries, as detailed further in Chapter 3, climate cases brought against governments (the majority of the Global South climate cases) will not relate specifically to domestic climate laws

[14] Lavanya Rajamani and Jacob Werksman, 'The Legal Character and Operational Relevance of the Paris Agreement's Temperature Goal' (2018) 376(2119) The Royal Society 1.

[15] ibid.

[16] Paris Agreement (n 5) art 4.3.

[17] Rajamani and Werksman (n 14).

[18] WGIII Report (n 4) SPM, E.3.3 (medium confidence).

[19] Sanja Bogojević, 'EU Climate Change Litigation, the Role of the European Courts, and the Importance of Legal Culture' (2013) 35(1) Law and Policy 184.

[20] Joana Setzer, Catherine Higham, and Emily Bradeen, 'Challenging Government Responses to Climate Change Through Framework Litigation' (*Grantham Research Institute on Climate Change and the Environment*, 7 September 2022) <www.lse.ac.uk/granthaminstitute/publication/challenging-government-responses-to-climate-change-through-framework-litigation/> accessed 7 May 2023. As of July 2022, 63 cases in Global North, with a 'significant minority' in Latin America (eight cases) and South Asia (seven cases).

(where they exist) or a country's NDC but more broadly to the implementation and enforcement of environmental and natural resources laws that contribute to climate mitigation and adaptation.

Viewed in this way, the capacity of climate litigation to shape international climate governance—for instance, in understanding the content of states' obligations under the Paris Agreement or how the responsibility of states to address climate change should be shared—would thus be left to international climate cases. Given the lack of an interstate enforcement mechanism under the Paris Agreement and the difficulties of applying other international law dispute settlement mechanisms to climate change disputes,[21] only a handful of such international climate cases exist, all of which have only emerged recently, as noted above.[22]

There are, however, other ways in which domestic climate cases may influence climate governance beyond states, giving rise to a transnational legal phenomenon.[23] Climate cases framed as part of a 'global' climate justice movement are often accompanied by campaigns that seek to speak to a wider national or international audience. These campaigns can help amplify the litigation's indirect or symbolic effects,[24] such as raising public awareness, mobilizing public sentiment, or shaping understanding of the problem, in ways that extend beyond the borders of a particular state. Good examples are the *Urgenda* and *Shell* cases, both brought in the Netherlands seeking, respectively, government and corporate accountability for contributing their 'fair share' to the global climate response.[25] As discussed further in Chapter 6, some Global South climate cases may also have these kinds of border-crossing effects, and several courts in the Global South are taking bold steps to compel action on climate change that may inspire similar approaches, both in the Global South and the Global North.

## 3. UNDERSTANDING OF THE 'GLOBAL SOUTH'

This book is focused on litigating climate change in developing countries across the three regions of Africa, Latin America and the Caribbean, and Asia and the

[21] Particularly due to the requirement for state consent and limited compulsory dispute settlement procedures in international law as well as other barriers to interstate actions.

[22] See Chapter 3. International Court of Justice 'Request for an Advisory Opinion of the International Court of Justice on the Obligations of States in Respect of Climate Change' (Press Release No 2023/21, 25 April 2023); Commission of Small Island States on Climate Change and International Law, 'Request for Advisory Opinion' (12 December 2022); Pablo Saavedra Alessandri, 'Request for an Advisory Opinion on Climate Emergency and Human Rights to the Inter-American Court of Human Rights form the Republic of Colombia and the Republic of Chile' (Petition, Inter-American Court of Human Rights, 9 January 2023).

[23] Peel and Lin (n 7).

[24] Concepts and types of impact from litigation are discussed further in Chapter 6.

[25] See Rajamani and Werksman (n 14) discussing fair shares.

Pacific.[26] Collectively these countries are referred to in the book as the 'Global South', distinguishing them from the developed countries of the 'Global North'. The concept of the 'Global South', however, is not one with a settled or uncontested meaning and does not map geographically to countries located in the Southern hemisphere.

In the law and development literature, the term 'Global South' has been used interchangeably with the terms 'Third World' and 'developing' to refer to states that lag behind the industrialized states of the 'Global North' in economic growth and prosperity, as well as political power and influence.[27] In the late 1990s, when economies such as Singapore, Hong Kong, and South Korea went '[f]rom Third to First' due to the adoption of successful export-oriented industrialization strategies, some commentators argued that the Third World no longer existed.[28] Rapid economic growth in China and India raises similar questions today. This points to growing diversity within and amongst developing states in terms of their economic development and global political influence. As Anand—a pioneer of the Third World Approaches to International Law (TWAIL) movement—observed of these states:

> [t]hey differ among themselves so greatly in economic promise that they are sometimes divided into 'third', 'fourth' and 'fifth' worlds ... In fact, for many purposes it is more misleading than illuminating to lump together countries of Asia, Africa, and Latin America. And yet, for a variety of purposes these countries perceive themselves as a group, a perception made more impressive because it overcomes an underlying, undeniable diversity.[29]

Amongst TWAIL scholars, it is acknowledged that the term 'Third World' may be problematic, but it is hard to do without the category. It provides the conceptual framing for counter-hegemonic discourse that unveils the close relationship between capitalism, imperialism, and international law, and explains why international law has always disadvantaged Third World peoples.[30]

Similar arguments have been made about the term 'the Global South'. Despite its widespread use in the social science and humanities literature, the concept of

[26] We do not include analysis of countries in the Middle East or Central Asia as no cases have yet emerged in these regions.

[27] Karin Mickelson, 'Rhetoric and Rage: Third World Voices in International Legal Discourse' (1998) 16(1) Wisconsin International Law Journal 353, 356; Usha Natarajan, 'TWAIL and the Environment: The State of Nature, the Nature of the State, and the Arab Spring' (2012) 14(1) Oregon Review of International Law 176, 179.

[28] See eg Mark T Berger, 'The End of the "Third World"?' (1994) 15 Third World Quarterly 257. The phrase 'From Third World to First' is borrowed from Lee Kuan Yew, *From Third World to First: The Singapore Story (1965–2000)* (Marshall Cavendish Editions 2000).

[29] RP Anand, 'International Law and the Developing Countries: Confrontation or Cooperation?' (Banyan Publications 1987).

[30] Philip Darby, 'Pursuing the Political: A Postcolonial Rethinking of Relations International' (2004) 33 Millennium: Journal International Studies 1. On the principal contentions of the TWAIL movement, see BS Chimni, 'Capitalism, Imperialism, and International Law in the Twenty-First Century' (2012) 14 Oregon Review of International Law 17.

the Global South and its content are acknowledged to be complex and contested.[31] Some commentators have argued that 'North' and 'South' are not useful analytical categories since there may be more differences between members of the South than between North and South.[32] Further, some contend that a North–South approach highlights differences and creates blocking coalitions in global politics.[33] However, as Gupta argues, even as the composition of the Global South and Global North changes, there will always be power disparities in governance processes, and a North-South analytical framework 'can help to focus more clearly on especially the equity issues and the views of the weakest actors which often become a casualty of efficient and patronizing processes'.[34] She also notes that it is wise to be wary of 'the possible persistence of neo-colonial approaches on the one hand and the psychology of third-worldism on the other hand'.[35]

In addition, the North–South divide continues to be a defining feature of global environmental politics, with most developing countries choosing to operate, even if nominally, under the banner of the collective 'Global South' in multilateral environmental negotiations.[36] Najam argues that the insistent choice to use the term 'South' reflects an important aspect of their collective identity in international politics.[37] As Dados and Connell put it:

> The term Global South functions as more than a metaphor for underdevelopment. It references an entire history of colonialism, neo-imperialism, and differential economic and social change through which large inequalities in living standards, life expectancy, and access to resources are maintained.[38]

In this book, while cognizant of the wider debate about use of the North–South categorization, our understanding of the Global South aligns with the developed/developing country distinction that is widely applied in international environmental and climate law scholarship.[39] This distinction also underpins the UNFCCC

[31] Elena Fiddian-Qasmiyeh and Patricia Daley, 'Conceptualising the Global South and South–South Encounters' in Elena Fiddian-Qasmiyeh and Patricia Daley (eds), *Routledge Handbook of South-South Relations* (Routledge 2018). Indeed, as the authors point out, some Global South scholars recognize the existence of 'multiple souths', including souths and southern voices within powerful metropoles.

[32] John Toye, 'Is the Third World Still There?' in Anna Farmar (ed), *The Developing World: An Introduction to Development Studies Through Selected Readings* (Development Education Support Centre 1988).

[33] Steinar Andresen and Jørgen Wettestad, 'International Resource Cooperation and the Greenhouse Problem' (1992) 2(4) Global Environmental Change 277; Richard E Benedick, 'Perspectives of a Negotiation Practitioner' in Gunnar Sjostedt (ed), *International Environmental Negotiation* (SAGE Publications 1993) 224.

[34] Joyeeta Gupta, 'Changing North-South Challenges in Global Environmental Politics' in Peter Dauvergne (ed), *Handbook of Global Environmental Politics* (2nd edn, Edward Elgar 2012) 97.

[35] ibid.

[36] Adil Najam, 'Why Environmental Politics Looks Different from the South' in Peter Dauvergne (ed), *Handbook of Global Environmental Politics* (Edward Elgar 2005) 111, 113.

[37] ibid.

[38] Nour Dados and Raewyn Connell, 'The Global South' (2012) 11(1) Contexts 12, 13.

[39] See eg Benjamin J Richardson and others, 'Introduction: Climate Law and Developing Countries' in Benjamin J Richardson and others (eds), *Climate Law and Developing Countries: Legal and Policy*

differentiation between Annex I parties—developed countries and former Soviet Union states—and non-Annex I parties—developing countries. In line with how these countries self-identify in the international climate change negotiations, as well as their classification as non-Annex I parties in the UNFCCC, major emerging economies such as China, India, Brazil, and South Africa are included in our analysis as part of the Global South.[40] Overall, of the 198 parties to the UNFCCC, 155 are categorized as non-Annex I parties.

Like others who have grappled with this categorization in the climate context, we acknowledge 'the irreducible heterogeneity of what is often referred to as the Global South and the highly complicated roles played by developing and developed nations in contributing and responding to the problem of climate change'.[41] For example, the place of China in the Global South/Global North mapping is often unclear given its exponential rise in recent years to become one of the world's largest economies, and the leading global emitter of greenhouse gases.[42] Nevertheless, we contend there is still value in the Global South grouping as the basis for an analysis of climate litigation, not least because it is these countries which face some of the most complex clean energy transition challenges and that are often among the most vulnerable to impacts from climate change.[43]

As we discuss further in Chapter 4, however, it is not all countries of the Global South in which we are seeing climate litigation emerge to date. This trend is

*Challenges for the World Economy* (Edward Elgar 2009) 1; Sumudu Atapattu and Carmen G Gonzalez, 'The North-South Divide in International Environmental Law: Framing the Issues' in Shawkat Alam and others (eds), *International Environmental Law And The Global South* (CUP 2015) 1.

[40] The UNFCCC website states: ';Developing countries generally work through the Group of 77 to establish common negotiating positions. The country holding the Chair of the G-77 . . . often speaks for the G-77 and China as a whole. However, because the G-77 and China is a diverse group with differing interests on climate change issues, individual developing countries also intervene in debates, as do groups within the G-77, such as the African States.' It should be noted that official statements refer to 'G-77 and China' because China is not formally a member of the G-77. See eg 'Statement on behalf of the Group of 77 and China by Ambassador Wael Aboulmagd, Chair of the G-77 and China for the Climate Change Process', Joint Opening Plenary of the 24th Session of the COP to the UNFCCC (COP24) Kotowice Poland, 2018. Although China firmly supports the G-77 agenda and stresses that 'even when China becomes stronger and more prosperous, it will remain a staunch member of the developing world because China and fellow developing countries have a similar past, common development tasks and ever expanding shared strategic interests', China is not a G-77 member because a core tenet of Chinese foreign policy is the principle of independence whereby China 'will not yield to any outside pressure or enter into alliance with any big power or group of countries': see Remarks by HE Foreign Minister Wang Yi of China at the 37th Foreign Ministers' Annual Meeting of the G-77, 'As a Member of the Developing World, China Will Always Speak for Developing Countries' (New York, 26 September 2013) <www.fmprc.gov.cn/mfa_eng//wjb_663304/wjbz_663308/2461_663310/t1081897.shtml> accessed 30 January 2023.

[41] Richardson and others (n 39) 5–6.

[42] Yuli Shan and others, 'China $CO_2$ emission accounts 1997–2015' (2018) 5 Scientific Data 17020. For up-to-date data on countries' emissions and climate action progress, see ClimateWatch <www.climatewatchdata.org/> accessed 1 December 2023.

[43] Hoesung Lee and others, *Climate Change 2023 Synthesis Report: Contribution of Working Groups I, II and III to the Sixth Assessment Report of the Intergovernmental Panel on Climate Change* (Intergovernmental Panel on Climate Change 2023) 184.

particularly concentrated in the developing countries classified as 'lower-middle income countries' (LMICs)[44] and 'upper-middle income countries' (UMICs)[45] according to the World Bank's system for classifying countries according to income level.[46]

Occupying the socio-economic middle ground between the low-income countries (LICs) of the Global South and the high-income countries (HICs) of the Global North, LMICs and UMICs—while a varied group socially, culturally, politically, and economically—nonetheless tend to confront a common set of challenges in realizing sustainable development. Moreover, although some MICs have improved institutional capacity, their institutions, including courts, are generally weaker and less well resourced than their counterparts in high-income countries. It is this combination of conflicts between sustainable economic growth and environmental degradation on the one hand, together with institutions that are 'good enough' to encourage a resort to dispute settlement on the other, that seems to be driving the predominance of climate litigation in these middle-income Global South countries. So far, of the 155 countries potentially encompassed by our understanding of the 'Global South', climate cases have been initiated in 20 of these countries, as summarized in Table 1.1 below.

## 4. WHAT IS CLIMATE LITIGATION?

Just as the term 'Global South' can have multiple meanings, different understandings of what constitutes 'climate litigation' prevail in the literature as well as in practice. As we discuss in Chapter 2, our goal in this book is not to resolve definitively the question of what counts as climate litigation.[47] Instead, since we are trying to provide a comprehensive assessment of a new climate legal phenomenon in the Global South, our interest is to focus on those cases which contribute in some meaningful way to building this jurisprudence.

Accordingly, in the book we rely on the distinctions between 'core', 'periphery', and 'incidental' cases that we have developed in previous scholarship.[48] Core cases describe those cases where climate change issues are at the centre of

[44] eg India, Pakistan, Indonesia, Nigeria, Papua New Guinea, and Vanuatu are included in the LMIC category (54 countries in total) <https://datahelpdesk.worldbank.org/knowledgebase/articles/906519-world-bank-country-and-lending-groups> accessed 30 July 2023.

[45] eg Argentina, Brazil, China, Colombia, Guyana, Malaysia, Mexico, South Africa, and Thailand are included in the UMIC category (54 countries in total) <https://datahelpdesk.worldbank.org/knowledgebase/articles/906519-world-bank-country-and-lending-groups> accessed 30 July 2023.

[46] ibid. Of the 28 low-income countries, only one—Uganda—has so far seen climate cases. Chile is another instance that does not fit the overall pattern, classified as high-income in the World Bank classification but as a developing country in the UNFCCC's non-Annex I grouping.

[47] See Chapter 2.

[48] Peel and Lin (n 7).

**Table 1.1** Countries of the Global South with climate cases

| Country | Income classification | No. of cases (as of May 2023) |
|---|---|---|
| India | LMIC | 18 |
| Indonesia | LMIC | 16 |
| Kenya | LMIC | 2 |
| Nigeria | LMIC | 1 |
| Pakistan | LMIC | 5 |
| Papua New Guinea | LMIC | 1 |
| Philippines | LMIC | 3 |
| Argentina | UMIC | 12 |
| Brazil | UMIC | 23 |
| China | UMIC | 1 |
| Colombia | UMIC | 3 |
| Guyana | UMIC | 3 |
| Mexico | UMIC | 14 |
| South Africa | UMIC | 11 |
| Thailand | UMIC | 1 |
| Chile | HIC | 7 |
| Uganda | LIC | 2 |
| Ecuador | UMIC | 3 |
| Peru | UMIC | 1 |
| Nepal | LMIC | 1 |

the legal argumentation in the case and/or of the resulting decision. Periphery cases, on the other hand, describe cases where climate change is one of many issues raised in a dispute which may be primarily 'about' other issues such as land use, water conservation, urban planning, and so on.[49] Further, we concentrate

[49] As described further in Section 5 below, the methodology we have applied in case selection, which includes a range of sources beyond the text of legal pleadings or judgments, has allowed us to identify additional climate cases where the litigants had a motivation related to addressing climate change in bringing the case.

on 'cases' or 'litigation' in the sense of disputes taken before a court or other quasi-judicial tribunal, rather than the broader set of 'legal interventions' that may precede formal litigation.[50] Within this category we include cases filed, pending decision, decided, or settled out of court. In this book, climate cases which are categorized as core or peripheral cases are included in our analysis and make up what we call the 'Global South docket'.[51]

This categorization means that we exclude cases that do not centrally or peripherally feature climate change issues. These cases are sometimes referred to as 'incidental' cases[52] and may include a mention of climate change but generally only in passing or without any actual engagement with the issue. As an example, we found a divorce proceedings case in Indonesia which cited climate change as one of the possible reasons for breakdown of the couple's relationship.[53] In the case, climate change was cited by the husband, an apple farmer, to explain his financial hardship in the divorce proceedings brought against him.[54] While there may be interesting studies that could be carried out about why incidental cases might include a passing mention of climate change, for our purposes we focused only on those cases that might make a meaningful contribution to climate governance, which is limited, in our view, to cases with climate issues at their core, or featured as part of a broader dispute about environmental degradation and sustainable development.

Finally, there are several 'core' and 'peripheral' climate change cases that include Global South plaintiffs or reference activities such as coal plants, mines, or other extractive projects or investments in the Global South, but which are brought before courts in Global North countries or, more recently, to regional and international courts and tribunals. These cases are not included in the Global South docket as they are not generally brought in courts in Global South countries and share much more in common with Global North climate litigation in their framing and legal claims. However, they represent an interesting transnational justice pathway for holding Global North actors accountable for their activities that take place in, or primarily impact, the Global South and so are considered further in Chapter 3 under the heading of 'transnational cases'.[55]

[50] The latter can include capacity-building activities with civil society or judges, complaints to regulators, letters issued to corporate actors to warn of potential litigation, or shareholder/investor activism and engagement preceding the taking of formal legal action.

[51] See further Chapter 3.

[52] In some scholarship 'incidental' is used with a different meaning. See eg Kim Bouwer, 'The Unsexy Future of Climate Change Litigation' (2018) 30(3) Journal of Environmental Law 232.

[53] Decision Number 440/Pdt.G/2016/PA.Mlg of the Religious Court of Malang.

[54] ibid.

[55] See further Chapter 3.

## 5. METHODOLOGY OF CASE SELECTION AND ANALYSIS

In locating cases falling within the Global South docket, we have applied a broad methodological approach. The starting point for our analysis was the major global databases of climate change cases maintained by the Sabin Center for Climate Change Law at Columbia University (Global Climate Change Litigation) and the London School of Economics (LSE) Grantham Research Institute on Climate Change and the Environment's Climate Laws of the World database. Building on an existing partnership, these two centres are working together to launch an enhanced version of the litigation database that will provide an integrated global resource.[56]

We undertook a systematic review of cases recorded in each Global South country category in both databases, including analysis of all reported judgments filed within these country categories to decide if they fell within the 'core' or 'peripheral' classifications described above, or ought to be excluded.[57]

For various reasons, including the more limited accessibility of judgments in languages other than English, as well as the incomplete electronic records of cases in several Global South countries,[58] the climate cases recorded in global databases did not capture all cases in the Global South docket. We therefore turned to other sources to help identify relevant cases including those at an early stage (eg cases filed but not decided or cases pending decision). These additional resources included newspaper reports, social media, conferences, interviews, and our own networks. Interviews were conducted with litigation participants including lawyers, environmental NGOs, local community groups, litigation funding organizations, and academic experts on climate law in Global South countries. The research was undertaken over the period of the Covid-19 pandemic which limited our capacity to undertake in-country field research as originally planned. However, through our networks and contacts obtained through interviews, we were able to identify further Global South cases litigating issues of climate change, beyond those included in the global databases.

As Chapter 3 of the book discusses further, cases included in the Global South docket were coded according to a set of variables in order to provide data on emerging trends across the jurisprudence as a whole. Our coding included analysing cases based on: (1) the identity of the plaintiffs (eg NGOs, local governments, individuals) and the defendants (eg governments, private corporations); (2) whether climate change was an issue at the core or at the periphery of the

[56] 'Climate Change Laws of the World' (Database, 2023) <https://climate-laws.org/faq#litigation-data> accessed 7 July 2023.

[57] For our results, see Chapter 3.

[58] As an example, the case law database of the National Green Tribunal in India is particularly difficult to use and may yield different results with similar search terms on different days depending on whether the database is experiencing technical difficulties.

lawsuit; (3) the nature of the legal avenue(s) pursued in the claim (eg whether environmental impact assessment was argued as the basis of the litigation, whether the public trust doctrine was invoked, or whether rights-based arguments were made by the parties); and (4) whether the court or the parties mentioned or relied on the Paris Agreement and national climate change laws in their decision and submissions respectively.

Other potential variables for coding the cases were considered but not included given the lack of meaningful data on these aspects. For instance, we have not coded the cases for whether they cited to IPCC reports or otherwise considered climate science arguments, nor whether the cases cross-referenced judgments from other jurisdictions, particularly other Global South countries. In respect of science, we found there was limited reference to and engagement with climate science in the cases, and where this material was present in the pleadings it seems to have been a consequence of litigants' working together with Global North organizations to bring the case. In fact, as Chapter 3 elaborates, this scientific or technical framing of climate issues is more prevalent in the Global North than in Global South climate cases, which have a stronger focus on issues of human rights and sustainable development.[59]

Likewise, while Global North courts often cite to other (Global North) cases in their climate judgments, this practice is not as common in the Global South. Some Global South courts, particularly in South Asia, make prolific references to international environmental principles, such as the precautionary principle, and may also cite—generally without much further engagement—the *Urgenda* case in the Netherlands.[60] But there does not seem to be much evidence, at least in the judgments issued in Global South climate cases, of the kind of 'transnational judicial dialogue' and cross-citation of case law from other countries seen in the Global North.[61] Potentially this is because much Global South climate case law has not been as 'visible' as that in the Global North, which is something that a book of this kind hopes to redress.

## 6. OUTLINE OF THE STRUCTURE OF THE BOOK

The remaining chapters of the book contain our analysis of the Global South docket and transnational cases involving Global South actors or activities, aiming to give a holistic view of what is now a considerable body of cases.

Chapter 2 revisits the issue of the definition of climate change litigation. The purpose of this discussion is to highlight the ways that conventional definitions

[59] See further Chapter 3.
[60] *Urgenda Foundation v State of the Netherlands* [2015] HAZA C/09/00456689.
[61] See eg *Gloucester Resources Ltd v Minister for Planning* [2017] NSWLEC 7.

have adopted an implicit Global North framing of what counts as a climate case. We argue that by adjusting the 'lens' through which we view climate litigation—in particular to include peripheral cases—we get a much broader and more inclusive picture of climate litigation and its occurrence in Global South countries. This chapter also considers distinctions drawn in the climate litigation literature between cases which are said to have a 'strategic' purpose and those which are labelled 'non-strategic',[62] as well as other categorizations such as 'framework' cases,[63] most of which apply more readily to a Global North context than a Global South one. Finally, the chapter examines the relationship of climate litigation to global climate governance and the implementation of the Paris Agreement, considering different pathways by which this relationship may be navigated.

Chapter 3, as mentioned earlier, contains the results of our identification and coding of the Global South docket and provides an analysis of emerging trends across this body of cases. Although the Global South climate litigation is still very 'young' there are clear trends that can be discerned, in particular around the use of constitutional rights as a pathway for climate litigation. We also highlight the ways that these broad trends differentiate across the three regions of Africa, Latin America and the Caribbean, and Asia and the Pacific, and offer some explanations for the reasons behind particular trends and their differences from the Global North.[64] The chapter also considers the developing body of transnational cases and the additional pathways they are providing to highlight climate justice issues facing communities in the Global South.[65]

Chapter 4 provides a more granular analysis of climate cases in three 'frontrunner' jurisdictions in Africa, Asia, and Latin America, namely South Africa, India, and Brazil. These middle-income countries currently have the highest number of climate cases in their respective regions. We use an in-depth analysis of the climate case law in these countries to provide insights into the conditions that encourage the development of climate litigation in Global South countries. A key finding is the promotive role played by institutional and socio-legal factors such as reasonable access to the courts to resolve disputes, a well-informed and progressive judiciary, the existence of climate laws and/or constitutional environmental rights, and an active civil society in the country—all important markers of a country's adherence to the environmental rule of law.[66]

Chapter 5 turns to an examination of the actors involved in Global South climate litigation and the roles they are playing. The landscape revealed by this analysis is

[62] See further Chapter 2.
[63] ibid.
[64] See Chapter 3.
[65] ibid.
[66] United Nations Environment Programme, *Environmental Rule of Law* (First Global Report, January 2019); United Nations Environment Programme, *Environmental Rule of Law: Tracking Progress and Charting Future Directions* (Second Global Report, November 2023).

a complex one. Cases have been brought mainly by local NGOs and community groups, but these groups often work in coalitions with an array of other actors, including individual lawyers and environmental public interest law firms, global NGOs, youth groups, Indigenous peoples, and philanthropic funders.[67] These actors are also heavily involved in Global North climate litigation. Some Global South jurisdictions, like Brazil and China, are seeing novel actors bringing climate cases, for instance, political parties and public prosecutors.[68] These different actors often bring various goals and expectations to climate litigation, which in turn may shape how claims are framed and what impact is sought from the litigation.

Chapter 6 turns to this question of the impact flowing from Global South climate litigation, defining impact in terms of the 'beyond courtroom' effects—direct and indirect, material and symbolic—that have been produced by the cases.[69] The novelty of Global South climate litigation, the 'youth' of emerging trends, as well as the fundamental complexity and context-dependence of impact assessments of litigation make this a challenging task. Our goal in this chapter is hence more modest. Using case studies of longer-running and/or high-profile Global South cases we aim to identify early impacts these cases have generated.[70] This analysis then leads to a discussion of areas for future development of the Global South climate case law, either to fill gaps in the jurisprudence so far, or to build on promising areas for this case law to make a distinctive and impactful contribution to multilevel climate governance.[71]

## 7. CONCLUSION

The aim of this book is by no means to have the 'last word' on Global South climate litigation but rather to start and continue a much-needed conversation on how such climate cases are emerging, their features and contributions in the context of Global South countries. Our approach in conducting this analysis has been to cast the net as widely as possible to capture the richness of legal development occurring in this space, but with a focus on core and peripheral climate cases that can be seen as part of a broader transnational effort to address climate change at multiple levels of governance.

The next chapter focuses in more closely on the questions of how climate litigation is understood and its potential governance role in the context of international treaty efforts seeking to mitigate and adapt to climate change.

67 See Chapter 5.
68 ibid.
69 See Chapter 6.
70 ibid.
71 ibid.

# 2
# Climate Change Litigation in the Era of the Paris Agreement

## 1. INTRODUCTION

As explained in Chapter 1, the focus of this book is on litigating climate change in the Global South. Climate change litigation in the Global South is part of a broader, global phenomenon in which environmental groups, individuals, and others have turned to courts to advance climate regulation and enforce relevant environmental laws.[1] While climate change litigation in the Global North has received significant (and growing) attention over the past decade,[2] only a relatively smaller number of studies have been devoted to Global South climate change litigation or climate cases in Global South countries.[3] A more holistic, inclusive view is warranted, however, especially as the number of climate cases in Global South countries continues to grow.[4] Including the Global South experience also better illuminates the progress of the quest for climate justice for vulnerable communities and groups, and may help buttress understanding of whether the goals of the Paris Agreement are being achieved.[5] This is particularly the case as climate change litigation emerges

[1] Joana Setzer and Catherine Higham, *Global Trends in Climate Change Litigation: 2022 Snapshot* (Policy Report, Grantham Research Institute on Climate Change and the Environment 2022) 6 (hereafter Setzer and Higham, *2022 Snapshot*). Setzer and Higham issued a further 2023 Snapshot update in June 2023: see Joana Setzer and Catherine Higham, *Global Trends in Climate Change Litigation: 2023 Snapshot* (Policy Report, Grantham Research Institute on Climate Change and the Environment 2023) (hereafter Setzer and Higham, *2023 Snapshot*).

[2] See literature reviews by Lisa Vanhala and Joana Setzer, 'Climate Change Litigation: A Review of Research on Courts and Litigants in Climate Governance' (2019) 10(3) WIREs Climate Change; and Jacqueline Peel and Hari Osofsky 2020 'Climate Change Litigation' (2020) 16(1) Annual Review of Law and Social Science 21, 38.

[3] See Jacqueline Peel and Jolene Lin, 'Transnational Climate Litigation: The Contribution of the Global South' (2019) 113(4) American Journal of International Law 679; Joana Setzer and Lisa Benjamin 'Climate Litigation in the Global South: Constraints and Innovations' (2020) 9(1) Transnational Environmental Law 77; Joana Setzer and Lisa Vanhala, 'Climate Change Litigation: A Review of Research on Courts and Litigants in Climate Governance' (2019) 10(4) WIREs Climate Change; Meredith Wilensky, 'Climate Change in the Courts: An Assessment of Non-U.S. Climate Litigation' (2015) 26 Duke Environmental Law and Policy Forum 131; Jutta Brunnée and others, *Climate Change Liability: Transnational Law and Practice* (CUP 2011).

[4] See further in Chapter 3, on the Global South docket.

[5] PR Shukla and others, 'Climate Change 2022: Mitigation of Climate Change (Contribution of Working Group III to the Sixth Assessment Report of the Intergovernmental Panel on Climate

*Litigating Climate Change in the Global South.* Jolene Lin and Jacqueline Peel, Oxford University Press.
 DOI: 10.1093/oso/9780191926525.003.0002

as an important mechanism for holding governments to account for the 'nationally determined contributions' (NDCs) to the global climate change response they have submitted under the Agreement.[6]

This chapter begins in Section 2 with a discussion of questions about the scope and definition of climate change litigation. Although this is a topic that has already been covered extensively in the literature, consensus remains elusive on an agreed understanding of climate change litigation.[7] Our purpose in this section of the chapter is not to resolve this continuing debate, but rather to illustrate how dominant definitions framed around climate change litigation emerging in the Global North do not adequately capture the realities and pathways that climate change litigation is taking in the Global South.

Section 3 then turns to the purposes for which climate change litigation may be brought, in other words, the ways that courts may function as tools of climate governance. We note a growing global trend of 'strategic' climate change litigation, where cases are brought with the aim of stimulating broader social and policy change, including raising the ambition of national and corporate climate action. These cases are sometimes distinguished from 'non-strategic' climate change litigation,[8] described as cases involving a more traditional purpose of interpreting and applying climate laws, as well as related environmental and natural resources laws. Climate justice narratives, calling for compensation to be provided to vulnerable communities for climate harms, are frequently presented by or on behalf of countries in the Global South.[9] But although several climate change cases emerging in the Global South or involving Global South plaintiffs have a strategic orientation and climate justice framing, many more cases are traditional public interest and enforcement claims, concerned with implementing existing environmental and natural resource management laws. This suggests a different 'mindset' underlying climate litigation in the Global South as opposed to that in the Global North; one more focused on applying existing laws and policies and overcoming flaws in their implementation, rather than pushing for more new or more ambitious regulation.[10]

Change)', Intergovernmental Panel on Climate Change (CUP April 2022) (hereafter WGIII Report); the WGIII Report examines for the first time the contribution of climate litigation to climate governance.

[6] Paris Agreement (opened for signature 22 April 2016, entered into force 4 November 2016) ATS 2016 24.

[7] See Peel and Osofsky (n 2).

[8] Setzer and Higham, *2022 Snapshot* (n 1).

[9] eg Sam Adelman, 'Climate Justice, Loss and Damage and Compensation for Small Island Developing States' (2016) 7(1) Journal of Human Rights and the Environment 32; Lavanya Rajamani, *Differential Treatment in International Environmental Law* (OUP 2006); and Shawkat Alam and others (eds), *International Environmental Law and the Global South* (CUP 2015).

[10] Asian Development Bank, *Climate Change, Coming Soon to a Court Near You: Climate Litigation in Asia and the Pacific and Beyond* (Report Series Purpose and Introduction to Climate Science, December 2020); Setzer and Benjamin (n 3); and Gitanjali Gill and Gopichandran Ramachandran 'Sustainability Transformations, Environmental Rule of Law and the Indian Judiciary: Connecting the Dots Through Climate Change Litigation' (2021) 23(3) Environmental Law Review 228.

Section 4 of the chapter examines the intersection between climate change litigation and efforts to ensure the accountability of governments for fulfilling their Paris Agreement commitments and NDCs. In the Global South, many governments have published comprehensive climate action plans and there are an increasing number of climate laws,[11] but these are often not being fully implemented. There are also many existing environmental and natural resources laws that are not properly enforced, which worsens the climate change problem. In this context, litigation is likely to be an important means of implementing and enforcing domestic climate and non-climate laws and policy regimes in Global South countries, which ultimately contributes to the implementation of the Paris Agreement. Given the bottom-up nature of the Paris Agreement and its emphasis on facilitative compliance (without resort to punitive enforcement mechanisms),[12] climate change litigation (in domestic, regional, and international courts) is well-placed to play an ever more important role in achieving accountability for climate action.

Section 5 concludes with a summary of ideas advanced in this chapter.

## 2. DEFINING CLIMATE CHANGE LITIGATION

Despite more than 20 years of scholarship on the topic of climate change litigation,[13] consensus on a definition of the phenomenon remains elusive.[14] In part, this reflects the breadth of the problem of climate change itself, which impacts many areas of society and intersects with other environmental and natural resource management issues. As the Geneva Association identified in a 2021 review, a key question in understanding what counts as climate change litigation has been what nexus is required between the litigation and climate change issues, or the '[e]xtent to which the case is about climate change'.[15] The Association identifies three different classifications in this regard:

(a) Climate change is central to the case: It is at the 'core' of the legal argument.

[11] London School of Economics, *Climate Change Laws of the World* (Database, January 2023) <https://climate-laws.org/> accessed 10 May 2023.

[12] Paris Agreement (n 6) art 15.

[13] Setzer and Vanhala (n 3); Jacqueline Peel and Hari Osofsky, 'Climate Change Litigation' (2020) 16(1) Annual Review of Law and Social Science 21; Jacqueline Peel, Alice Palmer, and Rebekkah Markey-Towler, 'Review of Literature on Impacts of Climate Litigation' (Report, 27 May 2022, Children's Investment Fund Foundation).

[14] eg Ivano Alogna, Christine Bakker, and Jean-Pierre Gauci, 'Climate Change Litigation: Global Perspectives—An Introduction' in Ivano Alogna, Christine Bakker, and Jean-Pierre Gauci (eds), *Climate Change Litigation: Global Perspectives* (Brill Nijhoff 2021) 20; noting that the edited collection chapters approach climate change litigation in a 'myriad of ways'.

[15] Maryam Golnaraghi and others, *Climate Change Litigation: Insights into the Evolving Global Landscape* (The Geneva Association 2021) 19.

(b) Climate change is peripheral to the case: There is explicit reference to climate change, but litigants rely on other grounds to call for climate-related behavioural change.
(c) Climate change is incidental to the case: This includes cases that make no specific reference to climate but have practical implications for climate change mitigation or adaptation.[16]

As we discuss further below, care needs to be taken with the Association's 'category (c) cases' which are generally 'peripheral' (rather than 'incidental') cases in our framing of climate change litigation, outlined in Chapter 1. In order to understand whether a category (c) case is actually 'about climate change', resort must usually be had to materials beyond the case record and judgment, such as interview data, social media, and NGO briefing materials on the litigation.

In this section of the chapter, we discuss some of the dominant definitions of climate change litigation that have emerged in the literature and in practice. These definitions have largely drawn from the experience of climate change litigation in developed countries of the Global North. We explain how these definitions do not always adequately capture the realities and pathways that climate change litigation is taking in the Global South. As the following chapter on the Global South climate change 'docket' discusses further, Global South countries where climate change litigation has emerged often have a high number of cases where climate issues are at the 'periphery' of the case.[17] To encompass this experience of climate change litigation alongside 'core' cases, we propose a broader, more inclusive approach.

## 2.1 Definitions of Climate Change Litigation from the Global North

Climate change litigation literature and practice have been shaped, until relatively recently, by the understandings that have emerged from jurisdictions with the highest number of cases and most notable case successes. The United States (US) remains the dominant climate litigation jurisdiction globally. The major global climate litigation databases recorded 1,590 cases filed in the US as of June 2023 compared with 751 cases filed in total in other countries and before international and regional courts and tribunals.[18] Outside of the US, Australia, the United Kingdom,

[16] ibid 6; Peel and Lin (n 3). As discussed further below, we make a similar distinction between core, peripheral, and incidental cases but understand the latter category differently to the Geneva Association as cases which, while they may contain a passing reference to climate change, have no meaningful engagement with the issue and hence are excluded from our understanding of 'climate change litigation' in this book.
[17] See further in Chapter 3.
[18] Setzer and Higham, *2023 Snapshot* 12 (n 1).

and the European Union remain the jurisdictions with the highest volume of cases.[19]

Cases raising climate change emerged in the US in the early 1990s, with the first example being *City of Los Angeles v National Highway Transportation Safety Administration (NHTSA)*, decided by the DC Circuit Court of Appeals in 1990.[20] The overwhelming majority of US cases have involved substantive challenges to agency permits and rules, and agency environmental impact assessments.[21] Following the 2007 decision of the US Supreme Court in *Massachusetts v EPA*—which established that the federal Environmental Protection Agency had a mandate to regulate greenhouse gas (GHG) emissions under the Clean Air Act[22]—climate change litigation expanded exponentially in the US. In the same year as the *Massachusetts v EPA* decision, Michael B Gerrard and J Cullen Howe from the law firm Arnold & Porter created a US Climate Litigation Chart to track this litigation. This database is now a joint project of Arnold & Porter and the Sabin Center for Climate Change Law at Columbia Law School (directed by Professor Michael Gerrard), maintaining records of US climate cases, as well as an additional chart encompassing global (non-US) climate change litigation.[23]

In compiling case records for its databases, the Sabin Center applies the following criteria:[24]

> First, cases must generally be brought before judicial bodies (though in some exemplary instances matters brought before administrative or investigatory bodies are also included). Historically, the term 'cases' in the U.S. database included more than judicial actions and proceedings. Other types of 'cases' formerly contained in the database included quasi-judicial administrative proceedings, rulemaking petitions, requests for reconsideration of regulations, notices of intent to sue (in situations where lawsuits were not subsequently filed), and subpoenas. Since 2018, these other types of cases have not been added to the U.S. database, and approximately 100 older such cases were removed from the database in November 2021.
>
> Second, climate change law, policy, or science must be a material issue of law or fact in the case. Cases that make only a passing reference to climate change, but

[19] ibid; Australia 130 cases, UK 102 cases, EU 67 cases.

[20] *City of Los Angeles v National Highway Transportation Safety Administration* (1990) 21 ELR 20170.

[21] David Markell and JB Ruhl, 'An Empirical Assessment of Climate Change in the Courts: A New Jurisprudence or Business as Usual?' (2012) 64 Florida Law Review 12, 38.

[22] *Massachusetts v EPA*, 549 US 497 (2007). Note the more recent Supreme Court case, *West Virginia v Environmental Protection Agency*, 597 US (2022), which casts some doubt on the extent of this mandate.

[23] Sabin Center for Climate Change Law, *Climate Case Chart—About* (January 2023) <http://climatecasechart.com/climate-change-litigation/about/> accessed 7 April 2023.

[24] ibid.

> do not address climate-relevant laws, policies, or actions in a meaningful way are not included.
>
> In general, cases that may have a direct impact on climate change, but do not explicitly raise climate issues, are also not included in the database. Examples of such cases may include challenges to government inaction on local air pollution or challenges to the development of fossil fuel infrastructure on the basis of other types of harm to human health and/or the environment. The intent of the litigants with regard to the climate-related consequences of such cases is not considered during the assessment process.

Similar understandings of climate change litigation have been applied in empirical reviews of the US climate litigation. For example, in their seminal review of this jurisprudence, published in 2012, Dave Markell and JB Ruhl defined climate change litigation 'as any piece of federal, state, tribal, or local administrative or judicial litigation in which the party filings or tribunal decisions directly and expressly raise an issue of fact or law regarding the substance or policy of climate change causes and impacts'.[25] In applying this definition in their empirical survey, the authors recognized that this excludes certain types of litigation, such as cases motivated by concern over climate change but not involving issues of law and fact directly relevant to questions of climate change law and policy.[26]

This narrower definition of climate change litigation has become influential in subsequent literature and efforts to catalogue and categorize climate change litigation, even as the phenomenon has spread beyond its epicentre in the US. For example, the *UN Environment Programme Global Climate Litigation Report: 2020 Status Review*, which was prepared by the United Nations Environment Programme in cooperation with the Sabin Center, considers 'climate change litigation' to 'include cases that raise material issues of law or fact relating to climate change mitigation, adaptation, or the science of climate change', echoing the approach of the Sabin Center's database.[27] Likewise, the UNEP's Status Review:

> excludes cases where the discussion of climate change is incidental or where a non-climate legal theory would guide the substantive outcome of the case. Thus, when climate change keywords are only used as a passing reference to the fact of climate change and those issues are not related to the laws, policies, or actions actually at issue, the case is excluded. Similarly, this report excludes cases that seek to accomplish goals arguably related to climate change adaptation or mitigation

[25] Markell and Ruhl (n 21) 27.

[26] ibid 26.

[27] United Nations Environment Programme, *Global Climate Litigation Report: 2020 Status Review* (Status Review 2020, Nairobi) 6. The 2023 Status Review update prepared by UNEP with the Sabin Center team continues this approach: see United Nations Environment Programme, *Global Climate Litigation Report: 2023 Status Review* (Status Review 2023, Nairobi).

> but that do not depend on the climate change dimensions of those goals. For example, lawsuits seeking to use human health regulations to limit air pollution from coal fired power plants may incidentally cause a court to compel that power plant to emit fewer greenhouse gases (GHGs). Such cases are not considered 'climate change litigation' for the purposes of this study.[28]

Collaborative efforts between the Sabin Center and the Grantham Research Institute at the London School of Economics to produce the Climate Change Laws of the World database, which includes a searchable global climate change litigation database, are also guided by the definition elaborated in the UNEP Status Review.[29] As the Grantham Research Institute explains, this 'narrower definition' helps 'facilitate the process of data collection and to emphasise the distinct nature and importance of cases that explicitly raise climate issues'.[30] While both cases on climate change mitigation and adaptation policies or efforts are included, this is only where climate change is 'a main or significant issue' in the litigation. Accordingly, there is exclusion of '[c]ases that make only a passing reference to climate change but do not address climate-relevant laws, policies or actions in a meaningful way . . . [or] cases that may have a direct impact on climate change, but do not explicitly raise climate issues'.[31]

Alternative approaches to defining climate change litigation have been suggested by other authors (although generally still in reflecting on the experience of cases in the Global North). For example, Francesco Sindico adopts a different methodological approach to compiling information on climate change litigation in the C2LI database that he established,[32] and in an edited collection on *Comparative Climate Litigation*. As Sindico describes it:

> C2LI is not a database of climate cases. It is also not an initiative for climate litigation; it is an initiative about climate litigation. C2LI looks at litigation through three specific scenarios: an individual challenging the State for ineffective climate action (public policy climate litigation); an individual challenging her government for authorising a specific project that leads to increased emissions or ineffective adaptation (public project based climate litigation); an individual challenging a private actor for operations that allegedly lead to more climate change (private climate litigation).[33]

[28] 2020 Status Review (n 27).
[29] ibid 7.
[30] Joana Setzer and Catherine Higham, *Global Trends in Climate Change Litigation: 2021 Snapshot*' (Policy Report, Grantham Research Institute on Climate Change and the Environment July 2021) 9.
[31] ibid.
[32] *Climate Change Litigation Initiative—C2LI* (Database, University of Strathclyde 2022) <www.strath.ac.uk/research/strathclydecentreenvironmentallawgovernance/ourwork/research/labsincubators/climatechangelitigationinitiative/> accessed 7 May 2023.
[33] ibid.

Other authors have looked to the way litigation is framed or the strategic motivations driving it in their classifications of climate change litigation. Chris Hilson argues that 'in order to count as climate litigation, cases should be framed as such', which places emphasis on the objectives of the litigants or the broader social movement supporting a case.[34] Similarly, in an early analysis of climate change litigation, Navraj Ghaleigh proposed a typology distinguishing different kinds of motivations for climate change litigation including 'promotive', 'defensive', 'boundary-testing', and 'perfecting' cases.[35]

Recognizing the differences that existed between scholars in 'seeking to give some shape to the notion of climate litigation in the literature', Jacqueline Peel and Hari Osofsky in their 2015 book, *Climate Change Litigation: Regulatory Pathways to Cleaner Energy*, proposed conceptualizing litigation on climate change as a series of concentric circles radiating out from 'core' cases where climate change was a central issue.[36] Cases where climate change was a peripheral issue, a motivation for the litigation, or where litigation had meaningful consequences for climate change mitigation or adaptation formed the successive outer rings of the circles (see Figure 2.1).[37] Even so, Peel and Osofsky's analysis was concentrated on US and Australian climate change litigation, seeking to determine via empirical methods, including interviews, the 'regulatory impact' of this litigation. As a consequence, their book is focused on cases with climate issues at their core.[38]

As these definitions and typologies illustrate, analyses of climate litigation with a focus on the Global North experience tend to emphasize cases where climate change is a central part of, or motivation for, the litigation.[39] Such cases can often be higher-profile because they attract significant academic focus (as novel climate claims) or due to the media attention that they generate.[40] In the literature addressing climate change litigation, concentrated in the Global North, it has been observed that this focus on high-profile cases can reduce analysis of lower-profile, more complex cases, or cases where climate change issues are more peripheral to the litigation (including where cases have an anti-regulatory orientation).[41] This issue of exclusion or neglect becomes problematic when we turn to cases emerging

[34] In Alogna, Bakker, and Gauci (eds) (n 14) 16.

[35] See Navraj Singh Ghaleigh, '"Six Honest Serving-Men": Climate Change Litigation as Legal Mobilization and the Utility of Typologies' (2010) 1(1) Climate Law 31.

[36] Jacqueline Peel and Hari Osofsky, *Climate Change Litigation: Regulatory Pathways to Cleaner Energy* (CUP 2015) 8.

[37] This figure appears in a later work by the authors: Jacqueline Peel and Hari Osofsky, 'Climate Change Litigation' (2020) 16(1) Annual Review of Law and Social Science 21–38, 24. The article is open access and its content licensed for unrestricted use, distribution, and reproduction in any medium, with appropriate acknowledgment, under a Creative Commons Attribution 4.0 International License.

[38] Peel and Osofsky (n 2) 8.

[39] Peel and Lin (n 3).

[40] Peel and Osofsky (n 2) 21.

[41] ibid; Kim Bouwer, 'The Unsexy Future of Climate Change Litigation' (2018) 30(3) Journal of Environmental Law 483.

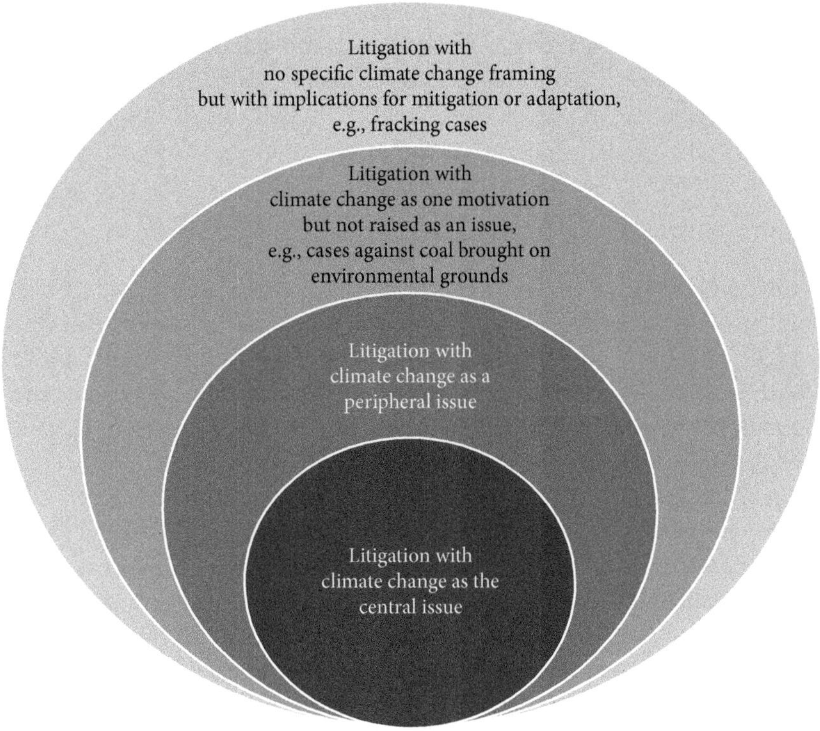


**Figure 2.1** Different notions of climate litigation
J Peel and HM Osofsky, 'Climate Change Litigation' 16 (2020) Annu Rev Law Soc Sci. 21

in the Global South which, as discussed further below, include many disputes that feature climate change issues only in a peripheral way.

Where reports or reviews have focused specifically on climate change litigation in Global South jurisdictions, a broader approach to understanding the scope of the phenomenon has been observed. A good example is the definition of climate change litigation adopted by the Asian Development Bank in its report on *Climate Litigation in Asia and the Pacific and Beyond*.[42] There the authors define climate change litigation 'broadly' as 'any case that is brought before judicial courts and administrative or specialized tribunals that (i) raises climate change as a central issue; (ii) raises climate change as a peripheral issue; or (iii) does not explicitly raise climate change but has ramifications for climate change mitigation or adaptation efforts, eg recognition of intergenerational responsibility'.[43] While the authors note

[42] Asian Development Bank (n 10).
[43] ibid 7. As for the Geneva Association definition discussed above, category (iii) of the ADB definition raises a question as to how 'ramifications for climate change mitigation or adaptation efforts' are

that this broader definition is most appropriate where looking at cases from countries in Asia and the Pacific, the need for a narrower definition is also recognized '[t]o prevent the scope of this report from becoming unwieldy' when looking beyond this region.[44]

## 2.2 Cases with Climate Change at the Periphery

As the Asian Development Bank report emphasizes, cases in regions of the Global South, such as Asia and the Pacific, may still include cases that raise climate change as a 'central issue'.[45] There are now several examples of these 'core' cases from Global South countries which centrally feature climate change arguments and fit conventional (Global North) definitions of climate change litigation. These case examples—like the Carbon Majors Inquiry carried out by the Philippines Human Rights Commission,[46] the Columbian case of *Future Generations v Ministry of the Environment and others*,[47] the decision of the Lahore High Court in Pakistan in *Leghari v Pakistan*,[48] and the Brazilian Federal Supreme Court case of *PSB and others v Brazil*[49]—have engaged closely with climate change issues and have achieved a high profile, with extensive discussion in the media and in academic commentary.[50] However, as Chapter 3's discussion and analysis of Global South climate change cases demonstrates, a substantial proportion of these cases do not feature climate change as a 'core' issue and are unlikely to attract the same degree of global visibility.

In previous work, we have described the latter category of cases as ones raising climate change as a 'peripheral issue'. 'Peripheral' climate change cases, as outlined in Chapter 1, are ones where climate change issues are subsidiary to other arguments raised in the litigation or one of many arguments or issues raised in a

determined if not mentioned in the case record. This evidence may be available from other sources beyond the case record.

[44] ibid 10.

[45] ibid 8.

[46] Commission on Human Rights of the Philippines, *National Inquiry on Climate Change* (Report, December 2022). This example is discussed further in Chapter 3.

[47] *Future Generations v Ministry of the Environment* (Colom Sup Ct, 11001-22-03-000-2018- 00319-01, 5 April 2018). This example is discussed further in Chapter 6.

[48] *Ashgar Leghari v Federation of Pakistan* (2015) Lahore High Court Green Bench WP No 25501/ 201. This example is discussed further in Chapter 6.

[49] *Partido Socialista Brasileiro (PSB), Partido Socialismo e Liberdade (PSOL), Partido dos Trabalhadores (PT) e Rede Sustentabilidade v União Federal*, Arguição de Descumprimento de Preceito Fundamental ADPF 708 filed 6 May 2020. This example is discussed further in Chapter 6.

[50] Emily Barritt, 'The Symbolic Value of Leghari v Federation of Pakistan: Climate Change Adjudication in the Global South' (2019) 30(2) King's Law Journal 203; Anam Gill, 'Farmer Sues Pakistan's Government to Demand Action on Climate Change' *Reuters* (14 November 2015); Joana Setzer and Délton Winter de Carvalho, 'Climate Litigation to Protect the Brazilian Amazon: Establishing a Constitutional Right to a Stable Climate' (2021) 30(2) Review of European, Comparative and International Law 197.

dispute. Another way of putting this is to say that while climate change litigation with climate as a central matter engages directly with climate change issues in the pleadings or judgment, 'peripheral' cases do so only indirectly.[51] Peripheral climate cases ought to be distinguished from cases where climate change is mentioned only in passing, without any direct or indirect engagement in the plaintiffs' arguments, case preparatory materials, or the decision of the court.[52] These latter cases—which we refer to as 'incidental' cases—generally make no meaningful contribution to advancing a specific climate jurisprudence and therefore fall outside the definition of climate litigation on which we base our analysis in this book.[53]

By contrast, including 'peripheral' climate change cases in climate litigation analyses, alongside higher-profile cases where climate change issues are 'core', yields a broader understanding of the emergence of climate litigation globally and trends across this jurisprudence. It also helps us to understand the ways Global South climate litigation might vary from that in the Global North and the different nature of its contribution to climate governance.

As we have discussed in previous work, a peripheral framing of climate change issues is likely to be more common in Global South countries for several reasons.[54] First, climate law frameworks, as well as associated avenues for access to justice, may be absent, less well-developed, or poorly implemented in the Global South, meaning that climate change cases in these jurisdictions need to draw on other laws that apply only indirectly to climate change.[55] Joana Setzer and Lisa Benjamin, in their analysis of Global South climate litigation, also point to poor implementation of laws as a particular source of challenges for climate change litigation in

[51] Peel and Lin (n 3).

[52] For example, there are numerous examples of Global South cases about projects with potential environmental impacts, such as roads or large infrastructure developments, that might have provided a forum for consideration of climate change impacts associated with transportation emissions or the removal of vegetation, but these matters were not raised or considered in any meaningful way in the judgments. See eg *Lahore Bachao Tehrik v Canal Road Project, Government of Punjab*, Lahore—SMC No 25/2009 [2011] PKSC 34 (15 September 2011) (involving widening of Canal Road and removal of the surrounding green belt areas); *Niwat v Electricity Generation Authority of Thailand* (Xayaburi Dam case), Decision of the Thai Supreme Administrative Court, 25 December 2015, unofficial translation available at <www.business-humanrights.org/sites/default/files/documents/CourtDecisionXayaburi25%2012%202015ENG.pdf> accessed 1 December 2023 (involving a major dam project on the Mekong River for electricity production). We have excluded these 'incidental' cases in our analysis, although recognizing that petitioners or judges in similar types of cases in the future may begin to engage with climate change issues.

[53] Peel and Lin (n 3) 703. In our analysis we include cases which broadly fit the ADB's third category described above, ie cases which do not explicitly raise climate change but have ramifications for climate change mitigation or adaptation efforts. In this instance it is our *empirical, interview-based methodology* that has allowed us to discover the climate change motivations of the cases concerned even if climate change is not mentioned in the case records. This is to be distinguished from the situation where cases *could* have climate change ramifications, eg cases to prosecute breaches of forestry laws, but there is no evidence that the case was motivated by climate change concerns on the part of litigants and there is no substantive discussion of climate issues in the judgment. Such a case would fall outside the definition of climate litigation which we have employed in this book.

[54] ibid.

[55] ibid 692.

Global South countries. They note that 'absence of climate frameworks in itself cannot explain the limited number of explicit climate change cases in the Global South',[56] and indeed, analyses of climate legislation show a growth in climate laws and policies across all countries, including those in the Global South.[57] However, Setzer and Benjamin point to 'well-known barriers' in Global South countries to implementing and enforcing laws and policies that are on the books, including weak and fragmented institutions, incomplete legal foundations, poor resourcing, and limited political will. In addition, in many Global South countries, climate policies and implementation efforts led by centralized ministries and departments may have little penetration into more remote or rural areas of the country.[58]

A second reason that climate change issues may feature less prominently in Global South cases is the lower public policy salience of such issues in many Global South countries, compared with more pressing social and policy concerns around economic development, poverty alleviation, or public health.[59] This can lead to climate change matters being packaged with a range of other issues in litigation, such as those pertaining to pollution, land use, forestry, natural resource conservation, disaster risk management, implementation of planning frameworks, or environmental justice and rights claims.[60] This framing aligns with the reality for many litigants in Global South countries, namely that climate change adds a layer of complexity to, or exacerbates, existing environmental challenges rather than presenting as a self-standing environmental issue.[61] Climate change is therefore a driver of further harm in addition to existing environmental problems like high levels of air pollution in cities, massive deforestation, land-clearing, and rapid biodiversity loss.

In most Global South countries, laws that respond to these non-climate environmental problems are generally already legislated, but limited implementation

[56] Setzer and Benjamin (n 3).

[57] London School of Economics, *Climate Change Laws of the World* (Database, January 2023) <https://climate-laws.org/> accessed 7 May 2023.

[58] Setzer and Benjamin (n 3) 58.

[59] Peel and Lin (n 3) 693. See eg Channel News Asia, 'As Living Costs Surge, Climate Change Takes a Backseat in Elections' (News Article, 29 April 2022) <www.reuters.com/article/climate-change-global-votes-idAFL5N2WA2BQ/> accessed 7 April 2023. This point was also raised by interviewees. For instance, in Interview No 7 conducted via Zoom on 20 April 2022, the interviewee, speaking about a case taken in Thailand, noted: 'we had floated the idea of raising the climate change issue in the litigation. But I think ultimately, the lawyers, you know, the Thai lawyers that were working on it, basically decided there were already enough issues with the EIA, that were issues that the community was actually concerned about... [N]o one from the community independently raised the concern about climate change ... [T]hey're more concerned about their farm, and the water that they get from the local stream which will be destroyed by the [coal] mine, and their livelihoods, that kind of thing."

[60] Peel and Lin (n 3) 694. See also Interview No 8 on 28 April 2022 regarding cases in Indonesia where the interviewee remarked: 'I don't think that it's about the climate change for the citizen yet at this point, because we know that the air pollution problem in Jakarta is real, I mean like we can see it with our eyes. But if we talk about the climate change I don't think that the public gets the idea that the air pollution is climate change. Because we, if we talk about the air pollution we always talk about the health impact at this time, we don't say about the climate change. I mean like we say about the climate change but most of the part is about the health impacts.'

[61] Interview No 7 conducted via Zoom on 20 April 2022.

or under-resourcing of enforcement contributes to ongoing problems, worsened by climate change impacts. Bringing cases within the scope of these laws may itself necessitate a non-climate framing (for example, casting the case as primarily about air pollution or enforcement of deforestation controls) in order to satisfy particular statutory requirements applicable to the bringing of a lawsuit.[62]

## 2.3 Towards a Broader Understanding of Climate Litigation

A more inclusive approach to understanding climate change litigation, which encompasses 'peripheral' cases, gives visibility to Global South cases and allows for a more holistic study of the phenomenon and its impacts. However, in the quest for greater inclusivity, care should be taken so as not to dilute the definition of climate change litigation to the point of being meaningless by incorporating cases where climate change is only mentioned incidentally or in passing but with no meaningful engagement with climate issues ('incidental' cases in our categorization approach outlined in Chapter 1).[63]

As Dave Markell and JB Ruhl noted in their early empirical analysis of the US climate case law, a 'holistic' study of climate change litigation can provide insights 'not only for how climate change policy is likely to evolve, and for who is likely to share it, but [also] more generally for the role of the courts in public policy governance'.[64] We expect that a broader understanding of climate change litigation that takes a more holistic perspective capturing the Global South experience will also be beneficial in assessing the contribution of such litigation to climate governance and implementation of the Paris Agreement. Indeed, Global South judges have often been highly creative in crafting legal remedies for climate inaction, as cases like the Pakistani case, *Leghari*,[65] and the Brazilian case, *PSB*, illustrate. In *PSB*, this creativity was demonstrated by the Court's approach to adopting a public forum

[62] Peel and Lin (n 3) 694. However, the cases we have included in our analysis still have climate concerns as a 'peripheral' issue or we have been able to identify a climate motivation for these cases based on interview data.

[63] Setzer and Higham, *2022 Snapshot* (n 1) 6. See eg *Prosecutor v Hairil* Case No 168/PID.B/LH/2018/PN Pbu (Pangkalan Bun District Court); *Prosecutor v Danar Bata* Case No 6/PID.SUS.TPK/2020/PN Gto (Gorontalo District Court); and *Prosecutor v PT Surya Panen Subur* Case No 54/PID.SUS/2014/PN.MBO (District Court of Meulaboh, in Aceh Province). For examples of other such cases in Indonesia, see Linda Yanti Sulistiawati 'Climate Change Litigation in Indonesia: The Dawn of a New Beginning?' (2023) (forthcoming, on file with the authors).

[64] Markell and Ruhl (n 20) 147.

[65] Emily Barritt and Boitumelo Sediti, 'The Symbolic Value of Leghari v Federation of Pakistan: Climate Change Adjudication in the Global South' (2019) 30(2) King's Law Journal 203; Waqqas Ahmad Wir, 'From Shehla Zia to Asghar Leghari: Pronouncing Unwritten Rights Is More Complex than a Celebratory Tale' in Jolene Lin and Douglas A Kysar (eds), *Climate Change Litigation in the Asia Pacific* (CUP 2020) 261. The *Leghari* case concerned a constitutional rights challenge to the Pakistani government's failure to implement its national climate policy framework. The case and its impacts are discussed further in Chapters 3 and 6.

so as to hear a much broader range of stakeholder views than would ordinarily be facilitated in a judicial setting, including 'scientists, environmentalists, Indigenous people, representatives from the agribusiness and financial sectors, economists, researchers, parliamentarians and representatives of the federal and state governments'.[66] As Joana Setzer notes, this approach marked a 'historic moment' for the Supreme Court.[67] Such creativity may ultimately provide a model for judiciaries both in the Global South and the Global North as domestic constituencies seek novel avenues to hold governments and other actors to account for their contribution to the global climate crisis.[68]

At a practical level, a broader understanding of climate change litigation, capable of including Global South cases, opens up alternative avenues for advocates and funders to explore opportunities for seeding or supporting climate cases in new jurisdictions. As Chapter 5 discusses further, greater understanding of this opportunity is already driving new pathways for climate change litigation in the Global South.[69] This may offer enhanced possibilities for achieving policy and regulatory impact, including on a regional or global scale, particularly where litigation is undertaken in Global South countries that are major emitters or particularly vulnerable to climate change impacts.[70]

In their study of climate change litigation in Asia and the Pacific, the authors of the Asian Development Bank study also support a broader approach.[71] They note that adopting a 'narrower definition' of climate change litigation would have limited the report's analysis of crucial issues for the region, such as the links between biodiversity and ecosystem resilience and climate change adaptation in Global South countries in Asia and the Pacific, as well as the wide-ranging effects of climate adaptation responses for all aspects of society. The authors conclude that 'across Asia and the Pacific, climate justice requires judges to consider whether the case before them has ramifications for climate change'.[72]

[66] Joana Setzer, *First Climate Case Reaches Brazil's Supreme Court* (Grantham Research Institute on Climate Change and the Environment 2020) <www.lse.ac.uk/granthaminstitute/news/first-climate-case-reaches-brazils-supreme-court/ accessed 10 January 2023.

[67] ibid.

[68] Peel and Lin (n 3) 700.

[69] See further the discussion of funders in Chapter 5.

[70] Peel and Lin (n 3) 700. We take up the question of measuring the impact of climate change litigation in Chapter 6.

[71] See also Daniel Hornung, Douglas A Kysar, and Jolene Lin, 'Introduction' in Jolene Lin and Douglas A Kysar (eds), *Climate Change Litigation in the Asia Pacific* (CUP 2020) 1, which discusses how editors did not want to restrict contributing authors to a certain definition because climate change litigation is only just emerging in this region and it would be unduly restrictive in such a fast-paced, developing space.

[72] Asian Development Bank (n 10). All this is not to suggest, however, that cases where climate change issues are only incidentally mentioned, or even cases which include several climate 'key terms' but do not deal with them in substance fall within the definition of climate litigation used in this book. This approach seeks to recognize the different concerns and circumstances manifesting in Global South litigation and the value of 'peripheral' cases, while not expanding the definition to such an extent that cases which do not have any material significance for climate issues are labelled as further examples of climate litigation.

## 3. PURPOSES OF CLIMATE CHANGE LITIGATION

With the growth of climate change cases, including their extension globally to countries of the Global South, there has been increased interest in the objectives being sought by litigants in this litigation. Existing and ongoing climate cases have been brought for a wide range of purposes, including challenging projects with climate change impacts, forcing disclosure of climate risk information, improving climate regulation, or recovery for climate-related loss and damage. At the same time—and particularly since the conclusion of the Paris Agreement—strategic uses of climate change litigation have increased, with litigants seeking to align their cases to particular climate mitigation or adaptation outcomes. This development parallels the trajectory of strategic human rights litigation and impact litigation more broadly, with potential lessons from these experiences for evolving climate change litigation.[73]

In this section of the chapter, we elaborate the notion of strategic climate change litigation, outlining how it is distinguished in the literature from 'non-strategic' uses of litigation for the purposes of application, interpretation, and enforcement of laws. On this classification, climate change cases in the Global South will often fall in the 'non-strategic' category of claims, focused on implementing existing environmental and natural resource management laws. There are signs that this trend is shifting as Global South climate change litigation attracts more attention (including from Northern funders) and litigants explore the potential for using litigation as a tool for enhanced governance and achieving climate justice. Nonetheless, it is likely that Global South climate change litigation, in the main, will continue to pursue different objectives and pathways than have been observed in the Global North.

### 3.1 'Strategic' versus 'Non-strategic' Climate Change Litigation

'Strategic' climate change litigation is increasingly recognized as a subset of general climate change litigation. But despite growing interest in strategic uses of climate change litigation, there is surprisingly little discussion in the literature of the notion of strategic climate change litigation or its distinction from so-called 'non-strategic' climate litigation. Most of the analysis of strategic climate change litigation also draws from the Global North experience, which raises questions about the extent to which this trend is being replicated in the Global South.

The term 'strategic litigation' is more frequently used in the human rights and social justice context, often interchangeably with similar terms like 'impact

[73] Ben Batros and Tessa Khan, 'Thinking Strategically About Climate Litigation' (2020) <https://ssrn.com/abstract=3564313> accessed 1 April 2023.

litigation', 'cause lawyering', 'public interest litigation', 'public policy litigation', and 'human rights litigation'. For example, the Open Society Justice initiative uses the term 'strategic human rights litigation' to mean:

> legal action in a court that is consciously aimed at achieving rights-related changes in law, policy, practice and/or public awareness above and beyond relief for the named plaintiffs.[74]

In their review of strategic litigation, Michael Ramsden and Kris Gledhill adopt a similar understanding of the concept. They identify four main characteristics of strategic litigation:

> (i) an extrinsic legacy component—it seeks outcomes with a long-term impact, going beyond the origins of the claimant's complaint; (ii) it is a method of advocacy rather than a cause, and therefore adaptable to a range of purposes; (iii) its objectives are similarly multi-faceted and go beyond creating effects within the court system; (iv) it views 'litigation' broadly, to include tribunals and international mechanisms of redress.[75]

Studies by socio-legal scholars also emphasize strategic litigation as a method of advocacy. According to these analyses, a feature of strategic litigation, including strategic climate change litigation, is that it often employs the law as a tool for promoting legal, policy, or social change, or for advancing social and economic justice.[76] It is common, in these circumstances, for discrete court actions to be combined with other non-legal methods, such as campaigns, protests, advocacy, and activism. This corresponds with political science theories of legal mobilization. In its narrowest application, 'legal mobilization' connotes 'high-profile litigation efforts for (or, arguably, against) social change'.[77] More broadly, the term describes 'any type of process by which individual or collective actions invoke legal norms, discourse or symbols to influence policy or behaviour'.[78] These actions generally form 'a larger, complex, dynamic, multistage process of disputing among various parties', of which court action is just one part or dimension.[79]

[74] Open Society Justice Initiative, *Global Human Rights Litigation Report* (Report, 2018) 25.

[75] Michael Ramsden and Kris Gledhill, 'Defining Strategic Litigation' (2019) 38(4) Civil Justice Quarterly 407.

[76] Susan Hansen, *Strategic Litigation* (The Atlantic Philanthropies 2018) 8.

[77] Lisa Vanhala and Jacqueline Kinghan, *Literature Review on the Use and Impact of Litigation* (Project Report 2018, The Public Law Project).

[78] ibid 5.

[79] Michael McCann, 'Litigation and Legal Mobilization' in Keith Whittington, Daniel Keleman, and Gregory Caldiera (eds), *The Oxford Handbook of Law and Litigation* (OUP 2008) 522, 525.

A trend of growing 'strategic' climate change litigation has been identified in several reviews of the field.[80] For example, in their 2022 snapshot of *Global Trends in Climate Change Litigation*, Joana Setzer and Catherine Higham found that the number of strategic cases continued to be 'dramatically on the rise' as was the case in 2021.[81] Of the non-US cases reviewed, close to 40% of cases brought in 2020 were classified as strategic cases, compared with less than 10% of cases a decade earlier in 2010 and no strategic cases two decades earlier in 2000.[82] Setzer and Higham describe 'strategic' cases as ones

> where the claimants' motives for bringing the cases go beyond the concerns of the individual litigant and aim to bring about some broader societal shift. For the most part, the goals of the claimants in such cases will include advancing climate policies, creating public awareness, or changing the behaviour of government or industry actors.[83]

By contrast, 'non-strategic' climate change cases would encompass those where the issues involved 'are primarily of relevance only to the parties involved'.[84] As an example of a non-strategic case, Setzer and Higham identify 'a dispute over planning permission for a new property in an area at risk from flooding or sea level rise, where climate change adaptation policies may be considered in the decision-making process'.[85] The authors acknowledge that their classification of cases as strategic or non-strategic 'entails a subjective assessment, often made on admittedly imperfect or incomplete information, about parties' intentions' and add that the classification does not imply that strategic cases are necessarily more important or more likely to achieve impact than non-strategic cases.[86] They also note that:

> Courts rarely have regard for the broader intentions of the parties when determining a case, meaning that cases brought with little or no strategic intent may

[80] In addition to Setzer and Higham, *2022 Snapshot* (n 1) see Kim Bouwer and Joana Setzer, *Climate Litigation as Climate Activism: What Works?* (British Academy, COP26 Briefings Series, December 2020) 3: 'Over the last few years, climate change litigation has been increasingly *used as a tool to influence policy outcomes and/or to change corporate and societal behaviour. This approach sees advocates using climate litigation to drive ambition in climate action*, taking a long view beyond the immediate success or failure of individual cases'; and Batros and Khan (n 73) 2: 'while many early efforts focused on challenging a particular fossil fuel-intensive project or harmful regulation, there has been a recent growth in "strategic cases" *which aim to produce ambitious and systemic outcomes*'.

[81] Setzer and Higham, *2022 Snapshot* (n 1) 4. Continuation of this trend was reinforced by Setzer and Higham's *2023 Snapshot*, 3–4.

[82] Setzer and Higham, *2022 Snapshot* (n 1). Note, however, that this does not necessarily suggest a corresponding drop in non-strategic cases. While strategic cases are likely to be captured in the database, non-strategic cases may be less likely to be captured.

[83] Setzer and Higham, *2022 Snapshot* (n 1) 12.

[84] ibid.

[85] ibid.

[86] ibid 13.

> nonetheless provide opportunities for courts to issue far-reaching judgments on novel legal issues.[87]

However, Setzer and Higham recognize that strategic climate change cases are more widely studied in academia and practice with a view to understanding their potential social and regulatory impact and effectiveness.[88] This often means that these cases receive more attention in assessments of the governance contributions of climate change litigation and in the deliberations of philanthropic funders supporting litigation.

Other discussions of strategic climate change litigation in the literature have focused on the outcomes of the litigation rather than the intent of the parties per se.[89] For instance, Jacqueline Peel, Hari Osofsky, and Anita Foerster see strategic climate change litigation as that which 'generally involves the use of the law and court action to advance beneficial outcomes for addressing climate change'.[90] The Geneva Association's 2021 review similarly describes strategic cases as

> designed to reach outcomes that go beyond obtaining results for the litigant bringing the case. These cases seek to advance climate policies, drive behavioural shifts in key actors, and/or create awareness and encourage public debate. Litigants bringing such cases make strategic decisions about who will bring the case, where and when the case will be filed, and what legal remedy will be sought.[91]

Because strategic climate change litigation is oriented to achieving outcomes beyond the courtroom, it often involves claims which are designed to capture public or media attention so as to increase the visibility and influence of resulting judgments.[92] Kim Bouwer and Joana Setzer identify three common forms of strategic

[87] ibid. See also Gill and Ramachandran (n 10) on climate cases before the Indian judiciary.

[88] Setzer and Higham, *2022 Snapshot* (n 1) 12. Citing Bouwer and Setzer (n 80); Peel and Osofsky (n 2). See also Geetanjali Ganguly, Joana Setzer, and Veerle Heyvaert, 'If at First You Don't Succeed: Suing Corporations for Climate Change' (2018) 38(4) Oxford Journal of Legal Studies 841, 842.

[89] Andrew Jackson, 'Ireland's Climate Action and Low Carbon Development Act 2015: Symbolic Legislation, Systemic Litigation, Stepping Stone?' (2020) University College Dublin Working Papers in Law, Criminology and Socio-Legal Studies Research Paper No 11/2020. Jackson uses different terminology of 'systemic' climate change litigation. This is picked up in the IPCC WGIII Report (n 5), which says that a distinction can be drawn between 'claims challenging the overall effort of a State or its organs to mitigate or adapt to climate change (sometimes referred to as "systemic climate litigation") and claims regarding authorisation of third-party activity', citing Bouwer (n 41); Michael B Gerrard, 'Climate Change Litigation in the United States: High Volume of Cases, Mostly About Statutes' in Alogna, Bakker, and Gauci (eds) (n 14) 33; and Navraj Singh Ghaleigh, 'Climate Constitutionalism of the UK Supreme Court' (2021) 33(2) Journal of Environmental Law 441.

[90] Jacqueline Peel, Hari Osofsky, and Anita Foerster, 'Shaping the "Next Generation" of Climate Change Litigation in Australia' (2017) 41(1) Melbourne University Law Review 793, 794.

[91] Maryam Golnaraghi and others, 'Climate Change Litigation: Insights into the Evolving Global Landscape' (The Geneva Association 2021) 19.

[92] Gill and Ramachandran (n 10).

climate change litigation used by activists, namely: litigation that targets high-carbon projects or policies (which they term 'hit the target litigation'), litigation that is part of a broader strategy to encourage further litigation or to ensure the viability of ongoing campaigns (termed 'stepping stone litigation'), and litigation that emphasizes inconsistencies between discourse and action (termed 'name and shame litigation').[93] As Gita Gill and Gopichandran Ramachandran point out, the kinds of legal arguments that emerge from these strategies comprise climate change cases that include 'human rights arguments . . . alternative strategies against "Carbon Majors" ranging from claims of nuisance to fraud and disclosure-related lawsuits . . . deceptive greenwashing marketing campaigns . . . and regulatory challenges to permits authorising high emitting projects'.[94]

A more recently identified category of 'strategic' climate change litigation is 'framework litigation' brought against governments, as was highlighted in Chapter 1.[95] Joana Setzer, Catherine Higham, and Emily Bradeen describe these cases as ones 'in which litigants challenge the ambition or implementation of a national or subnational government's overall policy response to climate change'.[96] Framework litigation according to these authors 'may involve (i) challenges to the overall level of ambition of the response; or (ii) failure to implement measures adequate to achieve the government's ambition'.[97] Like other forms of strategic litigation, Setzer, Higham, and Bradeen identify framework litigation as a feature predominantly of Global North climate litigation to date. As of 31 July 2022, they found at least 80 framework litigation cases had been filed against governments around the world but that the majority of these (63 cases) had been filed in or against Global North countries. In addition, of the framework litigation cases in the 'implementation' category, such cases are mostly found in Global South countries 'where, in attempting to overcome implementation constraints, litigation is often used as a last resort to compel governments to enforce existing policies for climate change mitigation and adaptation'.[98]

## 3.2 A Different Mindset in Global South Climate Change Litigation

As global reviews of climate change litigation have identified, not all climate change cases fit the category of strategic climate change litigation. In addition, it

[93] Kim Bouwer and Joana Setzer, *Climate Litigation as Climate Activism: What Works?* (British Academy, COP26 Briefings Series, December 2020) 14.

[94] Gill and Ramachandran (n 10), quoting Setzer and Higham, *Global Snapshot 2020* (n 8).

[95] This is a relatively recent trend with just under half of the 80 cases identified in 2022 filed in 2021 alone.

[96] Catherine Higham, Joana Setzer, and Emily Bradeen, *Challenging Government Responses to Climate Change Through Framework Litigation* (Policy Report, Grantham Research Institute on Climate Change and the Environment and Centre for Climate Change Economics and Policy 2022) 2..

[97] ibid.

[98] ibid 7.

appears that climate change cases in the Global South will often fall in the 'non-strategic' category of cases, particularly if they are 'peripheral' climate cases, as discussed above.

Gita Gill and Gopichandran Ramachandran argue that Global North climate litigation is driven by 'different mindsets, [comprehensive] legal and policy frameworks, and climate change challenges',[99] compared with the Global South, which is 'experiencing litigation change through the lexicon of "climate language, human and constitutional rights, disaster management, resource conservation, and enforcement of existing environmental and planning legislation"',[100] as well as what they term 'transformative dialogue' around sustainability.[101] Joana Setzer and Lisa Benjamin also emphasize the different motivations of Global South climate change litigation, with litigants generally not focused on 'eliciting new regulatory targets or instruments from governments on reducing emissions'.[102] They argue that instead litigants 'use existing legislative tools and human rights discourses to highlight the vulnerability of their populations to climate and protect their valuable ecosystems'.[103]

It is not clear from this analysis whether Global South climate change litigation is any less strategic than 'strategic climate change litigation', so termed, that is emerging in the Global North. It may be, for instance, that a peripheral framing and use of existing environmental laws is adopted to increase chances of success in the courts rather than because the litigants are solely concerned with the private interests of the litigants bringing the case.[104] In other words, some care is needed in assuming Global South cases are 'non-strategic' just because they do not employ similar legal arguments to high-profile cases in the Global North.[105] What is 'strategic' for tackling a given issue is therefore highly dependent upon the broader environmental, political, and legal context. For instance, focusing on preventing deforestation in a biodiversity-rich country such as Brazil can be considered highly

[99] Gill and Ramachandran (n 10), quoting Asian Development Bank, *Climate Change, Coming Soon to a Court Near You* (n 10).

[100] Quoting Shibani Ghosh, 'Climate Litigation in India' in Francesco Sindico and Makane Moïse Mbengue (eds), *Comparative Climate Change Litigation: Beyond the Usual Suspects* (Springer 2021) 347; Setzer and Benjamin (n 3); Shibani Ghosh, 'Litigating Climate Claims in India' (2019) 114 American Journal of International Law Unbound 45; Peel and Lin (n 3).

[101] Gill and Ramachandran (n 10).

[102] Setzer and Benjamin (n 3) 85.

[103] ibid.

[104] Asian Development Bank (n 10) xxv, noting that 'litigation preferences reflect domestic legal frameworks, with litigants looking for appropriate hooks to support their claims'.

[105] For instance, in Interview No 9 conducted via Zoom on 12 May 2022, an interviewee discussed the way air pollution litigation in Jakarta has been framed. The interviewee reflected that they 'did not really raise the argument regarding the climate change impact in that lawsuit, but yes we raised the argument regarding the preservation of the right to the environment ... based on the that historically ... this is a multi-year problem that Jakarta has a polluted air, that exceeding the national ambient air quality standard ... So we start with what is really happening in the face of the public, and air pollution is more imminent. Not talking about the science, but it's what is perceived in the public.'

'strategic' given the importance of protecting the Amazonian rainforest for keeping within the Paris Agreement temperature goals.

The Asian Development Bank report further highlights this point, noting the strong reliance on constitutional rights claims for climate change litigation in Asian countries given that these avenues: (1) have been used successfully in environmental litigation in the past; (2) allow direct access to superior courts; (3) provide a legal basis for a claim where the existing legal or policy framework is incomplete; and (4) present fewer hurdles for establishing standing to bring a case.[106]

Regardless of whether they are properly classified as strategic or non-strategic, there are numerous lawsuits on the Global South climate change 'docket'—discussed further in the next chapter—that do not ostensibly pursue aims or interests extending beyond those of the parties to the case. Examples in this vein include cases brought by governments to prosecute environmental breaches of the law (a trend particularly in countries such as Indonesia, China, and Brazil) and cases involving parties seeking to enforce their rights when these have been affected by the action or inaction of the defendants.[107] The Asian Development Bank study also identifies 'litigation seeking climate change adaptation' as a growing trend in Asian environmental lawsuits, which often 'emerges as a silent issue'. In this instance, climate change is not identified by the court or the parties as an issue, but the outcomes of the cases have clear co-benefits for building resilience to climate change impacts.[108] By contrast, there are signs that, just as the conclusion of the Paris Agreement prompted a rise in strategic climate cases in the Global North, there is also a similar trend of more 'strategic' cases (as per the definitions discussed above) arising in the Global South.[109]

The number of strategic climate change cases has increased since 2015 in Global South jurisdictions, a total of 88 cases as of May 2022.[110] A particular trend here is the growing number of cases brought on behalf of youth or future generations, which emphasize arguments of intergenerational climate justice. This jurisprudence includes the 2016 case of *Ali v Federation of Pakistan*, the 2017 case of *Pandey*

[106] Asian Development Bank (n 10).

[107] India, eg *Vimal Bhai v Minister of Environment and Forest* (2011) National Green Tribunal 5/11; *Punamchand v Union of India* (2013) High Court of Judicature of Bombay Bench; *Sher Singh v State of Himachal Pradesh* (2014) National Green Tribunal (CWPIL No 15 of 2010) ('*Rohtang Pass*' case). The Philippines, eg *Global Legal Action on Climate Change v The Government of Philippines* (2010) Philippines Supreme Court; and *Victoria Segovia v Climate Change Commission* (2017) Philippines Supreme Court (GR No 211010). See similar cases brought with anti-regulatory impact in Pakistan, eg *DG Khan Cement Company v Government of Punjab* (2021) Pakistan Supreme Court. Similarly in Argentina, eg *Barrick y Exploraciones Mineras Argentinas SA v Argentina* (2019) Corte Suprema de Justicia de la Nación CSJ 140/2011.

[108] Asian Development Bank (n 10) xxvii–xxviii. Noting again that in our study such a case would only be included within the 'climate change docket' if other evidence eg from interviews is available as to the climate-related motivations for the litigation.

[109] Particular Global South countries that seem to be following this trend include Pakistan, Mexico, Argentina, and Guyana, where existing climate cases appear to be almost entirely 'strategic' cases.

[110] Setzer and Higham, *2022 Snapshot* (n 1).

*v India*, the 2018 *Future Generations* case in Columbia, the 2018 *Asociación Civil por la Justicia Ambiental v Province of Entre Ríos and others.* case in Argentina, the 2019 cases of *Alvarez and others v Peru* and *Maria Khan and others v Federation of Pakistan and others*, the 2020 case of *Youth v Government of Mexico*, and the 2021 lawsuit in Brazil, *Six Youths v Minister of Environment and others.*[111]

Often more difficult to classify is the growing body of climate change cases in the Global South that are brought by citizens and non-government organizations to ensure better implementation and proper enforcement of existing policies and laws that could aid mitigation and adaptation efforts, or to prevent carbon-intensive development such as new coal-fired power plants and coal mines. These cases are not seeking new or better climate laws,[112] as is often the goal of strategic climate change litigation in the Global North, but may still have far-reaching effects for unsustainable forms of development. They include a range of project-based challenges, for example to fossil fuel developments, roads, and airports, as well as lawsuits challenging widespread deforestation practices, especially in the Amazon region.[113]

## 4. CLIMATE CHANGE LITIGATION AND THE PARIS AGREEMENT

While 'strategic' climate change litigation pushing for more ambitious climate laws and emissions reduction targets has not been a significant feature of Global South climate litigation in the same way as in the Global North,[114] the range of cases emerging in countries of the Global South—particularly those brought against governments—raises questions about the contribution of this litigation to the global climate change response and international climate governance. The Paris Agreement does not contain a binding international enforcement mechanism, and processes for reviewing the adequacy of NDCs, such as the global stocktake mechanism, have a collective focus.[115] This has opened space for domestic action, including litigation, to play a role in ensuring implementation of countries' Paris Agreement commitments and accountability for their delivery.

[111] *Ali v Federation of Pakistan* (Petition filed at Lahore Supreme Court, April 2016); *Pandey v India* Original Application No 187 of 2017 (National Green Tribunal of India); *Future Generations v Ministry of the Environment* (Colom Sup Ct, 11001-22-03-000-2018-00319-01, 5 April 2018); *Asociación Civil por la Justicia Ambiental c/ Entre Ríos, Provincia de* (Filed 2 July 2018); *Álvarez contra Perú* (Filed 16 December 2019); *Maria Khan v Federation of Pakistan* (Lahore High Court) No 8960 of 2019; *Jóvenes v Gobierno de México* (Filed 2 September 2020); *Six Youths v Minister of Environment and Others* (2021) Ação Popular nº 5008035-37.2021.4.03.6100. Further discussion of these cases can be found in Chapters 3 and 4.

[112] There are, however, exceptions, eg the Nepalese case of *Shrestha v Office of the Prime Minister* (2017) Supreme Court of Nepal 074-WO-028, which concerned the push for better climate laws.

[113] See the case examples discussed in Chapter 3.

[114] Higham, Setzer, and Bradeen (n 96).

[115] Paris Agreement (n 6) art 14.

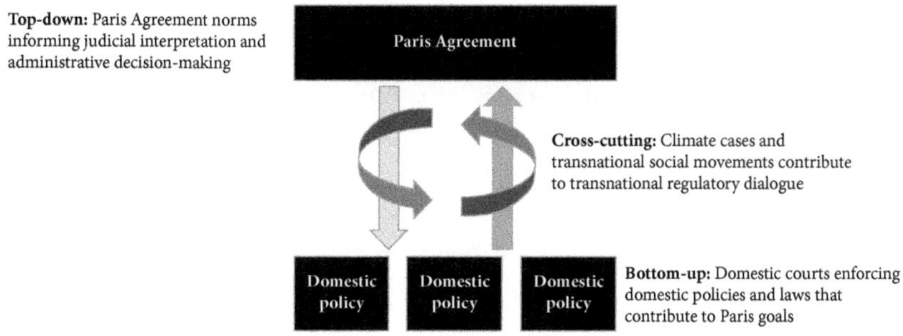


**Figure 2.2** Climate litigation intersections with the Paris Agreement

In this section of the chapter, we examine the developing intersection between climate change litigation and implementation of the Paris Agreement, and its implications for countries in the Global South. We discuss three key ways in which this intersection has been discussed and theorized in the literature, namely: (1) climate change litigation as a site for interpretation and application of Paris Agreement norms and principles ('top-down' pathway); (2) climate change litigation as a mechanism for enforcing domestic policies that contribute to the global climate change response ('bottom-up' pathway); and (3) climate change litigation and associated transnational social movements as a contributor to transnational climate governance ('cross-cutting' pathway). These three alternative pathways are represented in Figure 2.2.

As Chapter 3 discusses further, litigation appears to be emerging as a key mechanism in several Global South jurisdictions for implementing and enforcing climate and broader environmental and natural resource management regimes,[116] which may play a role in meeting their Paris Agreement commitments. It is also increasingly an aspect of transnational efforts to improve overall climate governance, including the accountability of nation states and major emitting corporations for their contributions to climate harms.

## 4.1 'Top-Down' Pathway: Climate Litigation and Interpretation of Paris Agreement Norms

A straightforward way in which domestic climate litigation may intersect with international climate governance is via international climate treaties such as the Paris Agreement informing the interpretation and application of domestic laws. This is described as a 'top-down' pathway in Figure 2.2. As Lennart Wegener notes,

[116] See further in Chapter 3.

the norms of the Paris Agreement are largely non-justiciable in domestic legal proceedings because they do not create specific, binding obligations of result for states parties. However, Wegener argues that Paris Agreement 'norms and mechanisms do find their way into domestic climate change jurisprudence as a means of interpretation and guidance, as well as significant indirect drivers of litigation itself'. In particular, he stresses that 'the ambitious goals of the present international climate change regime, states' NDCs and the mandatory updating can be highly valuable inputs into domestic litigation'.[117]

Similarly to Wegener, Chief Judge Brian Preston of the New South Wales Land and Environment Court in Australia looks at the influence of the Paris Agreement on national (and subnational) climate litigation, seeing a role for it to 'influence courts' application of its norms in domestic law and decision-making'.[118] According to His Honour, 'the Paris Agreement and its norms may provide a reference point and inform courts' interpretation, and application of, domestic laws and policies and fact-finding in litigation'.[119] Specific examples of norms of the Paris Agreement that have influenced domestic administrative and judicial decision-making are the temperature target (aiming to keep global average temperature rise well below 2°C while pursuing efforts to keep to a maximum 1.5°C rise), the provision in Article 4.1 of the Paris Agreement setting a collective obligation to reach net zero global emissions in the second half of the twenty-first century,[120] and what the Chief Judge refers to as the Paris Agreement's endorsement of a 'global norm to take action'.[121] In particular, His Honour sees the Paris Agreement as demonstrating global agreement on three key issues: (1) that increasing GHG emissions are causing climate change; (2) the IPCC's scientific consensus that climate change is caused by humans; and (3) that climate change will have dire consequences for the environment and human rights.[122]

Notably, these analyses of the 'top-down' interpretative role played by the Paris Agreement in climate change litigation are based on experiences from (mostly common law) countries situated in the Global North. Bearing in mind the different legal and socio-political contexts found in the Global South, these approaches to incorporating Paris Agreement norms in the interpretation of local laws and application of decision-making procedures may not have the same resonance. Indeed,

[117] Lennart Wegener, 'Can the Paris Agreement Help Climate Change Litigation and Vice Versa?' (2020) 9(1) Transnational Environmental Law 17, 23.

[118] Brian Preston, 'The Influence of the Paris Agreement on Climate Litigation: Legal Obligations and Norms (Part I)' (2021) 33(1) Journal of Environmental Law 1 (hereafter Preston, 'Legal Obligations and Norms').

[119] ibid 14.

[120] Paris Agreement (n 6) art 4.1.

[121] Preston, 'Legal Obligations and Norms' (n 118) 19.

[122] Brian Preston, 'The Influence of the Paris Agreement on Climate Litigation: Causation, Corporate Governance and Catalyst (Part II)' (2021) 33(2) Journal of Environmental Law 227, 229 (hereafter Preston, 'Causation, Corporate Governance and Catalyst (Part II))'.

in their analysis of constraints and limitations for climate litigation in the Global South, Joana Setzer and Lisa Benjamin suggest that rather than being a site for interpretation and application of international climate change norms, climate litigation is helping to 'bind together previously fragmented governance structures in the Global South to ensure that they focus on national climate goals, whether they be mitigation, adaptation, disaster management, or a combination of those aims'.[123] In the following chapter, we explore in more depth the extent to which arguments about, and interpretation of, the Paris Agreement provisions form part of the emerging 'Global South docket'.

## 4.2 'Bottom-Up Pathway': Climate Litigation and the Enforcement of Domestic Policies

Questions about the relationship between domestic climate change litigation and international climate governance are amplified in the context of the Paris Agreement due to its 'hybrid' structure.[124] Notably, the model of implementation adopted in the Paris Agreement is arguably one that depends for its effectiveness on domestic efforts of non-state and subnational actors to hold states parties accountable for their NDCs to the global climate change response[125]—NDCs which parties to the Paris Agreement pledge they 'inten[d] to achieve'.[126] Nationally Determined Contributions are the Paris Agreement's central mechanism for achieving the Agreement's objective to 'strengthen the global response to the threat of climate change, in the context of sustainable development and efforts to eradicate poverty' by:

> (i) holding the increase in global average temperature to well below 2°C above pre-industrial levels and to pursue efforts to limit the temperature increase to 1.5°C above pre-industrial levels; (ii) increasing the ability to adapt to adverse impacts of climate change; and (iii) making finance flows consistent with a pathway towards low greenhouse gas emissions ('GHG') and 'climate-resilient development'.[127]

[123] Setzer and Benjamin (n 3) 85.

[124] Daniel Bodansky, 'The Legal Character of the Paris Agreement' (2016) 25(2) Review of European, Comparative and International Environmental Law 142.

[125] Thomas Hale, 'Catalytic Cooperation' (2020) 20(4) Global Environmental Politics 73.

[126] Peel and Lin (n 5), 698; Paris Agreement (n 6).

[127] This objective is a 'collective goal, framed in aspirational language, and as such does not impose binding obligations in State Parties, despite suggestions to the contrary' per Geert Van Calster and Leonie Reins, 'Introduction' in Geert van Calster and Leonie Reins (eds), *The Paris Agreement on Climate Change: A Commentary* (Edward Elgar 2021) 1.

States parties have the obligation to formulate their own mitigation, adaptation, and finance commitments in NDCs, and to update these regularly towards the achievement of more ambitious outcomes, but these requirements are procedural 'obligations of conduct' rather than substantive 'obligations of result'.[128] The Paris Agreement seeks to promote ratcheting in the ambition of parties' NDCs over time to meet its goals through its processes of regular submission of new NDCs, assessment of their collective progress through the global stocktake, and its requirements for reporting and review through the Agreement's enhanced transparency mechanism.[129] Nonetheless, a gap potentially remains between what parties pledge through their NDCs and actions that are taken to address the global climate change challenge. This gap is emphasized by assessments of current NDCs which stress that they do not 'add up' to the collective ambition needed to meet the Paris Agreement's temperature goal.[130]

The Paris Agreement itself offers no enforcement mechanism to bridge this gap.[131] The treaty contains a facilitative compliance mechanism that is intended to be 'non-adversarial and non-punitive'.[132] The Paris Agreement, like the UN Framework Convention on Climate Change, preserves the capacity for states to bring claims against one another in international dispute settlement procedures but these procedures generally only operate if both states consent to submit the claim to an international court or tribunal.[133] Reflecting these procedural hurdles, there have been no international climate change cases determined as yet and the prospects for interstate dispute settlement seem slim,[134] although there remains considerable academic commentary on the prospect of international climate cases in different international dispute settlement forums.[135] While prospects for international climate change jurisprudence have improved with climate-related advisory opinion requests to the International Court of Justice, the International Tribunal for the Law of the Sea, and the Inter-American Court

[128] Lavanya Rajamani and Jacob Werksman, 'The Legal Character and Operational Relevance of the Paris Agreement's Temperature Goal' (2018) 376(2119) The Royal Society.

[129] Harro van Asselt and Kati Kulovesi, 'Enhanced Transparency Framework for Action and Support' in van Calster and Reins (eds) (n 127) 304.

[130] WGIII Report (n 5).

[131] Alexander Zahar, 'A Bottom-Up Compliance Mechanism for the Paris Agreement' (2017) 1(1) Chinese Journal of Environmental Law 1.

[132] Paris Agreement (n 6) art 15(2).

[133] Bodansky, 'The Legal Character of the Paris Agreement' (n 124).

[134] Peel and Lin (n 3), 695.

[135] Daniel Bodansky, 'The Role of the International Court of Justice in Addressing Climate Change: Some Preliminary Reflections' (2017) 49(1) Arizona State Law Journal 689; Rozemarijn J Roland Holst, 'Taking the Current When it Serves: Prospects and Challenges for an ITLOS Advisory Opinion on Oceans and Climate Change' [2022] Special Issue Article, Review of European, Comparative and International Environmental Law 1. International economic law forums, such as the WTO dispute settlement system or investor-state arbitrations, may also emerge as international climate dispute settlement sites of the future, with already a number of cases concerning renewable energy installations: see Sandrine Maljean-Dubois and others, 'International Investment Law and Climate Change: Introduction to the Special Issue' (2022) 23(5–6) Special Issue, The Journal of World Investment & Trade 737.

of Human Rights,[136] in such instances courts would only provide general interpretations of relevant international law principles and obligations rather than determining the accountability of any state(s) for alleged non-compliance with their international climate and other commitments.

Domestic climate change cases may help fill this accountability gap under the Paris Agreement by linking its international obligations of conduct with national obligations of result.[137] This is particularly evident in some of the Brazilian cases, including *Laboratorio de Observatorio do Clima v Minister of Environment and Brazil*[138] and *Six Youths v Minister of Environment and others*,[139] both of which are discussed in more detail in Chapter 4. In *Six Youths*, it was alleged that the Federal Government had violated the Paris Agreement and the Brazilian Constitution by updating Brazil's NDC to be significantly less ambitious than its previous one.[140]

It is the case of *PSB and others v Brazil*, however, which is particularly noteworthy.[141] This case was brought by four political parties including the Brazilian Socialist Party (PSB), who were in opposition to then President, Jair Bolsonaro's, Federal Union. The plaintiffs sought an injunction that would compel the government to reactivate the Brazil's latent National Climate Change Fund (Fundo Clima), which was established in 2009 to finance mitigation and adaptation projects. In their arguments, the plaintiffs called upon the precautionary principle as well as Brazil's commitments under the Paris Agreement. They claimed that a failure to implement the Climate Fund was one of many examples of how Brazil is disrespecting its 'legal obligation to implement measures that are compatible with the achievement of these goals'.[142] In addition to this international focus, the

[136] Bodansky (n 135); Holst (n 135). International Court of Justice 'Request for an Advisory Opinion of the International Court of Justice on the Obligations of States in Respect of Climate Change' (Press Release No 2023/21, 25 April 2023); Commission of Small Island States on Climate Change and International Law, 'Request for Advisory Opinion' (12 December 2022); Pablo Saavedra Alessandri, 'Request for an Advisory Opinion on Climate Emergency and Human Rights to the Inter-American Court of Human Rights form the Republic of Colombia and the Republic of Chile' (Petition, Inter American Court of Human Rights, 9 January 2023).

[137] Anna-Julia Saiger, 'Domestic Courts and the Paris Agreement's Climate Goals: The Need for a Comparative Approach' (2020) 9(1) Transnational Environmental Law 37, 38. See also Lennart Wegener, 'Can the Paris Agreement Help Climate Change Litigation and Vice Versa?' (2020) 9(1) Transnational Environmental Law 17, 34: a shortcoming in the Paris Architecture is 'loose legal links between the collective overall goal, and individual efforts'. National courts can therefore provide an incentive to enhance the ambition of NDCs.

[138] *Laboratorio de Observatorio do Clima v Minister of Environment and Brazil* (2021) Ação Civil Pública N° 1027282-96.2021.4.01.3200.

[139] *Six Youths v Minister of Environment* (2021) Ação Popular n° 5008035-37.2021.4.03.6100.

[140] Maria Antonia Tigre, 'A New Climate Litigation Claim in Brazil Raises the Pressure for Increased Climate Action and Protection of the Amazon Rainforest' *Sabin Center for Climate Change Law* (Columbia Law School 2021) https://blogs.law.columbia.edu/climatechange/2021/12/01/a-new-climate-litigation-claim-in-brazil-raises-the-pressure-for-increased-climate-action-and-protection-of-the-amazon-rainforest/ accessed 1 December 2023.

[141] *Partido Socialista Brasileiro (PSB) v União Federal* (n 49).

[142] Joana Setzer, *First Climate Case Reaches Brazil's Supreme Court, News and Commentary* (Commentary, Grantham Research Institute on Climate Change and the Environment 30 September 2020) <www.lse.ac.uk/granthaminstitute/news/first-climate-case-reaches-brazils-supreme-court/> accessed 1 December 2023.

plaintiffs relied upon Article 225 of the Brazilian Constitution, which provides that all government bodies must 'protect the environment and fight pollution in any of its forms', and 'to preserve forests, fauna and flora', as well as recognizing the right to a healthy environment.[143]

In its ruling, the Court adopted a highly innovative, rights-based approach where it held that (1) the Paris Agreement is a human rights treaty, and for that reason enjoys '*supralegal* status'; and (2) based on Article 225, the separation of powers doctrine, and its international obligations, that the Brazilian Government is constitutionally obliged to combat climate change and prevent regression of environmental protection.[144]

Cases such as these demonstrate a second linkage between climate change litigation and international climate governance focused on the capacity of the former to serve as a 'bottom-up' mechanism for enforcing domestic climate policies that contribute to fulfilling Paris NDCs (see Figure 2.2). As Lord Carnwath explains:

> National legislatures bear the primary responsibility to give legal effect to the commitments undertaken by states under the Paris agreement. However, the courts will also have an important role in holding their governments to account, and, so far as possible within the constraints of their individual legal systems, in ensuring that those commitments are given practical and enforceable effect.[145]

It is important to appreciate, nonetheless, that the role of climate change litigation in this instance is not generally as a tool for directly enforcing the Paris Agreement in domestic law. Rather, as Kim Bouwer emphasizes, it involves 'testing domestic policies enacted to give effect to international commitments'.[146] This makes an appreciation of 'domestic contexts' and applicable national procedural and substantive laws necessary for understanding the role climate change litigation plays in international climate governance and 'requires a comparative approach'.[147] Anna-Julia Saiger argues that a comparative approach enables exploration of

> the contexts that allow domestic courts to link international obligations of conduct with national obligations of result. In this way international legal scholarship can make sense of domestic judicial decisions in the climate change regime and, at the same time, acknowledge the importance of domestic contexts.[148]

[143] ibid.

[144] Unofficial Decision Translation, *Partido Socialista Brasileiro (PSB) v União Federal* (n 49).

[145] Lord Carnwath, 'Climate Change Adjudication After Paris' (2016) 28(1) Journal of Environmental Law 5, 9.

[146] Bouwer (n 41) 492.

[147] Saiger (n 137) 37.

[148] ibid 52.

She suggests that elements to consider in this assessment include: constitutional and administrative provisions at the national level, procedural law and access to justice requirements, legal culture and institutional backgrounds, the analyst's scholarly standpoint, and courts' motivations and self-conceptions.[149] Given the diversity of domestic legal contexts arising in the Global South, this call for attention to context is particularly important in examining how climate cases are interacting with international climate obligations.

### 4.3 'Cross-cutting Pathway': Climate Litigation and Transnational Climate Governance

As outlined in Chapter 1, on one view, the climate change litigation that has emerged globally—given its domestic focus—may have little to contribute to international climate governance and implementation of the Paris Agreement. It is certainly the case that climate cases are strongly shaped by the applicable laws and legal systems in any particular country and are targeted to the resolution of local conflicts.[150] However, the impacts of climate change litigation can also be seen to be 'greater than the sum of their parts', particularly where similar arguments are pursued across different countries or where there are connections to a wider transnational social movement, focused, for example, on arguing for science-based action or ensuring climate justice.

In some cases, advocates and practitioners may see their climate litigation work as contributing to the global effort to address climate change, and cases may be accompanied by campaigns that seek to appeal to a wider national, or international, audience. Scholars of climate change litigation (although mostly based in the Global North) equally have tended to discuss climate change case law in terms that stress its broader impacts and cross-cutting role in multilevel climate governance.[151] In this context, climate change litigation becomes 'part of the transnational regulatory dialogue over climate change' that helps shape climate governance through the case law's broader effects on governmental regulatory decision-making, corporate behaviour, and public understanding of the problem of climate change.[152] This is represented as a 'cross-cutting' approach in Figure 2.2 where climate change cases and associated social movements help to develop common narratives or pathways that generate shared ways of understanding or dealing with the climate change problem.

[149] ibid 53.
[150] Peel and Lin (n 3) 695.
[151] ibid 696.
[152] ibid 699, quoting Hari Osofsky, 'The Continuing Importance of Climate Change Litigation' (2010) 1(1) *Climate Law* 3.

Chief Judge Preston emphasizes this view in his discussion of 'the ripple effect' of climate change litigation.[153] His Honour argues that 'certain landmark judicial decisions have had an influential transnational impact on global climate litigation' through inspiring other climate change litigation in the same or other countries.[154] Cinnamon Carlarne puts a similar view in her chapter in 'Debating Climate Law'.[155] She argues that 'climate litigation is playing an incontrovertibly important role in shaping the rule of law around climate change'.[156]

As we noted in Chapter 1, the Intergovernmental Panel on Climate Change (IPCC), in its 2022 Working Group III report, includes an assessment of the governance contribution of climate change litigation that also highlights its broader role in shaping such 'transnational regulatory dialogue'. The IPCC describes climate change litigation as 'an attempt to control, order or influence the behaviour of others in relation to climate governance' which 'has been used by a wide variety of litigants (governments, private actors, civil society and individuals) at multiple scales (local, regional, national and international)'.[157] In respect of climate change cases brought against governments (which still comprise the majority of cases), the IPCC report distinguishes between 'claims challenging the overall effort of a State or its organs to mitigate or adapt to climate change', such as the *Urgenda* and *PSB* cases, and 'claims regarding authorisation of third-party activity', such as challenges to coal power projects which have been a feature of the climate change jurisprudence in Global South countries such as South Africa.[158] The IPCC concludes that the 'outcomes of climate litigation can affect the stringency and ambitiousness of climate governance' and that there is 'now increasing academic agreement that climate litigation has become a powerful force in climate governance'.[159]

A similar sentiment informs a new thread of scholarship in the climate change litigation literature which collects together under the label of studies on 'just transition litigation', although it is not clear that this term is being used consistently by different authors. For instance, Annalisa Savaresi and Joana Setzer see 'just transition litigation' as a subset of rights-based claims involving 'lawsuits in which human rights law and remedies are used to challenge measures and projects designed to deliver climate change adaptation and/or mitigation'.[160] Framed in this

[153] Preston, 'Causation, Corporate Governance and Catalyst (Part II)' (n 122).
[154] ibid 247.
[155] Cinnamon Piñon Carlarne, 'The Essential Role of Climate Litigation and the Courts in Averting Climate Crisis' in Benoit Mayer and Alexander Zahar (eds), *Debating Climate Law* (CUP 2021) 111.
[156] ibid 113. Carlarne also argues strongly in favour of courts playing this role, see especially at 127. For an alternative view, stressing institutional limits of courts to fill any governance gap created by inaction of governments, see the following chapter by Guy Dwyer, 'Climate Litigation: A Red Herring Among Climate Mitigation Tools' at 128.
[157] WGIII Report (n 5) s 13.4.2, 1375.
[158] See further the case examples discussed in Chapters 3 and 4.
[159] WGIII Report (n 5) s 13.4.2, 1377.
[160] Annalisa Savaresi and Joana Setzer, 'A First Global Mapping of Rights-Based Climate Litigation Reveals a Need to Explore Just Transition Cases in More Depth' (Grantham Research Institute on

way, 'just transition litigation' may have an anti- rather than a pro-climate impact by objecting to the way certain climate-favourable projects (eg measures to reduce forest emissions or to construct renewable energy sources such as hydroelectric dams or windfarms) are carried out and their impacts on human rights or local communities. Savaresi and Setzer advocate for greater attention to these kinds of cases (for which they cite examples both in the Global North and in the Global South) in order 'to appreciate the tensions associated with a transition towards zero carbon societies, and the ways in which they may be resolved through the adoption of a rights-based approach to climate change decision-making'.[161]

By contrast, a 2023 working paper by Maria Antonia Tigre and co-authors on 'Just Transition Litigation in Latin America' is focused on 20 cases all filed in Global South countries since 2016.[162] Tigre et al define 'just transition litigation' in this context as claims that question:

> the distribution of the benefits and burdens of transition policies and activities towards net zero-emission and climate resilient societies amongst local communities and affected stakeholders. Litigators typically invoke human rights but may rely on other national and international laws and principles. Just transition cases are not necessarily pro- or anti-climate action. Instead, they demand that climate action is undertaken in a just and inclusive manner with attention to those in vulnerable situations.[163]

The cases they consider, discussed further in Chapter 3, reflect the complex justice questions raised by decarbonization efforts taken in pursuit of the goals of the Paris Agreement. While not 'core' climate cases given their focus on the socio-environmental impacts of mining and energy projects, 'they peripherally relate to climate change with a broad decarbonization strategy for an energy transition at its core'.[164] Tigre et al consequently find that the majority of these cases reference the Paris Agreement and/or countries' NDCs as part of an argument by petitioners in favour of a 'just energy transition'. However, they also note that the cases are used to address both fossil fuel activities and projects, such as the impacts of

Climate Change and the Environment 29 March 2022) <www.lse.ac.uk/granthaminstitute/news/a-first-global-mapping-of-rights-based-climate-litigation-reveals-a-need-to-explore-just-transition-cases-in-more-depth/> accessed 10 February 2023.

161 ibid.

162 The cases come from Brazil, Mexico, Chile, and Colombia. See Maria Antonia Tigre and others, 'Just Transition Litigation in Latin America: An Initial Categorization of Climate Litigation Cases Amid the Energy Transition' *Sabin Center for Climate Change Law* (Columbia Law School 2023) <https://scholarship.law.columbia.edu/cgi/viewcontent.cgi?article=1198&context=sabin_climate_change> accessed 7 February 2023. On the Sabin Center's Global Climate Litigation Chart, 10 'Just transition' cases are currently listed from the same four countries.

163 ibid.

164 ibid.

decommissioning coal-fired power plants on workers, local communities, and local environments, as well as new renewable energy-related activities, projects, or policies that support clean energy transition:

> In this sense, litigation may address burdens and benefits of the construction of new wind parks or the licensing or operation of rare minerals mines (e.g., lithium), both of which are necessary for scaling up the production of new renewable energy and relevant new renewable energy technologies (e.g., lithium).[165]

This analysis suggests that 'cross-cutting pathways' of intersection reflected in climate litigation concerning Paris Agreement requirements and the domestic implementation of decarbonization policies are likely to operate in a two-way direction. On the one hand, Paris Agreement obligations are incentivizing the formulation of local implementation policies for mitigation and climate resilient development, but at the same time the litigation is also providing a site to challenge how those policies are given effect, in turn shaping the global understanding of the requirements of a 'just transition'.[166] Global South countries may provide a particularly rich context for examining these cross-cutting intersections given the rapid energy transition occurring in a number of these countries, coupled with significant natural and mineral resources that will be a major contributor to global efforts to mount an effective response to the climate change problem.

## 5. CONCLUSION

This chapter has discussed common understandings of climate change litigation and its potential to be used as a (strategic) tool for influencing climate change governance, including by creating links between the norms and objectives of the Paris Agreement and domestic policy and legal actions. In every case, the dominant conceptions that have emerged are shaped by experiences of climate change litigation in the Global North, which is where, to date, most climate change litigation has taken place and around which most scholarly analysis of climate change litigation is framed. There has been a strong trend in the climate change litigation scholarship overall towards classification and typologies. These approaches tend to rule certain cases in or out of the realm of 'climate change litigation' or classify claims according to their 'strategic' or 'non-strategic' purpose. As contributors to this literature, we have also participated in this exercise although we recognize its limitations.

[165] ibid.
[166] Paris Agreement (n 6) preamble.

However, what is lost in this type of analysis is often a capacity to perceive different pathways and purposes for climate cases such as those emerging across many countries of the Global South. Widening the lens of analysis to include cases with climate change issues at the periphery as well as those with climate at their core, and to recognize the different 'mindset' informing climate litigation beyond countries of the Global North, allows us to capture a more comprehensive picture of global climate litigation developments. This also permits a better understanding of the value of litigation as a pathway for global decarbonization and adaptation to the adverse impacts of climate change in pursuit of the goals of the Paris Agreement.

In line with this commitment to a more holistic and context-sensitive approach, the following chapters consider more closely the experience of climate change litigation emerging in the Global South. Chapter 3 first presents a survey of the expanding 'Global South docket', drawing out key trends while also acknowledging that the individual trajectory of climate litigation in each Global South jurisdiction will differ according to its particular social context, political settings, and legal culture.[167] Chapter 4 then takes a deeper dive into the jurisprudence of the 'frontrunner jurisdictions' of India, Brazil, and South Africa where most of this litigation is emerging in an effort to discover what factors and conditions underlie the growth of climate cases in these countries.

[167] Peel and Lin (n 3) 718.

# 3
# The Global South Climate Litigation Docket

## 1. INTRODUCTION

Apart from a few high-profile cases such as the *Carbon Majors Petition* in the Philippines, the *Leghari* case in Pakistan, and *Future Generations v Ministry of the Environment and Others* in Colombia, climate litigation in the Global South has generally received less attention than climate litigation in the Global North.[1] To most observers, it would appear that there are few developments worth noting about climate litigation in these parts of the world. However, as argued in Chapter 2, a key approach underpinning this book is an adjustment of the 'lens' through which we view climate litigation, which is likely to reveal notable case law developments in the Global South. Further, an understanding of the characteristics of climate litigation in the Global South leads to a richer, more nuanced, and complex understanding of transnational climate litigation and its role in addressing the climate emergency.

In order to develop an understanding of climate litigation in the Global South and its contribution to climate governance and implementing the Paris Agreement, this chapter provides a comprehensive examination of cases in the 'Global South docket'. In total, we analyse 128 cases, of which over half are filed (pending decision) or decided in Latin America and the Caribbean (66 cases), 46 in Asia and the Pacific, and 16 in Africa.[2] As discussed in Chapter 1, part of the difficulty of analysing Global South climate case law lies in the fact that not all cases are included in existing global databases. In selecting cases for analysis, our first step was to review the cases that are presently included in the two leading climate change case-law databases, the Sabin Center for Climate Change Law's Non-US Climate Change Litigation database,[3] and the London School of Economics' Grantham

[1] Commission on Human Rights of the Philippines, *National Inquiry on Climate Change* (Report, December 2022); *Ashgar Leghari v Federation of Pakistan* (2015) Lahore High Court Green Bench WP No 25501/201; Tribunal Superior del Distrito Judicial de Bogotá, Sala Civil Especializada en Restitución de Tierras, abril 5, 2018, Luis Armando Tolosa Villabona, 11001-22-03-000-2018-00319-01.

[2] Our analysis of Africa encompassed the Middle East, but no cases were found in that region.

[3] Sabin Center for Climate Change Law, 'Global Climate Change Litigation' Database (*Sabin Center for Climate Change Law*) <http://climatecasechart.com/non-us-climate-change-litigation/> accessed 28 June 2023.

*Litigating Climate Change in the Global South*. Jolene Lin and Jacqueline Peel, Oxford University Press.
 DOI: 10.1093/oso/9780191926525.003.0003

Research Institute Climate Change Laws of the World database.[4] We included cases where climate change is 'at the periphery' of the argument or decision but excluded cases where climate change (and related) issues are mentioned incidentally but are not otherwise considered in a meaningful way in the judgments.[5] This involved a systematic review of cases recorded in each Global South country category in both databases, including analysis of all reported judgments filed within these country categories to decide if they fell within the 'core' or 'peripheral' classifications, or ought to be excluded.

This case-law data was supplemented by the use of other resources to identify cases that are not recorded in the Sabin Center and LSE databases for a variety of reasons, including gaps in the reporting of cases from non-English speaking jurisdictions. These additional resources included newspaper reports, social media, conferences, interviews, and our own networks. In total, as of May 2023, 128 cases were identified as making up the Global South docket.

The selected cases were coded for the following variables: (1) the identity of the plaintiffs (eg NGOs, local governments, individuals) and the defendants (eg governments, private corporations); (2) whether climate change was an issue at the core or at the periphery of the lawsuit; (3) the nature of the legal avenue(s) pursued in the claim (eg whether environmental impact assessment was argued as the basis of the litigation, whether the public trust doctrine was invoked, or whether rights-based arguments were made by the parties); (4) whether the court or the parties mentioned or relied on the Paris Agreement and national climate change laws in their decision and submissions respectively.

Our compilation of the cases which are decided, are pending decision, or are emerging (ie filed but court or tribunal hearings are yet to take place) in Global South jurisdictions can be found online.[6] The compilation of 'transnational' cases, also accessible online, sets out information about cases that have been decided, are pending decision, or are emerging in Global North jurisdictions that involve Global South plaintiffs and/or projects.[7] These cases are not included in the Global South docket as the fact that they are being filed and decided in courts in the Global North means they are not true Global South cases. It is also notable that the framing of arguments and approach taken in these cases share more in common with Global North climate jurisprudence than cases in the Global South docket. However, we have included a summary of these cases in our online compilation and will also discuss some of these cases in this chapter because they represent

[4] Grantham Research Institute on Climate Change and the Environment, 'Climate Change Laws of the World' Database (*Grantham Research Institute on Climate Change and the Environment*) <https://climate-laws.org/> accessed 28 June 2023.

[5] See Chapter 2, Section 2 for detailed discussion of the definition of climate litigation used in this book.

[6] Insert URL for online database of cases [Note: this will be inserted later after discussion with OUP]

[7] ibid.

two related trends of Global South inhabitants seeking climate justice in the Global North courts, and efforts to hold corporations and governments of the Global North legally accountable for their activities in the Global South.

Following this Introduction, Section 2 presents our analysis of the data whereby we seek to answer questions such as the relative proportion of 'core' versus 'peripheral' cases in the Global South docket and to highlight the key trends that emerge from our analysis. These trends include the more pronounced use of rights-based arguments in cases in the Global South compared to the Global North and the relatively low number of 'framework litigation' cases in the Global South.[8] We note that a large majority of the cases are filed by non-governmental organizations (NGOs) against governments.

In Section 3, we consider how climate change law in the Global South, analysed in Section 2, is characterized by some features that distinguish it from Global North climate cases. We consider three characteristics to be fairly distinctive of the Global South climate docket and argue that paying more attention to these features allows us to reframe our understanding of climate litigation and how it is unfolding in different but connected ways globally. First, a distinctive feature of the Global South docket is the prevalence of climate constitutional cases. In such cases, claimants argue that their constitutional rights (which often include the right to a clean and healthy environment) are violated by governmental failure to reduce greenhouse gas emissions (GHGs) adequately and to adapt to the impacts of climate change. Secondly, the majority of Global South climate cases demonstrate a preference for putting policies and laws already 'on the books' into action and ensuring proper enforcement, rather than pushing for new or better climate laws. In bringing suits that focus on seeking judicial intervention to compel the proper enforcement of existing laws, Global South plaintiffs often rely on fairly well-established legal arguments and domestic jurisprudence. Thirdly, many climate cases in the Global South adopt what might be termed a 'stealthy' strategy, which dilutes the political potency of climate issues by packaging them together with less controversial claims.

Rather than suggesting that these characteristics are firmly associated with Global South climate litigation, we think of these characteristics as a series of spectrums, with Global South cases presently concentrated at one end and Global North cases at the other. Further, within the Global South docket itself, we see some regional variations along these spectrums too. These spectrums are by no means determinative but serve to provide some intellectual scaffolding to help us identify the distinctive nature of the contribution made by Global South cases to transnational climate governance.

[8] See discussion on framework litigation in Section 2.3 of this chapter.

Section 4 examines the 'transnational' cases, ie cases that have been decided, are pending decision, or are emerging in Global North jurisdictions that involve Global South plaintiffs and/or projects. Such cases can be the basis for creating truly transnational coalitions of actors seeking accountability from Global North governments and corporations for their failure to reduce GHGs and decarbonize their economic activities. The inclusion of Global South plaintiffs supports submissions to the court that the failure to address climate change has far-reaching adverse consequences for vulnerable populations that make a negligible contribution towards causing climate change. This is a variation of the dominant narrative in many Global North cases, which has tended to frame climate change impacts in terms of statistics and causative probabilities rather than accounts of human suffering.[9] It is also notable that major fossil fuel projects and investments in the Global South that are predominantly financed by developed countries and their financial institutions are being challenged in courts of Global North jurisdictions, providing an interesting pathway of holding developed countries and financial institutions accountable for aligning their investments with the Paris Agreement. Section 5 draws some conclusions about the arguments raised in this chapter.

## 2. TRENDS IN THE GLOBAL SOUTH DOCKET

In this section, we provide an overview of the trends and legal developments that can be observed of the Global South's experience of climate litigation. It is noteworthy that since the publication of our earlier work on Global South climate litigation in 2019, there has been a significant increase in the number of cases filed (and decided). Cases have also been filed in some jurisdictions for the first time. In this regard, Thailand, Guyana, and Papua New Guinea are notable examples.[10] The

[9] Laura Gyte, Violeta Barrera, and Lucy Singer, 'The Story of Our Lives—Narrative Change Strategies in Climate Litigation' in César Rodríguez-Garavito (ed), *Litigating the Climate Emergency* (CUP 2022).

[10] In *Saonu and Morobe Provincial Government v Minister for Environment and Conservation and Climate Change* OS (JR) of 2021, Court of Justice at Waigani, the plaintiff provincial authorities of Saonu and Morobe sought to invalidate a mining permit issued by the national environmental authority, arguing that it would violate Papua New Guinea's pledges to act on climate change. In 2022, after several years of struggling to stop coal mining on lands forcibly taken from them, 50 representatives of the Kaboedin Village in the Omkoi district in Thailand filed a case in the Chiang Mai Administrative Court to challenge the validity of the environmental impact assessment that had been conducted by the project developers (Chiang Mai Administrative Court, Black Case No Sor 1/2565, 4 April 2022). Between 2020 and 2022, three climate lawsuits were filed in Guyana. In *Thomas v Environmental Protection Agency* 2020-HC-DEM-CIV-FDA-460, Guyanese scientist Dr Troy Thomas alleged that the 23-year permits that the agency had issued to Esso (a subsidiary of ExxonMobil) for oil exploration violated reg 19 of subsidiary legislation passed pursuant to the Environmental Protection Act that limited the time span of such permits to five years. The plaintiff, the defendant agency, and Esso (who was added as an interested party) reached a settlement after three preliminary hearings. The permit was shortened to five years as required by reg 19, requiring Esso to obtain a new environmental permit if it wanted to continue its petroleum operations after 1 June 2022. As Melinda Janki writes, '[t]he case marks the first time a litigant has relied on climate change and the Paris Agreement in support of a claim for a judicial remedy in Guyanese court': Melinda Janki, 'Guyana: A First Step Towards Climate Litigation' (*Climate*

body of scholarship on climate litigation continues to grow rapidly and, as pointed out in Chapter 2, many experts have sought to shed light on the phenomenon by identifying various categories and sub-categories of climate lawsuits. In the analysis below, we draw upon this body of scholarship to complement our observations and discussion of the Global South docket of climate cases.

## 2.1 Core versus Periphery?

In this book, it is understood that a case with climate change at its 'core' is one where arguments raising issues of climate law or climate science are central to the pleadings or judgment. By contrast, 'peripheral' climate cases are ones where climate issues are subsidiary to other arguments, or one of many arguments or issues raised in a dispute. To put it another way, 'core' climate cases engage directly with climate change issues whereas 'peripheral' cases do so indirectly. However, where climate change is mentioned only incidentally or in passing, without any direct or indirect engagement in the plaintiffs' arguments, case preparatory materials or the decision of the court, these cases make no meaningful contribution to advancing a specific climate jurisprudence and hence have been excluded.

Just slightly over half (54.7%) of the cases in the Global South docket feature climate change at the periphery rather than at the core of the case (see Table 3.1 below). It is notable that there is significant variation across the three regions. Of the cases in Asia 73% feature climate change 'at the periphery' while this is case for only 39% of cases in Latin America. In Africa, 62.5% of the cases feature climate change 'at the periphery'.

Several core cases in the Global South docket have been filed in the past three years. Some examples will be discussed here. At the time of writing, there have been six cases in Mexico challenging the policies and laws governing the energy sector. For example, *Greenpeace Mexico v Ministry of Energy and Others (on the Energy Sector Program)* was filed in 2020 by Greenpeace which sought to challenge the government's energy policy proposals on the basis that the proposals over-emphasized the use of coal-powered sources and did not adequately support the development of renewable energy.[11] The proposals therefore were, according to the plaintiff's submissions, unconstitutional and not aligned with Mexico's obligations pursuant to the Paris Agreement. A second example is *Greenpeace v Mexico (Budget reduction for combating climate change).*[12] In 2019, Greenpeace sought

*Law*, 29 December 2020) <https://blogs.law.columbia.edu/climatechange/2020/12/29/guyana-a-first-step-towards-climate-litigation/> accessed on 13 July 2023.

[11] Amparo indirecto 104/2020, Greenpeace Mexico, Judicial de la Federación.

[12] Amparo en Revisión 526/2020, Greenpeace México, Suprema Corte de Justicia de la Nación.

**Table 3.1** Overview of the Global South docket—core versus periphery cases

| Region | Total number of cases | Number of cases with climate change 'at the periphery' | Number of cases with climate change 'at the core' | Percentage of 'peripheral' cases |
|---|---|---|---|---|
| Asia | 46 | 34 | 12 | 73.91% |
| Africa | 16 | 10 | 6 | 62.5% |
| Latin America | 66 | 26 | 40 | 39.39% |
| Pacific Islands | 0 | 0 | 0 | 0% |
| TOTAL | **128** | **70** | **58** | **54.69%** |

to challenge the Mexican House of Representatives' decision to reduce the size of the national budget dedicated to climate change mitigation and adaptation from previous years in the 2019 Annual Expenditure Bill. The decision by the Supreme Court in this case had little meaningful impact in reality. The decision was handed down after the 2019 Annual Expenditure Bill was no longer in force.

*Iten ELC Petition No. 007 of 2022—Legal Advice Centre T/K Kituo cha Sheria & Another v Attorney General and 7 Others* was filed in the Iten Environment and Land Court in Kenya in 2022.[13] The plaintiffs are victims of climate change-related flooding which has caused displacement, and loss of life and property. Arguing that the harm they have suffered gave rise to a violation of several of their constitutional rights, they seek judicial enforcement of the duties of public officials pursuant to Kenya's Climate Change Act. The case is pending decision at the time of writing. As will be discussed in Chapter 4, there have been several core cases filed in Brazil that seek to hold the government accountable for their non-compliance with national and international climate change law.[14]

As indicated above, roughly half (54.7%) of the Global South docket comprises of 'periphery' cases. It is easy to think that 'core' cases are more important and bear more significance for climate governance because of their central focus on climate change policies and laws. However, that would be too glib a conclusion to draw. There are several 'periphery' cases which have contributed to climate jurisprudence and led to significant direct and indirect effects. It is therefore worth noting that the terms 'core' and 'periphery' do not connote evaluations or judgements of impact or importance but are simply terminology we have adopted to distinguish

[13] ELC Petition No 7 of 2022.
[14] See Chapter 4, Section 2.3.

categories of cases. Further, as we will discuss in Section 3, there are a number of reasons why lawyers and litigants decide to advance case theories that lend themselves to a 'peripheral' framing of climate change. For one, it may well be a strategic choice in some Global South jurisdictions where there is moderate judicial receptivity to climate change-related arguments in court.

An example of a 'periphery' case in the Global South docket that has contributed to the development of climate jurisprudence is *DG Khan Cement Company v Government of Punjab*, a decision by the Supreme Court of Pakistan.[15] A cement company challenged a new regulation by the provincial government of Punjab barring the construction of new cement plants or the expansion of existing cement plants in environmentally fragile zones called 'Negative Areas'. The plaintiff company challenged the regulation on the grounds that it violated its constitutional right to freedom of trade, business, and profession under Article 18 of the Constitution, and that the government acted hastily by issuing the regulation without full consideration of scientific impacts.

The Supreme Court of Pakistan upheld the government's regulation. It took into account that the government had been guided by expert advice that new or expanded cement plants could cause further depletion of groundwater and other harmful environmental impacts.[16] In the decision, the court emphasized the need for consideration of climate change in government decisions and the impact of climate change on water resources. The court wrote: '[o]nly by devising and implementing appropriate adaptation measures will it be possible to ensure water, food and energy security for the country . . . The Notification, in the current facts of the case, is a climate resilient measure and in step with the National Climate Change Policy and the Constitution.'[17]

The quality of the legal reasoning, within the context of climate change being one of the grounds for decision (a 'periphery' case), is noteworthy. First, addressing climate change necessarily requires an integrated approach.[18] By engaging with the issue of integrating climate change considerations and relevant policies in government decisions across various policy areas, the court helped to clarify what effective regulation and policymaking in the face of climate change needs to be.

[15] *DG Khan Cement Company Ltd v Government of Punjab through its Chief Secretary, Lahore, etc*, CP 1290-L/2019.
[16] ibid [14].
[17] *DG Khan* (n 15) [18]. For discussion, see Parvez Hassan, 'The D.G. Khan Cement Case: Pakistan Judiciary Again Leads the Way' (IUCN, 23 April 2021) <www.iucn.org/news/world-commission-environmental-law/202104/dg-khan-cement-case-pakistan-judiciary-again-leads-way> accessed 11 July 2023.
[18] Eloise Scotford and Stephen Minas, 'Probing the Hidden Depths of Climate Law: Analysing National Climate Change Legislation' (2018) 28(1) Review of European, Comparative and International Environmental Law 67, 70–71<https://onlinelibrary.wiley.com/doi/10.1111/reel.12259> accessed 10 July 2023.

Secondly, climate change can require decisions to be made in conditions of scientific uncertainty. 'While scientific understandings of climate change are relatively settled, assessments of future climate change are not straightforward', for example.[19] Justice Syed Mansoor Ali Shah applied the precautionary principle (citing Principle 10 of the 1992 Rio Declaration) as well as the emerging environmental principle of *in dubio pro natura*. This principle is included as Principle Five of the International Union for Conservation of Nature (IUCN) World Declaration on the Environmental Rule of Law and states:

> [I]n cases of doubt, all matters before courts, administrative agencies, and other decision-makers shall be resolved in a way most likely to favour the protection and conservation of the environment, with preference to be given to alternatives that are least harmful to the environment. Actions shall not be undertaken when their potential adverse impacts on the environment are disproportionate or excessive in relation to the benefits derived therefrom.[20]

Justice Shah held that '[on] the facts of [this] case, the Provincial Government was obliged to take a precautionary approach and act in accordance with the principle of *in dubio pro natura* until, inter alia, a detailed hydrological study assessing the potential of groundwater resources for industrial purposes of the project area is carried out'.[21]

## 2.2 Who Are the Plaintiffs?

Non-governmental organizations play a leading role in climate litigation in both the Global North and the Global South. The plaintiff in the vast majority of Global South cases is an environmental NGO. At the time of writing, NGOs filed (either on their own or as part of a coalition) 39.84% of the cases in the Global South docket. Of Global South cases, 6.25% have been filed by global environmental NGOs in their own name or as lead plaintiff. Examples include *Greenpeace Argentina and Others v Argentina*,[22] *Greenpeace v Instituto Nacional de Ecología y Cambio*

[19] Elizabeth Fisher, Eloise Scotford, and Emily Barritt, 'The Legally Disruptive Nature of Climate Change' (2017) 80(2) Modern Law Review 173, 179 <https://onlinelibrary.wiley.com/doi/abs/10.1111/1468-2230.12251> accessed 1 December 2023.

[20] 'Environmental Rule of Law' (IUCN) <www.iucn.org/our-union/commissions/world-commission-environmental-law/our-work/history/foundational-documents-4> accessed 3 July 2023.

[21] *DG Khan* (n 15) [16].

[22] Judicial de la Nación [C.Fed] [Federal Court of Appeals of Mar del Plata], 5/12/2022, 'Fundación Greenpeace Argentina y ostros c/ Estado Nacional de la República Argentina y ostros s/ Amparo Ambiental' (2022).

*Climático and Others*,[23] and *Greenpeace Indonesia and Others v Bali Provincial Governor*.[24] Domestic environmental NGOs filed 36.72% of Global South cases.

The Ministério Público, literally translated as the Public Ministry but more usefully translated for English language readers as the public prosecution office, is a unique independent body that has been described by the constitutional lawyer and former President of Brazil, Michel Temer, as the 'fourth branch' of government in Brazil.[25] The prosecutors play a unique role in bringing climate change-related cases in Brazil and represent a notable pathway for climate litigation that is rarely seen in the Global North. In the Brazilian docket, the public prosecution service is named the plaintiff in 17.39% of the cases.[26] It has been noted that, given the impressive track record of Brazilian prosecutors in pursuing environmental litigation, one would expect the prosecutors to be responsible for pursuing more climate lawsuits.[27] However, the heavy workloads that Brazilian prosecutors have been facing has meant that they work closely with other actors to facilitate climate litigation but are not actively filing many cases.[28] Nonetheless, as discussed in Chapter 5, the prosecutors in Brazil are creating a unique pathway for climate litigation that is not seen in the Global North and which the Chinese prosecution service is seeking to replicate (albeit with significant modifications to adapt the approach to the Chinese legal system and political economy). A fuller discussion of this and the various actors who are filing lawsuits, supporting litigation as part of broader climate advocacy efforts, as well as funding litigation can be found in Chapter 5.

## 2.3 Who Are the Defendants?

Of the Global South docket, 79.69% comprise of legal challenges against governments. These cases include: (1) administrative law challenges to government decision-making under planning and environmental legislation; (2) proceedings that challenge the alleged lack of ambition in the state's national climate change policies and/or the state's Paris Agreement Nationally Determined Contribution (NDC); and (3) legal proceedings that seek judicial intervention to compel the

[23] Amparo en Revisión 159/2022, Decimo Primer Tribunal Colegiado en Materia Administrativa del Primer Circuito, *Greenpeace Mexico v Instituto Naciónal de Ecología y Cambio Climático*.

[24] Decision No 2/G/LH/2018/PTUN.DPS (Administrative Court of Denpasar).

[25] 'MPF' (Ministerio Publico Federal) <www.mpf.mp.br/> accessed 11 July 2023; Temer, 'Temer: Public Ministry is Fourth Power of the Republic' (Association of the Public Ministry of the Federal District and Territories) <https://amp-df.jusbrasil.com.br/noticias/2112040/temer-ministerio-publico-e-quarto-poder-da-republica> accessed 11 July 2023.

[26] For further discussion, see Chapter 4 (Section 2.3 about Brazil) and Chapter 5 (Section 2.2 on prosecutors and regulatory agencies).

[27] Discussion during by-invitation only workshop, *Climate Change Litigation in the Global South* (National University of Singapore, Faculty of Law, 22–23 May 2023).

[28] ibid.

government to properly implement climate change policies and programmes that have been formulated but not acted upon.

The first category of administrative law challenges is a fairly established pathway of climate change litigation, particularly in Australia, and is often also referred to as the 'first generation' of climate change cases.[29] The core of first generation climate litigation comprises of lawsuits that have focused on individual, GHG emissions-intensive projects and have been brought under environmental statutes.[30] *Earthlife Africa Johannesburg v The Minister for Environmental Affairs* is a 'first generation' climate lawsuit that is well known for being the first 'core' climate lawsuit in South Africa. It has since inspired similar lawsuits across the jurisdiction, as will be discussed in Chapter 4 of this book.

Environmental permitting has also been the subject of administrative law challenges on the basis that, inter alia, the government in question had awarded the permit without consideration of climate change impacts and/or the impact of the permitted activity on climate change. In March 2023, Indigenous people in Indonesia's easternmost province sued the government for issuing palm oil plantation permits to a company that will allow it to plant oil palm on 39,190 hectares of land, which, the plaintiffs argue, are their ancestral lands.[31] The plaintiffs argue that the award of the permit was procedurally flawed because there was no proper consultation of the Indigenous owners of the land.[32] Further, the award of the permit went against Indonesia's legal commitments under the Paris Agreement to reduce its GHG emissions as the clearing of the land to create plantations will increase deforestation.[33]

'Government framework litigation' has been described as being a subset of climate litigation whereby, '[r]ather than challenging specific projects or individual policies that result in increased greenhouse gas emissions or that fail to account for climate-relevant impacts, these cases challenge broad, whole-of-government policies and decisions'.[34] Framework litigation can be distinguished into two

[29] Climate change litigation in Australia has been described as having a 'long and proud history', with the first case, *Greenpeace Australia Ltd v Redbank Power Pty Ltd* [1994] 86 LGERA 143, decided in 1994. This case has served as a model for much of the ensuing 'first generation' of climate litigation in Australia; Peel, Osofsky and Anita Foerster, 'Shaping the "Next Generation" of Climate Change Litigation in Australia' (2017) 41 Melbourne University Law Review 793; Jacqueline Peel, Hari M Ofosky, and Anita Foerster, '"Next Generation" Climate Change Litigation in Australia' in Jolene Lin and Douglas A Kysar (eds), *Climate Change Litigation in the Asia Pacific* (CUP 2020).

[30] Peel, Osofsky, and Foerster, '"Next Generation" Climate Change Litigation' (n 29) 181.

[31] Greenpeace Indonesia, 'West Papuan Indigenous Defender Files Lawsuit Over Palm Oil Company Forestland Grab' (*Greenpeace*, 13 March 2023) <www.greenpeace.org/southeastasia/press/58630/west-papuan-indigenous-defender-files-lawsuit-over-palm-oil-company-forestland-grab/?ref=the-wave> accessed 10 July 2023.

[32] ibid.

[33] ibid.

[34] Catherine Higham, Joana Setzer, and Emily Bradeen, 'Challenging Government Responses to Climate Change through Framework Litigation' (*Grantham Research Institute on Climate Change and the Environment*, September 2022) 2 <www.lse.ac.uk/granthaminstitute/publication/challenging-government-responses-to-climate-change-through-framework-litigation/> accessed 19 January 2024.

broad types: 'ambition cases' and 'implementation cases'.[35] The second category of cases mentioned above largely comprises of 'ambition cases' which challenge the absence or inadequacy of a government's policy response to climate change, for example, by adopting climate targets that are not aligned with the objective of achieving net zero emissions by the middle of this century. A target is, simply speaking, the end result that one wishes to attain. In the climate change law and policy context, a target is often expressed in terms of a cut in GHG emissions by a certain percentage (or to net zero) by a specific year (eg 2030 or 2050).[36] The Paris Agreement states a temperature target in Article 2(1) and it states in Article 4(1) the related target of '[achieving] a balance between anthropogenic emissions by sources and removal by sinks of greenhouse gases in the second half of this century'.

Most government framework litigation has been filed in Global North jurisdictions. Examples include *Urgenda v the Netherlands*,[37] *The People's Climate Case*,[38] *Friends of the Irish Environment v Ireland*,[39] *ENVironnement JEUnesse v Procureur General du Canada*,[40] and *Do-Hyun Kim and Others v South*

[35] ibid 6.

[36] For discussion about climate targets, see Chris Hilson, 'Hitting the Target? Analysing the Use of Targets in Climate Law' (2020) 32(2) Journal of Environmental Law 195.

[37] [2015] HAZA C/09/00456689. For discussions, see Margaretha Wewerinke-Singh and Ashleigh McCoach, 'The State of Netherlands v Urgenda Foundation: Distilling Best Practice and Lessons Learnt for Future Rights-Based Climate Litigation' (2021) 30(2) Review of European, Comparative & International Environmental Law 275; Josephine van Zeben, 'Establishing a Governmental Duty of Care for Climate Change Mitigation: Will Urgenda Turn the Tide?' (2015) 4(2) Transnational Environmental Law 339.

[38] In Case T–330/18 *Carvalho v European Parliament and Council*, which is also popularly referred to as '*The People's Climate Case*', 10 families from Portugal, France, Germany, Kenya, Fiji, Italy, Romania, and the Swedish Sami Youth Association brought an action in the European Union (EU) General Court seeking to compel the EU to adopt more stringent GHG emissions reduction targets. The General Court dismissed the case on the basis that the plaintiffs did not have sufficient standing to bring the case. On appeal, the European Court of Justice upheld the General Court's decision. See recording of Gerd Winter's discussion of this case: Gerd Winter, 'Armando Carvalho et al Versus European Union, called People's Climate Case: Litigation between Law, Science and Policy' (Duke Nicholas Institute for Energy, Environment & Sustainability, 2020) <https://nicholasinstitute.duke.edu/media/gerd-winter-armando-carvalho-et-al-versus-european-union-called-peoples-climate-case> accessed 17 July 2023; Gökşen Şahin, 'People's Climate Case—Families and Youth Take the EU to Court Over its Failure to Address the Climate Crisis' in Claude Henry, Johan Rockström, and Nicholas Stern (eds), *Standing Up for a Sustainable World* (Edward Elgar, 2020).

[39] 2017 No 793 JR. For discussion, see Orla Kelleher, 'The Supreme Court of Ireland's decision in Friends of the Irish Environment v Government of Ireland ('Climate Case Ireland')' *(EJIL:Talk!*, 9 September 2020) <www.ejiltalk.org/the-supreme-court-of-irelands-decision-in-friends-of-the-irish-environment-v-government-of-ireland-climate-case-ireland/> accessed 11 July 2023; Sanita Van Wyk, 'Climate Change Litigation: Determinations of the Supreme Court of Ireland on the National Mitigation Plan' (University of Oregon, 3 May 2022) <https://scholarsbank.uoregon.edu/xmlui/handle/1794/27112> accessed 12 July 2023.

[40] 2021 QCCA 1871. Also see Environnement Jeunesse, 'Youth vs Canada' (Environnement Jeunesse) <https://enjeu.qc.ca/justice-eng/> accessed 10 July 2023. For discussion and analysis, see Camille Cameron and Riley Weyman, 'Recent Youth-Led and Rights-Based Climate Change Litigation in Canada: Reconciling Justiciability, Charter Claims and Procedural Choices' (2021) 34(1) Journal of Environmental Law 195.

*Korea*.[41] In the Global South, *Maria Khan and Others v Pakistan* was filed in 2019 by a group of women against the Pakistani government for violating their right to equal protection under the law by systematically refusing to address climate change and its disproportionate impacts on women.[42] Their petition's central pleading is for the government to reduce its GHG emissions by reducing the use of fossil fuels and promoting renewable energy sources. The case is pending decision at the time of writing.

The third category of cases mentioned above (ie legal proceedings that seek judicial intervention to compel the government to properly implement climate change policies and programmes that have been formulated but not acted upon) is mostly found in Global South jurisdictions where litigation is often a last resort to compel governments to enforce existing climate mitigation and adaptation policies.[43] The best-known example is the case of *Ashgar Leghari v Pakistan*.[44] Leghari, an agriculturalist and lawyer, sued the Pakistani government because it did not implement its National Climate Change Policy 2012 nor its Framework for Implementation of Climate Change Policy 2013. Leghari argued that the government should pursue climate change mitigation and adaptation efforts and its failure to meet its climate change adaptation targets resulted in immediate impacts on Pakistan's water, food, and energy security. The court ruled in favour of Leghari and concluded that there had been no real progress with implementation.[45] To oversee the execution of policy, the court constituted the Climate Change Commission which issued its final report in 2018 stating that around 60% of the priority tasks within the implementation framework had been completed.[46] The commission was dissolved and the court set up a Standing Committee on Climate Change to serve as a continuing channel of communication between the court and the executive 'to facilitate future work on climate change'.[47] The *Leghari* case has been described as having mainstreamed a 'climate change narrative' in Pakistan's environmental law jurisprudence and introduced the issue of climate change to the judiciary such that

[41] Constitutional Court, *Case No 2020 Heonma 389*, a pending case filed on 13 March 2020. See Sejong Youn, 'Youth Climate Litigation in Korea: Constitutional Complaint Case No. 2020 Heonma 389' (*Asia-Pacific Centre for Environmental Law*, 28 October 2020) <https://law.nus.edu.sg/apcel/media/book-launch-climate-change-litigation-in-the-asia-pacific/> accessed 17 July 2023.

[42] *Maria Khan v Pakistan*, Writ Petition No 8960 of 2019, High Court of Lahore.

[43] Higham, Setzer, and Bradeen (n 34) 7.

[44] *Leghari v Federation of Pakistan*, PLD 2018 Lahore 364. See Asghar Leghari's presentation at the International Seminar on Climate Litigation as Governance Tool (23–24 October 2020, online): Asghar Leghari, 'From Environmental Justice to Climate Justice: Pakistan's Experience' (*Duke Nicholas Institute for Energy, Environment & Sustainability*) <https://nicholasinstitute.duke.edu/events/international-seminar-climate-litigation-governance-tool/videos-panel-3> accessed 17 July 2023.

[45] *Leghari v Federation of Pakistan* 2018 PLD (n 44).

[46] Parvez Hassan, 'Judicial Commissions and Climate Justice in Pakistan' (Asia Pacific Judicial Colloquium on Climate Change, Lahore, 26–27 February 2018) 9–13 <www.ajne.org/sites/default/files/event/7219/session-materials/apac-judicial-colloquium-cc-session-1-4-judicial-commissions-climate-justice-pakistan-parvez-hassan.pdf> accessed 12 July 2023.

[47] ibid 13.

Pakistan's judges are now familiar with the issue of climate change and the concept of climate justice.[48] This has led to judges coming up with innovative remedies for plaintiffs.[49] For instance, judges have ordered the government to impose penalties for acts of deforestation and to increase public awareness of the importance of tree-planting.[50]

There are a small handful of cases that have been filed against companies or corporate actors in the Global South. The Indonesian government (through the Ministry of Environment and Forestry (MoEF)) has filed cases against timber companies and oil palm concession holders for illegal logging and causing peatland fires, with the plaintiff government seeking compensation for the environmental damage resulting from the defendants' illegal conduct as well as costs to restore the natural resource to its prior state.[51] The climate change dimension lies in the MoEF's claim requiring the defendant to pay the costs of restoring the damaged carbon sink such as a forest or a mangrove area (restoration costs) or compensation for the release of GHGs due to the destruction of the carbon sink (compensation costs).

## 2.4 Rights-Based Litigation

Rights-based climate litigation refers to cases that rely on constitutional rights or human rights claims, including alleged violations of the right to life or environmental rights.[52] Overall, the Global South has seen a substantial number of rights-based cases (see Figure 3.1 below), making up 62.5% of the Global South docket. Rights-based arguments were advanced in 12 out of 16 cases that have been filed or adjudicated in Africa (two in Kenya, one in Nigeria, seven in South Africa, and two in Uganda). In South Africa, the courts are required (by section 39(2) of the Constitution) to promote the purpose and spirit of the Bill of Rights when

[48] Justice Ayesha Malik, Supreme Court of Pakistan, speaking on a webinar 'Climate Change Litigation and Human Rights' (Sabin Center-UNEP Annual Conference on Global Climate Litigation, 19 April 2022, 10.30am to 12 noon (EST)). For further discussion of the case's impacts, see Chapter 6.
[49] ibid.
[50] ibid.
[51] *Minister of Environment v PT Selatnasik Indokwarsa and PT Simpang Pesak Indokwarsa*, Decision No 105/PDT/G/2009/PN.JKT.UT (District Court of North Jakarta, 2010); *Minister of Environment v PT Kalista Alam*, Decision No 12/PDT.G/2012/PN.MBO (District Court of Meulaboh, 2013); *Minister of Environment v PT Merbau Pelalawan Lestari*, Decision No 157/PDT.G/2013/PN.PBR (District Court of Pekanbaru, 2014); *Minister of Environment v PT Bumi Mekar Hijau*, Decision No 51/PDT/2016/PT.PLG (High Court of Palembang, 2016); *Minister of Environment v PT Jatim Jaya Perkasa*, Decision No 108/PDT.G/2015/PN.JKT.UTR (District Court of North Jakarta, 2016); *Minister of Environment v PT Waringin Agro Jaya*, Decision No 456/PDT.G-LH/2016/PN.JKT.SEL (District Court of South Jakarta, 2017).
[52] David Boyd has traced the diffusion of environmental rights in national constitutions: David R Boyd, *The Environmental Rights Revolution: A Global Study of Constitutions, Human Rights, and the Environment* (UBC Press 2011).

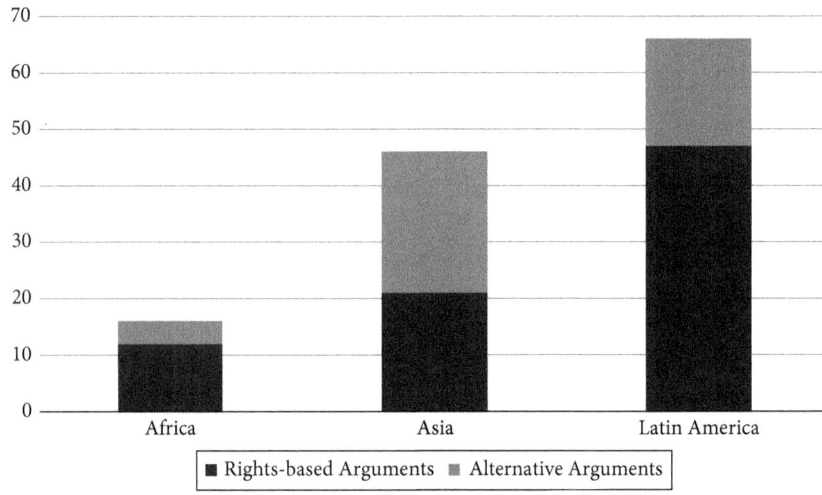


**Figure 3.1** Advancing rights-based arguments

interpreting statutory legislation. The Bill of Rights includes the right to a healthy environment, which was taken into consideration when determining whether the EIA ought to include climate change assessment in the *Earthlife* case.[53]

In total, 21 out of 46 cases in Asia advanced rights-based arguments. In the Philippines, the Constitution provides that '[t]he State shall protect and advance the right of the people to a balanced and healthful ecology in accordance with the rhythm and harmony of nature'.[54] In addition, the Philippine Supreme Court in 2009 introduced rules and mechanisms to facilitate protection of the constitutionally enshrined rights to life and a healthy environment.[55] In two out of the three Filipino cases, the plaintiffs relied on these mechanisms, known as the 'writ of kalikasan', to compel the government to act on climate change. As for the third case, the *National Inquiry on the Impact of Climate Change on the Human Rights of the Filipino People* (otherwise popularly known as the *Carbon Majors Inquiry*) this concerned a petition to the Human Rights Commission of the Philippines to

[53] South African Constitution, s 24; *Earthlife Africa Johannesburg v Minister of Environmental Affairs* ZAGPPHC 58; [2017] 2 All SA 519 (GP).

[54] constitution, art II, s 16.

[55] This was the result of an initiative led by Chief Justice Puno of the Philippines Supreme Court to promote public interest litigation in the quest for environmental justice. A copy of these Rules can be found at: Rules of Procedure for Environmental Cases, AM No 09-6-8-SC, 2010 (Republic of the Philippines) <www.lawphil.net/courts/supreme/am/am_09-6-8-sc_2010.html> accessed 13 July 2023. For discussion, see Gloria Estenzo Ramos, 'Innovative Procedural Rules on Environmental Cases in the Philippines: Ushering in a Golden Era for Environmental Rights Protection' (2011) 1 IUCN AEL Journal of Environmental Law <www.iucnael.org/en/e-journal/previous-issues/157-issue-20111.html> accessed 12 July 2023.

investigate the impacts of climate change on human rights.[56] It is a well-known global precedent as it marks the first time that a human rights body has articulated the responsibility of transnational corporations for human rights violations resulting from climate change impacts.[57]

Pakistan's Constitution does not recognize environmental rights such as the right to a healthy environment and the right to clean water. However, over the years, Pakistan's superior courts (the High Courts and the Supreme Court) have expanded the scope of the Fundamental Rights and judicially interpreted Article 9 (right to life) and Article 14 (dignity of man being inviolable) to include a right to a clean and healthy environment. Article 9 states that no person shall be deprived of life or liberty except in accordance with law. While this textually encapsulates a negative obligation of the state, Article 9 has been interpreted by the courts to place positive obligations on the state.[58] This has led to a significant rise in public interest litigation petitions relying on the constitution as a basis to raise environment-related claims.[59]

In Latin America, there has been a marked increase in the number of rights-based cases filed and decided by the courts in the past few years. Forty-seven out of 66 cases in Latin America included rights-based arguments. An example is *Herrera Carrion and Others v Ministry of the Environment and Others (Caso Mecheros)*.[60] In February 2020, nine girls filed for a constitutional injunction against the government of Ecuador, arguing that the state's permitting of gas flaring as a common practice caused serious adverse impacts on the environment and human health to the extent that it rose to the level of violation of their constitutional rights to health, water, and a healthy environment as well as the rights of Nature. The case was dismissed at first instance but, on appeal, the Provincial Court of Justice of Sucumbio ruled in favour of the youth plaintiffs. The court held that the practice of gas flaring

[56] Commission on Human Rights of the Philippines, *National Inquiry on Climate Change* (Report, December 2022) <https://chr.gov.ph/wp-content/uploads/2022/12/CHRP_National-Inquiry-on-Climate-Change-Report.pdf> accessed 10 July 2023.

[57] Greenpeace, 'The Climate Change & Human Rights Inquiry Archive' (*Greenpeace Philippines*, 2022) <www.greenpeace.org/philippines/the-climate-change-human-rights-inquiry-archive/> accessed 13 July 2023. See also Annalisa Savaresi and Jacques Hartmann, 'Using Human Rights Law to Address the Impacts of Climate Change: Early Reflections on the Carbon Majors Inquiry' in Lin and Kysar (eds) (n 29); Annalisa Savaresi and Juan Auz, 'Climate Change Litigation and Human Rights: Pushing the Boundaries' (2019) 9(3) Climate Law 244.

[58] This process began in India and has been replicated in Pakistan. See Waqqas Ahmad Mir, 'From Shehla Zia to Asghar Leghari: Pronouncing Unwritten Rights Is More Complex than a Celebratory Tale' in Lin and Kysar (eds) (n 29) 263.

[59] This is not surprising. A constitution represents the highest or supreme law of a jurisdiction. The embedding of environmental rights expands the legal stock for those seeking to mobilize the law. Sociolegal scholars' reference to 'legal stock' in relation to environmental and climate change law itself is to 'the body of constitutional provisions, laws, statutes, and for some, soft law measures enacted to govern social-ecological systems'; Lisa Vanhala, 'Environmental Legal Mobilization' (2022) 18(2) Annual Review of Law and Social Science 1, 5.

[60] Judgment No 21201-2020-00170, 29 June 2021, Multi-Competent Chamber of the Provincial Court of Justice of Sucumbíos.

disregarded Ecuador's Paris Agreement NDC and was also unconstitutional. The court ordered the phased elimination of gas flaring by December 2030.

There have been recent efforts to adapt environmental constitutionalism—the practice of incorporating environmental rights and obligations in national constitutions—to meet the challenges of tackling climate change.[61] As discussed above, environmental rights in national constitutions have been used in climate litigation that seeks to establish credible and long-lasting climate action. In a novel case, *IEA v Brazil*, the plaintiff NGO sought the recognition of a fundamental right to a stable climate which it argued is implicitly entrenched in the Brazilian Constitution.[62] Specifically, IEA argued that the magnitude of the climate emergency and the adverse impacts of climate change on so many fundamental rights warranted the recognition of a new and autonomous right that guarantees a stable climate system that is capable of sustaining human life.[63] At the time of writing, the case is pending final decision by the courts.

In their mapping exercise, Ghaleigh, Setzer, and Welikala identified 11 jurisdictions that have dedicated climate constitutional provisions or 'climate clauses'.[64] These include: Algeria, Bolivia, Côte d'Ivoire, Cuba, Dominican Republic, Ecuador, Thailand, Tunisia, Venezuela, Vietnam, and Zambia.[65] The climate clauses do not recognize the right to a stable climate per se, or the temperature targets of the Paris Agreement or reports by the Intergovernmental Panel on Climate Change (IPCC).[66] Instead, the climate constitutional provisions that have been identified are fairly open textured, broadly stating a commitment to tackling climate change. Tunisia's Constitution appears most advanced in its articulation of the connection between environmental rights and climate action. Article 45 of the Tunisian Constitution proclaims '[t]he state guarantees the right to a healthy and balanced environment and the right to participate in the protection of the climate'. As Ghaleigh, Setzer, and Welikala put it: '[t]hese climate clauses are first movers in the articulation of climate considerations in the language of constitutional

[61] Navraj Singh Ghaleigh, Joana Setzer, and Asanga Welikala. 'The Complexities of Comparative Climate Constitutionalism' (2022) 34(3) Journal of Environmental Law 517, 522.

[62] Ação Civil Pública 5048951-39.2020.4.04.7000, Judicial Section of Paraná, 11th Federal Court of Curitiba, 5 April 2022.

[63] It is noteworthy that the legal team in *IEA v Brazil* sought to adapt arguments used in *Juliana v the United States* to the Brazilian context; Joana Setzer and Délton Winter de Carvalho, 'Climate Litigation to Protect the Brazilian Amazon: Establishing a Constitutional to a Stable Climate' (2021) 30(2) Review of European, Comparative & International Environmental Law 197, 203.

[64] Ghaleigh, Setzer, and Welikala (n 61) 523.

[65] ibid.

[66] ibid. For discussion about the right to a stable climate, see eg James R May and Erin Daly, 'Can the U.S. Constitution Encompass a Right to a Stable Climate? (Yes, It Can.)' (2021) 39 (1) UCLA Journal of Environmental Law and Policy 45; Carolyn Kormann, 'The Right to a Stable Climate Is the Constitutional Question of the Twenty-First Century' *The New Yorker* (15 June 2019) <www.newyorker.com/news/daily-comment/the-right-to-a-stable-climate-is-the-constitutional-question-of-the-twenty-first-century> accessed 13 July 2023; Alessandra Lehmen, 'Advancing Strategic Climate Litigation in Brazil' (2021) 22(8) German Law Journal 1471.

provisions.'[67] We can expect them to feature in future climate litigation as citizens seek to hold their governments accountable for the fulfilment of these constitutional protections.

## 2.5 Environmental Impact Assessment

Environmental Impact Assessment (EIA) is 'the most commonly known, used, and globally widespread, environmental planning and management [tool]'.[68] Most countries around the world have laws mandating the use of EIA to assess the environmental implications of proposed projects and provide decision-makers with adequate information to decide whether a specific activity or project should proceed, and, if so, how it should be undertaken.[69] Some jurisdictions (eg the European Union) have issued guidance on the integration of climate change and biodiversity in EIA.[70] Jurisdictions which have not integrated climate change consideration into their EIA laws have been suitable venues for filing climate lawsuits that refer to or use an EIA requirement as a ground of action. Such cases typically involve judicial or administrative review of a regulatory decision to approve the EIA conducted for a proposed project (eg a coal mine) on the basis that, inter alia, the decision is flawed because the EIA process did not adequately take climate change into consideration. Environmental impact assessment cases continue to contribute significantly to the development of jurisprudence in climate litigation 'hotspots' like the United States and Australia.[71]

Thirty-nine out of the 128 cases in the Global South docket referred to an EIA requirement or used this requirement as one of the grounds of action. Due to the success of the *Earthlife* case in South Africa challenging the validity of the EIA approval for the proposed development of a coal-fired power plant on the basis that the EIA was flawed because it did not include a climate change assessment, several cases in South Africa have used an EIA requirement as a ground of

[67] Ghaleigh, Joana Setzer, and Asanga Welikala (n 61) 525.

[68] UN Environment Law Division and UN Environment World Conservation Monitoring Centre, *Assessing Environmental Impacts—A Global Review of Legislation* (Report, 2018) <www.unep.org/resources/assessment/assessing-environmental-impacts-global-review-legislation> accessed 13 July 2023.

[69] It is noteworthy that Principle 17 of the Rio Declaration on Environment and Development (1992) states: '[e]nvironmental impact assessment, as a national instrument, shall be undertaken for proposed activities that are likely to have a significant adverse impact on the environment and are subject to a decision of a competent national authority.'

[70] Climate-ADAPT, 'Guidance on Integrating Climate Change and Biodiversity into Environmental Impact Assessment' <https://climate-adapt.eea.europa.eu/en/metadata/guidances/guidance-on-integrating-climate-change-and-biodiversity-into-environmental-impact-assessment> accessed 7 July 2023.

[71] Jacqueline Peel and Hari M Ofosky, *Change Litigation—Regulatory Pathways to Cleaner Energy* (CUP 2015); Jacqueline Peel, Hari M Ofosky, and Anita Foerster, ''Next Generation' Climate Change Litigation in Australia' in Lin and Kysar (eds) (n 29).

action.[72] Environmental impact assessment climate cases make up 10 of the 16 cases in Africa.

The record on the use of EIA as a ground of action is more mixed in the climate case law emerging from Asia. Fourteen out of the 46 cases in Asia relied on EIA as a ground of action. Notable examples include *Wijana, Sufarlan, Astawa and Greenpeace Indonesia v Governor of Bali*,[73] and the *Omkoi coal mining* case.

In *Wijana, Sufarlan, Astawa and Greenpeace Indonesia v Governor of Bali*, a group of citizens and Greenpeace Indonesia brought an action before the Denpasar Administrative Court challenging the issuance of an environmental permit to construct a plant in Celukan Bawang, a subdistrict in northern Bali. The citizen plaintiffs were mainly concerned about the potential impacts of the power plant on their livelihoods while Greenpeace was concerned with the environmental and climate change impacts of the project.[74] The petitioners argued that the permit should be quashed because the EIA for the proposed project did not take GHG emissions into account. The petitioners made extensive references to similar cases in other jurisdictions wherein the court ruled that an EIA that did not take GHG emissions into account would be deemed flawed.[75] Nonetheless, the court was silent on the link between GHG emissions and EIA and also did not consider the issue of whether a project proponent is under a legal obligation to take into account likely GHG emissions when applying for the relevant permit to operate. Instead, the Administrative Court dismissed the lawsuit principally on the ground that there was a lack of evidence regarding the alleged pollution and environmental damage.[76] As the plaintiffs had yet to sustain any injury, they lacked standing in the court's view. This dismissal was upheld on appeal by both the Court of Appeal and the Supreme Court.

The plaintiffs in *Residents of Omkoi v Expert Committee on EIA Consideration and the Office of Natural Resources and Environmental Policy and Planning* are members of an ethic Karen community that lives in the remote mountains of the Omkoi district of Chiang Mai Province, Thailand.[77] They sought to challenge the opening of an open-pit lignite coal mine on land that they sold to the 99 Thuwanon

[72] See discussion of these cases in Chapter 4.

[73] *Greenpeace Indonesia v Governor of Bali* (n 24).

[74] ibid.

[75] See Amicus Brief submitted by a coalition of Indonesian and international environmental groups: Indonesian Centre for Environmental Law and others, Amici Curiae Brief' (*Climate Case Charts*, 2018) <http://climatecasechart.com/non-us-case/greenpeace-indonesia-and-others-v-bali-provincial-governor/> accessed 12 July 2023.

[76] *Greenpeace Indonesia v Governor of Bali* (n 24) 148–49. For discussion about climate change litigation in Indonesia more broadly, see Andri G Wibisana and Conrado M Cornelius, 'Climate Change Litigation in Indonesia' in Lin and Kysar (eds) (n 29). Also see discussion in the section on NGOs in Chapter 5 of this book.

[77] For more details, see Earthrights International, 'Silencing Indigenous Communities: The Case of a Lignite Coal Mine In Omkoi District, Thailand' (*Earthrights International*, March 2022) <https://earthrights.org/wp-content/uploads/Omkoi-Case-Study-2022.pdf> accessed 1 March 2023; Climate Case Chart, 'Residents of Omkoi v Expert Committee on EIA Consideration and the Office of Natural Resources and Environmental Policy and Planning' (*Climate Case Chart*) <http://climatecasechart.

Company and their agents under duress and threats. Specifically, the plaintiffs challenged the validity of the EIA report that was submitted to and received approval from the relevant authority on the basis that the report was deeply flawed. The report, the plaintiffs alleged, contained factual errors and misinformation including forged signatures and falsified data. In addition, there was no public participation and the plaintiffs' community was not consulted as required by law. The Chiang Mai Administrative Court issued an injunction to suspend the activities pending final judgement. In their order, issued on 23 September 2022, the court reaffirmed the right of meaningful participation in the EIA process.[78]

Fifteen out of the 66 cases in Latin America (22.73% of the cases) relied on EIA as a ground of action. Environmental impact assessment is a foundational pillar of Chile's environmental legal and regulatory framework.[79] It plays an important role in managing the environmental and social externalities of Chile's mining industry which is one of the largest in the world. In 2021, Chile was the world's largest producer of copper, the second-largest producer of molybdenum, and the third-largest producer of silver.[80] Since 2017, three cases have been filed on the basis that the EIA conducted for the mining (and major infrastructure) project was flawed because it failed to take climate change impacts and Chile's legal commitments to reduce GHG emissions into account.[81] In the vast majority of EIA cases, the legal challenge is mounted to halt the authorization of *proposed* projects. However, in a landmark case decided in April 2022, the Chilean Supreme Court ruled that an *existing* power plant needed to undergo a new EIA to include properly modified requirements to account for the plant's impacts on climate change.

In *Mejillones Tourist Service Association and Others v the Environmental Evaluation Service (SEA) of Antofagasta*, local tour operators and local community organizations brought a constitutional lawsuit against the SEA of Antofagosta.[82] The plaintiffs challenged the agency's refusal to include climate

com/non-us-case/fifty-representatives-of-the-residents-of-omkoi-v-expert-committee-on-eia-consideration-in-the-mining-and-extracting-industry-and-the-office-of-natural-resources-and-environmental-policy-and-planning/> accessed 13 July 2023.

[78] Climate Case Chart (n 77).

[79] OECD, *Environmental Governance and Management* (OECD Environmental Performance Reviews: Chile, 2016) ch 2 <www.oecd-ilibrary.org/sites/9789264252615-9-en/index.html?itemId=/content/component/9789264252615-9-en> accessed 13 July 2023.

[80] For an insightful and concise account of how Chile became one of the world's largest recipients of foreign capital in the mining sector at the expense of the environment and human health, see Johanna Järvelä and Lotta Aho, 'Going against the Tide: Towards Binding Environmental Regulation of Mining in Chile' (*Centre for Business and Development Studies*, 12 May 2022) <https://cbds.cbs.dk/going-against-the-tide-towards-binding-environmental-regulation-of-mining-in-chile/> accessed 13 July 2023.

[81] The three cases are *Jara Alarcon Luis v Environmental Assessment Service*, Res Ex No 1317 of 15 November 2016; *Grez v Environmental Evaluation Service of Chile*, R-77-2018; *Mejillones Tourist Service with the Environmental Evaluation Service of Antofagasta*, Case Number 6930-2021, Case Number 71.628-2021.

[82] Chile Ministry of Justice, 'Corte Supreme ordena revision extraordinaria de impacto Ambiental de central termoeléctrica en Mejillones' (*Poder Judicial Republica de Chile*, 21 April 2022) <www.pjud.cl/prensa-y-comunicaciones/noticias-del-poder-judicial/72641> accessed 14 July 2023.

change considerations in the modification of the environmental authorization it gave in 2007 for the construction of the Angamos coal-fired power station which serves the world's largest copper mine (Escondida).[83] The plaintiffs argued that the agency's refusal to update the environmental authorization (which will compel the power plant to improve its environmental performance and reduce its impacts on climate change) was arbitrary and violated their constitutional rights, including the right to life and the right to live in an environment free of pollution.[84] The Court of Appeal of Antofagasta rejected the plaintiff's claims.

However, the Chilean Supreme Court overturned the Court of Appeal's ruling on appeal. The Supreme Court ruled there was a sound legal basis (in Article 25 of Law No 19.300) for allowing revisions to environmental authorizations to take into account environmental considerations that were not considered in the earlier EIA.[85] The Supreme Court ruled that such revisions were necessary to respond to changing environmental circumstances, including climate change. The court also expressly referred to Chile's legal obligations pursuant to the UN Framework Convention on Climate Change (UNFCCC) as underpinning its decision to order the agency to undertake a new EIA for the Angamos power plant.[86] This decision has potentially far-reaching consequences. If agencies are required to modify the climate change requirements for existing EIA authorizations and, furthermore, to order new EIAs if needed, this will amplify the utility of EIA to decarbonize projects and drive the transition towards a low-carbon economy. For instance, a new EIA might lead to the conclusion that the project is no longer profitable because of the introduction of climate mitigation and adaptation requirements, driving investors and project developers to reconsider the business case for extending the lifespan of fossil fuel projects.

## 2.6 References to the Paris Agreement

In Chapter 2, we discussed three key ways in which the intersection between climate change litigation and the Paris Agreement have been conceptualized, namely: (1) the cross-cutting pathway whereby climate litigation and related transnational social movements contribute to transnational climate change governance; (2) climate change litigation as a mechanism for enforcing domestic policies and laws (the bottom-up pathway); and (3) the top-down pathway which situates climate litigation as a site for the interpretation and application of the Paris Agreement norms

[83] Bnamericas, 'AES Gener Launches Construction at Angamos Coal-fired Plant' (*Bnamericas*, 28 August 2023) <www.bnamericas.com/en/news/aes-gener-launches-construction-at-angamos-coal-fired-plant> accessed 14 July 2023.

[84] *Mejillones Tourist Service* petition (n 81).

[85] ibid judgment.

[86] ibid.

and principles.[87] Since almost all jurisdictions in the Global South have ratified the Paris Agreement, goals to achieve carbon neutrality and reduce GHG emissions are part of each jurisdiction's overall climate change policies.

As a starting point for considering more deeply how climate change litigation in the Global South contributes to international climate law and governance, we sought to ascertain how many cases refer to the long-term temperature goal, the NDCs, and other norms of the Paris Agreement. In a significant proportion of cases (about 46% of the total number of cases under consideration), the plaintiffs or the court referred to the country's Paris Agreement NDC or, more broadly, the country's international legal commitments pursuant to the Paris Agreement to support more concerted climate action across governmental policy implementation and law enforcement.

In some notable cases, plaintiffs have sought to: (1) challenge the ambition of a country's NDC; and (2) challenge domestic laws for their alleged incompatibility with the Paris Agreement norms and principles. *Six Youths v Minister of Environment and Others* is an archetype of the first approach. As will be discussed further in Chapter 4, this case was filed by six youths who sought an injunction against the Brazilian government's 2020 NDC on the basis that it was significantly less ambitious and therefore in contravention of the Paris Agreement. While the Brazilian court summarily dismissed the initial complaint and the decision went up on appeal, in a similar case filed in Mexico, the appellate court granted the injunction against the updated NDC coming into force on the basis that it was a violation of the principle of non-regression in human rights law.[88] Subsequently, the Eleventh Collegiate Court of the First Circuit in Administrative Matters ruled against the plaintiff NGO. It did so on the basis that the plaintiff had not provided enough evidence to prove that the NDC was regressive. Nor did the court accept that the plaintiff had made out its claim that Mexico could emit up to 14 million tons of carbon dioxide equivalent ($CO_2e$) more than it would otherwise have if it had adhered to its more stringent 2015 NDC.[89] The case was therefore dismissed.

Cases in the Global South docket which sought to challenge domestic laws for their alleged incompatibility with the Paris Agreement norms and principles include *Greenpeace Mexico v Ministry of Energy and Others (on the Energy Sector Program)* and *Society for Protection of Environment and Biodiversity v Union of India*.[90] It is notable that, in these two cases, the plaintiffs are not challenging the lack of implementation of domestic environmental or climate change laws but rather challenging proposed amendments to existing laws that would have the effect

[87] This is visually captured in Figure 2.2 in Chapter 2 of this book.

[88] Amparo en Revisión 159/2022 (n 23).

[89] ibid.

[90] *Pandey v Union of India*, Original Application No 187/2017; *Society for Protection of Environment & Biodiversity v Union of India*, Application No 677 of 2016 (MA No 148 of 2017); Amparo indirecto 104/2020 (n 11).

of worsening climate change. In the *Greenpeace* case, the proposed legal amendments would lead to greater production of electricity from coal. In the *Society for Protection of Environment and Biodiversity* case, the proposed amendment to the EIA law would exempt certain construction works from the EIA process (or a full application of it). This was likely to have the intended consequence of fast-tracking infrastructure development, which would be at the expense of the environment, including climate change.

## 2.7 Public Trust Doctrine

The public trust doctrine is a common law doctrine, which has its roots in Roman law.[91] According to the public trust doctrine, the state holds natural resources on trust and, as a trustee, is required to utilize these resources wisely and for the benefit of the citizens.[92] In the words of the court in *MC Mehta v Kamal Nath*, 'the State is the trustee of all natural resources which are by nature meant for public use and enjoyment'.[93] A key feature of the doctrine is that the public often has a right to sue to enforce it.[94] This feature makes the public trust doctrine a potentially powerful pathway to conserve the environment and the climate.

The landmark United States (US) constitutional climate lawsuit, *Juliana v United States*, is a case in point.[95] In this case, the youth plaintiffs assert that, through the government's actions that cause climate change, it has violated their constitutional rights to life, liberty, and property, as well as failed to protect essential public trust resources.[96] The suit was first filed in the US District Court for the District of Oregon in 2015. The district judge denied the motion to dismiss, holding that the public trust and other theories could legally support the

[91] Hope M Babcock, 'The Public Trust Doctrine: What a Tall Tale They Tell' (2009) 61 South Carolina Law Review 393, 396. Justinian codified the doctrine in *Corpus Juris Civilis* (about 529 BCE) as follows: 'by the law of nature these things are common to all mankind, the air, running water, the sea and consequently the shores of the sea': David Takacs, 'The Public Trust Doctrine, Environmental Human Rights, and the Future of Private Property' (2008) 16 NYU Environmental Law Journal 711, 713. Charles Wilkinson has found references to the public trust doctrine in Chinese water law (249–207 BCE), Islamic water law, Nigerian traditional customs, and the laws of medieval Spain and France: Charles F Wilkinson, 'The Headwaters of the Public Trust: Some Thoughts on the Source and Scope of the Traditional Doctrine' (1989) 19 Environmental Law 425, 426. See also Anne Richardson Oakes, 'Judicial Resources and the Public Trust Doctrine: A Powerful Tool of Environmental Protection?' (2018) 7(3) Transnational Environmental Law 468.

[92] Richardson Oakes (n 91).

[93] *MC Mehta v Kamal Nath* (1997) 1 SCC 388.

[94] In American jurisprudence, see eg *Paepcke v Public Building Commission*, 263 NE2d 11,18 (Ill 1970) (finding that the public trust doctrine provides grounds for standing to the public); Mary Turnipseed and others, 'The Silver Anniversary of the United States' Exclusive Economic Zone; Twenty-Five Years of Ocean Use and Abuse, and the Possibility of a Blue Water Public Trust' (2009) 36 Ecology Law Quarterly 1, 19 (describing the public trust doctrine as presumptive grounds for standing).

[95] *Juliana v United States*, 217 F Supp 3d 1224 (D Or 2016).

[96] ibid.

claims.[97] The Ninth Circuit eventually dismissed the case on standing but the lower court's decision remains of significance. As one commentator has put it, '[J]uliana signals a federal court using the public trust doctrine to strike a strong blow in the climate change fight. And it stands as an example that other courts could follow.'[98]

In the Global South docket, we found that the public trust doctrine featured in 12 cases. Nine out of 12 of these cases were brought in India and Pakistan where the public trust doctrine is well established in the environmental jurisprudence.[99] In the *Rohtang Pass* case in India, the court ordered the government to take immediate action to address the pollution crisis because, inter alia, the public trust doctrine required it to do so.[100] Plaintiffs also raised the public trust doctrine in *Pandey v India*, *Ali v Pakistan*, the *Carbon Majors Petition*, and the *Ugandan Youths* case.[101] It is notable that these four cases were filed with the support of Our Children's Trust, the non-profit public interest law firm that represented the plaintiffs in *Juliana v United States* and has formed partnerships with youth and lawyers in the Global South to support their efforts to bring lawsuits compelling government action on climate change.[102]

While the public trust doctrine features in the background of judicial reasoning or as a normative concept to support the court's decision, none of the 12 decisions turned on the issue of whether the state had violated the obligations imposed by the public trust doctrine. In fact, the scope of the public trust doctrine remains relatively broad and, in most cases, appears to be aspirational rather than a legal doctrine that is applied by the courts to adjudicate the matter at hand. Nonetheless, it can be argued that the public trust doctrine is underutilized and there is potentially more receptiveness to arguments based on the doctrine in Global South jurisdictions compared to the Global North.

[97] *Juliana v United States*, 947 F 3d 1159 (9th Cir 2020) 1263.

[98] Joseph Regalia, 'The Public Trust Doctrine and the Climate Crisis: Panacea or Platitude' (2021) 11(1) Michigan Journal of Environmental and Administrative Law 1, 13. *In re Hawai'i Electric Light Co* (SCOT-22-0000418), the Supreme Court of Hawai'i issued a landmark decision rejecting a power purchase agreement between the Hawai'i Electric Light Company and an energy company. The majority opinion emphasized the state's constitutional duty to protect the right to a clean and healthy environment. The concurring opinion of Justice Wilson went further, finding that the right to a clean and healthy environment is part of the due process right to 'life, liberty, and property' as well as in the public trust doctrine, which requires conserving and protecting natural resources for the benefit of present and future generations.

[99] Muhammad Wajid Munir, 'Putting Public Trust Doctrine to Work: A Study of Judicial Intervention in Environmental Justice' (2018) 4(1) Lahore University of Management Studies Law Journal 96.

[100] *Court on its own motion v State of Himachal Pradesh*, Original Application No 237 (THC) of 2013 (CWPIL No 15 of 2010).

[101] *Pandey v India* (n 90); *Rabab Ali v Federation of Pakistan*, Constitutional Petition No 1 of 2016; *Carbon Majors* (n 56), *Nisi Mbabazi v The National Environmental Management Authority and Another*, Civil Suit No 283 of 2012.

[102] Our Children's Trust, 'Home' (*Our Children's Trust*) <www.ourchildrenstrust.org/> accessed 10 July 2023.

As mentioned earlier, the doctrine is well rooted in the jurisprudence of a number of Global South jurisdictions, including India and Pakistan. In *Sugathapala Mendis v Chandrika Kumaratunga and Others (Waters Edge* case), the Supreme Court of Sri Lanka upheld the right of any citizen to seek redress for a violation of the public trust.[103] In this case, the petitioners challenged the government's decision to sell land that had been set aside to serve a public purpose (ie to protect surrounding urban areas from flooding) to a private company. The company had plans to use the land to create an exclusive private golf resort. The court ruled that the petitioners had standing to sue the government officials involved in approving the transfer of land ownership for violating the public trust (and the fundamental right to equality before the law guaranteed by the Constitution). The court stated that 'it is to be noted for our purposes that all facets of the country—its land, economic opportunities or other assets—are to be handled and administered under the stringent limitations of the trusteeship posed by the Public Trust Doctrine and must be used in a manner for economic growth and always for the benefit of the entirety of the citizenry of the country'.[104] Accordingly, the court ruled that, inter alia, the land would be restored as far as possible to its original purpose of flood prevention. The relevant agencies were ordered to present the court with a land restoration plan within three months of the date of the judgment.[105]

The above example illustrates the potential for advocates and judges to rely on existing environmental jurisprudence for the extension of the public trust doctrine to protecting the climate. Further, while the public trust doctrine is most recognized in common law countries, the concept of the public trust is well established in the civil law tradition and is codified in law in a number of civil law countries. For example, Article 33(3) of the Constitution of the Republic of Indonesia states that '[t]he land, the waters and the natural resources within shall be under the powers of the state and shall be used for greatest benefit of the people'. In addition, several Indonesian natural resource management laws contain a similar provision that the resource in question is 'controlled by the state' but 'should be used for the benefit and welfare of the people'.[106] There is potential for arguments premised on such provisions in civil legal codes to provide a promising avenue for the further development of climate case law in the Global South.

[103] 2008 2 Sri LR 339.

[104] ibid 374.

[105] ibid 388.

[106] See eg the preambles of the Agrarian Law Act, 1960, No 5; the Water Resources Act, 2004, No 7; The Management of Coastal Areas Act, 2007, No 27; and Islands and Law No. 1 of 2014 amending Law No. 27 of 2007 on the Management of Coastal Zone and Small Islands; Forestry Law Act, 1999, No 41; and the Spatial Planning Law Act, 2007, No 26.

## 2.8 Just Transition Litigation

A just transition for communities as the global economy responds to climate change was one of the elements recognized in the Paris Agreement.[107] At the Twenty-Fourth Conference of the Parties (COP24) in Katowice, parties to the UNFCCC signed the Solidarity and Just Transition Silesia Declaration which called for more progress in addressing labour market vulnerabilities in carbon-intensive sectors that face transition risks.[108] The Glasgow Just Transition Declaration, signed by more than 30 countries at the Twenty-Sixth Conference of the Parties (COP26), declares support for workers, communities, and regions that are particularly affected by moves away from carbon-intensive economic activities.[109] Some states, including Chile and South Africa, have included notions of just transition in their Paris Agreement NDCs.[110] The importance of a just transition continues to gain recognition as governments and businesses grapple with the challenges of scaling up decarbonization efforts, and nowhere are these challenges more acute than in the Global South where nearly 750 million people live without electricity.[111]

Just transition litigation is a small subset of climate cases and has been defined as follows:

> Just transition litigation questions the distribution of the benefits and burdens of transition policies and activities towards net zero-emission and climate-resilient societies amongst local communities and affected stakeholders. Litigators typically invoke human rights but may rely on other national and international laws and principles. Just transition cases are not necessarily pro- or anti-climate

[107] The Agreement's preamble states: '[t]he Parties to this Agreement . . . [take] into account the imperatives of a just transition of the workforce and the creation of decent work and quality jobs in accordance with nationally defined development priorities'.

[108] International Organisation of Employers, 'Silesa Declaration' (*International Organisation of Employers*) <www.ioe-emp.org/index.php?eID=dumpFile&t=f&f=134978&token=91237abd5b4e38c1e7c2e4364b2b8e7095d8e0fd> accessed 14 July 2023.

[109] 'Supporting the Conditions for a Just Transition Internationally' (*The National Archives*, 4 November 2021) <https://ukcop26.org/supporting-the-conditions-for-a-just-transition-internationally/.> accessed 14 July 2023. The declaration calls for social dialogue amongst workers, governments, and other groups affected by the transition to green economies and the creation of decent jobs in new supply chains.

[110] UNFCCC, 'South Africa—First Nationally Determined Contribution under the Paris Agreement Updated September 2021' (*NDC Registry*, September 2021) 5 <https://unfccc.int/sites/default/files/NDC/2022-06/South%20Africa%20updated%20first%20NDC%20September%202021.pdf> accessed on 14 July 2023; UNFCCC, 'Chile's Nationally Determined Contribution Update 2020' (*NDC Registry*, 2020) 23–27 <https://unfccc.int/sites/default/files/NDC/2022-06/Chile%27s_NDC_2020_english.pdf> accessed 14 July 2023.

[111] Laura Cozzi and others, 'For the First Time in Decades, the Number of People without Access to Electricity is Set to Increase in 2022' (*International Energy Agency*, 3 November 2022) <www.iea.org/commentaries/for-the-first-time-in-decades-the-number-of-people-without-access-to-electricity-is-set-to-increase-in-2022> accessed 14 July 2023.

> action. Instead, they demand that climate action is undertaken in a just and inclusive manner with attention to those in vulnerable situations.[112]

Just transition litigation has also been described more broadly to refer to rights-based cases that 'do not object to climate action in and of itself, but rather to the way in which it is carried out and/or to its impacts on the enjoyment of human rights'.[113]

Our analysis of the Global South docket shows that there are, at the time of writing, 18 just transition lawsuits in Latin America, and none in Africa and Asia. As just transition cases are responding to how communities are affected by decarbonization policy developments (eg increasing renewable energy in the country's energy mix), it is not surprising that cases have yet to be filed in Africa and Asia where the development of decarbonization policies generally lags behind Latin America.[114] However, as decarbonization policy processes get underway in these regions, we can expect just transition litigation to emerge as Indigenous communities, NGOs, and perhaps even trade unions question how they are included in political negotiations and processes, as well as how the benefits of the energy transition are distributed amongst affected parties.

*Regional Government of Atacama v Ministry of Mining and Others* is an interesting case that has arisen in the context of the Chilean government's strategic plans to exploit its lithium resources.[115] Lithium is a critical component of the batteries that power electric vehicles. Lithium mining in the Atacama has already caused extensive damage to its fragile ecosystems, depleted water resources, and harmed the way of the life of the Indigenous communities.[116] In this case, Atacama's regional

[112] Maria Antonia Tigre and others, 'Just Transition Litigation in Latin America: An Initial Categorization of Climate Litigation Cases Amid the Energy Transition' (*Sabin Center for Climate Change Law*, January 2023) 13 <https://scholarship.law.columbia.edu/sabin_climate_change/197/> accessed 14 July 2023.

[113] Annalisa Savaresi and Joana Setzer, 'Rights-Based Litigation in the Climate Emergency: Mapping the Landscape and New Knowledge Frontiers' (2022) 13(1) Journal of Human Rights and the Environment 7, 29.

[114] The highest number of cases in Mexico and Chile is attributable to geopolitical developments. In Mexico, several controversial energy industry reforms have provided conditions ripe for the emergence of just transition litigation. In Chile, ambitious plans to develop the lithium mining sector and renewable energy have been implemented without the participation of affected stakeholders, including local governments and Indigenous communities: Maria Antonia Tigre and others (n 112) 18.

[115] Corte de Apelaciones de Copiapó, 30 January 2022, 'Gobierno Regional de Atacama con Ministerio de Mineria y Otro', Rol de la causa: Rol Protección 9-2022. It should be noted that in April 2023, Chile's President Gabriel Boric unveiled plans to bring the lithium industry under state control. He also said that stricter environmental controls will be imposed and that future projects would only proceed after consultations with local communities and should use water-efficient technologies: Michael Stott and Kenza Bryan, 'Chile's President Moves to Bring Lithium under State Control' *Financial Times* (21 April 2023) <www.ft.com/content/ebd48bbc-1390-4679-99fe-682975bbdba8> accessed 14 July 2023.

[116] The Atacama's spectacular salt flats contain one of the world's largest deposits of lithium where most of Chile's lithium extraction occurs; Amanda Maxwell, 'Exhausted: How We Can Stop Lithium Mining from Depleting Water Resources, Draining Wetlands, and Harming Communities in South

government sought to challenge the national mining authority's bidding process to award more exploration permits on the basis that the process was carried out without assessment of the environmental impacts and without public participation and consultation. While the plaintiff government recognized that increasing lithium production is a response to global demand and is critical for the global transition towards low-carbon development, it sought greater accountability and assessment of the adverse environmental and social impacts of lithium mining. On 14 January 2022, the Court of Appeals of Copiapo granted an order to suspend contracts that had already been awarded to two companies while the writ of protection (a remedy for the protection of constitutional rights) is pending.

## 2.9 Adaptation Litigation

Compared with GHG emissions mitigation that tends to be the focus of international and national regulation, climate change adaptation is an issue that is suited to consideration at more local levels of governance. For a start, variations in the manifestation of climate change impacts across different areas, even within the same country or region, militate in favour of responses that are localized and tailored to the needs of the local community. Across the world, climate litigation tends to focus on mitigation.[117] There is a need for greater consideration of how the law can help societies cope with climate change impacts, particularly in the Global South where the burden of adaptation is greatest.

Our analysis of the Global South docket shows that, at the time of writing, there are a handful of climate adaptation cases: three in Africa, four in Asia, and three in Latin America. This set of cases suggests that while adaptation litigation in the Global North has tended to focus on planning laws and infrastructure development, litigation in the Global South has addressed a broader range of factors that contribute to adaptive capacity. These factors include protection of fragile ecosystems such as glaciers and rainforests, lack of implementation of climate adaptation policies, poor implementation of existing land management and environmental laws which has the effect of reducing adaptive capacity, and disaster management.

America' (*NRDC*, 26 April 2022) <www.nrdc.org/resources/exhausted-how-we-can-stop-lithium-mining-depleting-water-resources-draining-wetlands-and> accessed 14 July 2023; Dave Sherwood, 'Chile Judge Calls for Water Supply on "Fragile" Lithium-Rich Atacama Salt Flat' *Reuters* (27 July 2020) <www.reuters.com/article/uk-chile-lithium-idUKKCN24S14U> accessed 14 July 2023.

[117] Joana Setzer and Catherine Higham, *Global Trends in Climate Litigation: 2021 Snapshot* (Policy Report, Grantham Research Institute on Climate Change and the Environment 2 July 2021) <www.lse.ac.uk/granthaminstitute/wp-content/uploads/2021/07/Global-trends-in-climate-change-litigation_2021-snapshot.pdf> accessed 14 July 2023. Setzer and Highman found that out of the more than 1800 cases worldwide, only 180 touch on the issue of adaptation.

*Leghari v Pakistan* is a prime example of systemic adaptation litigation, which has been defined as 'cases that aim to advance adaptation action nationwide, especially by compelling States to adopt, improve and implement national or subnational legal and policy frameworks on adaptation.'[118] Given that Global South jurisdictions are more vulnerable and have less adaptive capacity, the replication of *Leghari v Pakistan* across the Global South to compel states to adopt and implement adaptation policies would be useful to advance adaptation. Another noteworthy case is the *Colombian Youths* case. The Supreme Court of Justice ordered, inter alia, the government to work with affected communities to develop an 'intergenerational pact for the Colombian Amazon.'[119] Along with measures to reduce deforestation and GHG emissions, the plan had to include strategies directed towards climate change adaptation.[120] It has been argued that this case type can be strategically used to protect other ecosystems that are particularly vulnerable to climate change (eg marine coral reefs), and that the right to a healthy environment and rights of nature can constitute the primary legal basis for this type of case.[121]

## 3. FEATURES OF GLOBAL SOUTH CLIMATE CASES

A central message of this book is that an appreciation of the Global South's experience with climate litigation and its case law goes a long way towards a fuller picture of transnational climate litigation and its contribution towards global climate governance. In this section, we seek to distil some characteristics of climate cases in the Global South as a basis for reframing and broadening understandings of the phenomenon of transnational climate litigation. Rather than suggesting that these characteristics are firmly associated with Global South climate litigation, we think of these characteristics as a series of spectrums (see Figure 3.2, Figure 3.3, and Figure 3.4 below), with Global South cases presently concentrated at one end and Global North cases at the other. Further, within the Global South docket itself, we see some regional variations along these spectrums too. These spectrums are by no means determinative but help elucidate the distinctive nature of the contribution made by Global South cases to transnational climate litigation.

118 Riccardo Luporini, 'Strategic Litigation at the Domestic and International Levels as a Tool to Advance Climate Change Adaptation?' (2023) 4(1) Yearbook of International Disaster Law Online 202, 225.
119 *Future Generations* (n 1) 49.
120 ibid.
121 Luporini (n 118) 224.

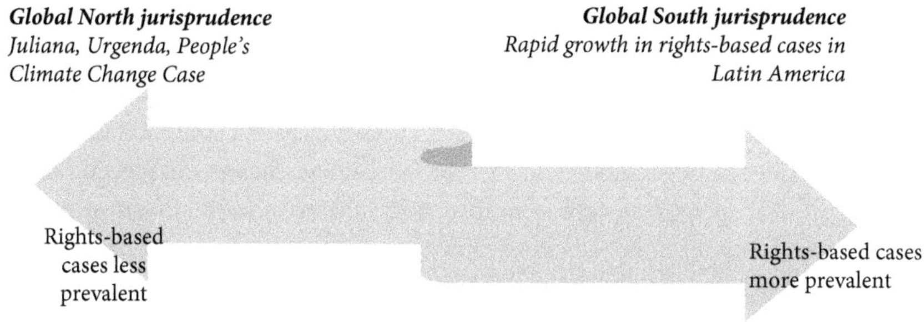


**Figure 3.2** Spectrum of rights-based litigation

## 3.1 Prevalence of Rights-Based Claims

First, a distinctive feature of the Global South docket is the high number of human rights or (environmental) constitutional rights claims asserting that failures of mitigation or adaptation violate rights protections (Figure 3.2).[122] As shown in the earlier discussion about rights-based litigation, the relatively greater prevalence of rights-based claims in the Global South docket can be explained, at least in part, by the fact that many of the national constitutions of Global South jurisdictions contain the right to live in a healthy environment. On 28 July 2022, the United Nations (UN) Human Rights Council passed a resolution recognizing 'the right to a clean, healthy and sustainable environment' as a 'human right that is important for the enjoyment of human rights'.[123] Although such resolutions are not legally binding on the UN Member States, this recognition of the right to a healthy environment is expected to be a catalyst that will prompt more countries to enshrine the right to a

[122] In general, rights-based climate litigation is being conducted both in the Global North and the Global South, and there is a rising trend in the Global North following the success of high-profile cases including *Urgenda v the Netherlands* and *Neubauer v Germany*. The point being advanced in this chapter is that, as a pathway of litigation, rights-based claims make up a majority of Global South cases. It should also be noted that prior to 2015, only a handful of rights-based climate cases had been filed anywhere in the world. Between 2015 and 2021, 148 rights-based climate lawsuits against states (and, to a far lesser extent, against corporations) have been filed in 30 jurisdictions and in 8 international judicial or quasi-judicial bodies. Europe is the most active region with respect to rights-based climate litigation (34% of the total number of rights-based cases worldwide), followed by Latin America and North America at roughly the same levels (18% of the total number of rights-based cases worldwide); see Cesar Rodriguez-Garavito, 'Litigating the Climate Emergency: The Global Rise of Human Rights-Based Litigation for Climate Action' in César Rodríguez-Garavito (ed), *Litigating the Climate Emergency* (CUP 2022) 12.

[123] UN General Assembly, Resolution on the Human Right to a Clean, Healthy and Sustainable Environment, Doc A/76/L.75 (26 July 2022); UNEP, 'In Historic Move, UN Declares Healthy Environment a Human Right' (*UNEP*, 28 July 2022) <www.unep.org/news-and-stories/story/historic-move-un-declares-healthy-environment-human-right> accessed 15 July 2023; UN, 'UN General Assembly Declares Access to Clean and Healthy Environment a Universal Human Right' *UN News* (28 July 2022) <https://news.un.org/en/story/2022/07/1123482> accessed 14 July 2023.

healthy environment in their national constitutions. It also has significant normative value and can be used to bolster legal arguments in court.

Vilchez and Savaresi have found that out of 112 rights-based climate cases, 42 cases invoked the right to a healthy environment.[124] These 42 lawsuits preponderantly targeted states (39 out of 42 cases) and focused on climate change mitigation (32 out of 42 cases). All except one case were filed with national judicial or quasi-judicial bodies.[125] As many rights-based climate litigation cases remain pending, comparing outcomes can be challenging. However, Vilchez and Savaresi's assessment of decided cases shows that, on balance, rights-based cases which invoked the right to a healthy environment tended to be more successful than those which did not.

There is also generally more scope, compared to the Global North, to base claims on constitutional rights for climate protection in the Global South. Rights-based cases in Colombia, India, Pakistan, and the Philippines have been able to draw on jurisprudence and remedies developed over the years in the context of claims challenging the violation of socio-economic rights such as the right to housing and the right to health. Indigenous rights litigation has also provided valuable jurisprudence relevant to climate litigation, including jurisprudence on the requirement for free, prior, and informed consultation and consent of affected communities.[126] As Rodriguez-Garavito has argued:

> this Southern route to climate litigation is not serendipitous, or the result of the absence of specialized climate change legislation that litigants would otherwise have used in framing their cases. Instead, it is a route whose tracks were firmly laid over the last three decades through public interest law practice, research, and judicial activism regarding constitutional rights in general and socioeconomic rights (SERs) in particular.[127]

We see significant potential for the growth of rights-based climate litigation in the Global South, particularly as litigants seek to replicate the success of landmark cases and in jurisdictions where the preconditions for such litigation are in place. Such preconditions will include the existence of constitutional jurisprudence that is protective of the environment, and which could be readily expanded to encompass climate-change related issues such as deforestation.[128] There is also significant

[124] Pau de Vilchez and Annalisa Savaresi, 'The Right to a Healthy Environment and Climate Litigation: A Game Changer?' (2021) 32(1) Yearbook of International Environmental Law 3.

[125] ibid 5.

[126] See Chapter 5 for further discussion.

[127] César Rodríguez-Garavito, 'Human Rights: The Global South's Route to Climate Litigation' (2020) 114 American Journal of International Law Unbound 40–44.

[128] See Chapter 4 for further discussion. It should also be noted that context is, of course, all important. Rights-based litigation will never emerge in China, for example, because the Chinese Constitution is not justiciable; Fu Hualing and Zhai Xiaobo, in highlighting that China's Constitution is not justiciable, note that '[a] constitution that is non-justiciable is remote from the people's daily lives

potential for rights-based litigation to advance climate adaptation. There is currently a dearth of cases on climate adaptation and yet, human rights law lends itself quite easily to litigating adaptation measures designed to protect individuals and communities from the effects of climate change including forced displacement and adverse health impacts.[129]

It is noteworthy that while no actual rights-based climate change litigation has yet to materialize in the Pacific Islands, the legal systems of all the Pacific Island states are amenable to such litigation.[130] They have written constitutions that entrench a range of civil and political rights including the right to life. Fiji's Constitution also enshrines the right to a clean and healthy environment.[131] It has been argued that while a lawsuit against the state may be futile as well as contrary to climate justice (given that other states are responsible for much larger shares of global emissions and also have greater capacities to decarbonize their economies), an alternative option would be for Pacific Island plaintiffs to sue the 'carbon majors'.[132] The 'carbon majors' refer to the 90 oil, gas, coal, and cement companies that have contributed most to causing anthropogenic climate change.[133]

## 3.2 Preference for Enforcement of Existing Laws

A notable feature of many cases in the Global South docket is that they involve lawsuits brought by citizens and NGOs to compel their governments to put mitigation and adaptation policies into action, or to stop building new coal-fired power plants and mining coal (see Figure 3.3). The South African cases, for example, are all focused on halting the construction of new coal-fired power plants by challenging the adequacy of the EIA undertaken. Equally, the Global South climate case law features a significant number of enforcement lawsuits brought by the government against companies in the extractive sector (eg timber logging and oil palm cultivation). This is particularly the case in Indonesia and Brazil. The outcome in the majority of these law enforcement proceedings is that the defendant company was found guilty and had to pay compensation for the carbon released into the atmosphere due to its illegal activities or was ordered to alter its carbon-intensive commercial practices such as setting fires to clear land for commercial agriculture.[134]

and opaque even for scholars'. See Fu Hualing and Zhai Xiaobo, 'What Makes the Chinese Constitution Socialist?' (2018) 16(2) International Journal of Constitutional Law 655.

[129] Rodriguez-Garavito (n 122) 34.

[130] Margaretha Wewerinke-Singh, 'Litigating Human Rights Violations Related to the Adverse Effects of Climate Change in the Pacific Islands' in Lin and Kysar (eds) (n 29) 99.

[131] Constitution of the Republic of Fiji 2013, art 40.

[132] Wewerinke-Singh (n 130) 100.

[133] Climate Accountability Institute, 'Carbon Majors' (*Climate Accountability Institute*, 2020) <https://climateaccountability.org/carbonmajors.html> accessed 15 July 2023.

[134] See Chapters 4 and 5 for further discussion.

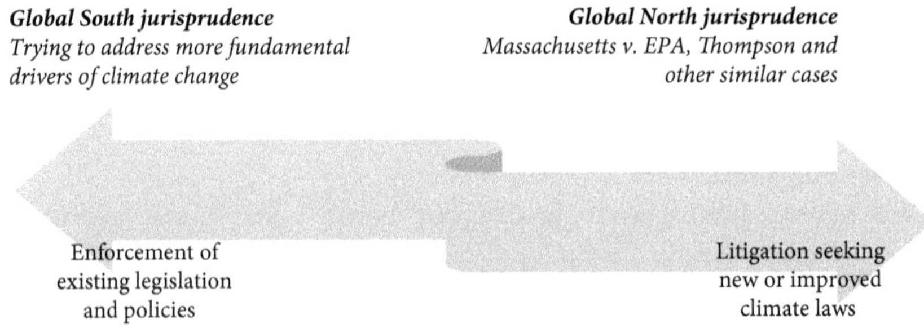


**Figure 3.3** Spectrum of enforcement-focused litigation

This case law suggests a different pathway for Global South climate claims than that which has frequently been the norm in the Global North. In the United States and Australia, for example, climate litigation has often been brought with the goal of exerting pressure on governments to fill a regulatory gap.[135] The *Massachusetts v EPA* case decided by the US Supreme Court sought to have the EPA implement a mandate under the Clean Air Act to regulate greenhouse gases as an air pollutant.[136] The high-profile *Urgenda* case in the Netherlands,[137] and other Global North cases inspired by it such as the *Thompson* case in New Zealand,[138] also have pursued a climate law reform rationale, arguing for the adoption of stricter emissions reduction targets by governments. In the European Union, analysts of climate litigation there have highlighted that a significant portion of the case law is aimed at 'perfecting' climate law measures, such as the European Emissions Trading Scheme, to make them more effective as a mitigation tool.[139]

By contrast, the Global South climate cases demonstrate an apparent preference for putting policies and laws already 'on the books' into action and ensuring proper enforcement, rather than pushing for new or better climate laws. In bringing suits that focus on seeking judicial intervention to compel the proper enforcement of existing laws, Global South plaintiffs are able to rely on fairly well-established legal arguments, thereby avoiding the risks of judicial reluctance to address climate change directly. For Global South advocates, there is also likely to be a strategic choice made to prefer tried-and-tested methods because of perceived higher chances of success. This is an important consideration when Global South organizations have

[135] Peel and Osofsky, *Change Litigation* (n 71) 28–53.

[136] *Massachusetts v Environmental Protection Agency*, 549 US 497 (2007).

[137] *Urgenda v Netherlands*, Case No 200.178.245/01, Hague Court of Appeal, 9 October 2018. An unofficial translation of the case from the Dutch original is available at <www.urgenda.nl/wp-content/uploads/ECLI_NL_GHDHA_2018_2610.pdf> accessed 14 July 2023.

[138] *Thompson v New Zealand* (Minister for Climate Change) CIV 2015-485-919 (2017) NZHC 733 <www.nzlii.org/cgi-bin/download.cgi/cgi-bin/download.cgi/download/nz/cases/NZHC/2017/733.pdf> accessed 14 July 2023.

[139] Navraj Singh Ghaleigh, 'Economics and International Climate Change Law' in Kevin R Gray, Richard Tarasofsky, and Cinnamon Carlarne (eds), *The Oxford Handbook of International Climate*

fewer financial resources to bring test cases. Tried-and-tested arguments and established legal precedents in most Global South countries are likely to be drawn from broader environmental or other non-specific climate laws and be applied to climate change by analogy or to address climate change indirectly.[140]

Further, in seeking enforcement of these laws, plaintiffs in Global South countries are often trying to address what they perceive to be more fundamental drivers of climate change. For example, in *Pandey v Union of India*, the nine-year-old claimant sought the proper enforcement of the national forestry law, air pollution control law, and environmental impact assessment (EIA) law on the basis that the non-implementation of these existing laws 'has led to adverse impacts of climate change across the country'.[141]

## 3.3 'Stealthy' Climate Litigation

Many Global South cases evince what might be termed a 'stealthy' strategy, which dilutes the political potency of climate issues by packaging them together with less controversial claims (see Figure 3.4). An important reason why litigants in some Global South countries may prefer to pursue climate cases in a more indirect fashion is because of traditions of judicial restraint and limited judicial review operating in these jurisdictions. This is a relevant factor in a number of Southeast Asian jurisdictions, which eschew notions of an activist judiciary present in other common law jurisdictions, including some South Asian countries such as India and Pakistan.[142]

More generally, there is often a tailoring of legal claims in Global South climate cases to what is seen to be the most important policy issue in the jurisdiction, which is not always climate change. For example, in China, the key concern may be air pollution, whereas in other countries forestry and burning practices may be more salient. While we have not yet seen a substantial number of climate cases emerge in the Pacific, the prevalence of disputes over land tenure or resettlement rights in many Pacific Island countries suggests that these kinds of claims might be a more likely vehicle for including climate-related arguments.[143]

*Change Law* (OUP 2016); Sanja Bogojević, 'EU Climate Change Litigation, the Role of the EU Courts, and the Importance of Legal Culture' (2013) 35 Law & Policy 184–207.

[140] See the various country studies and analytical chapters in Lin and Kysar (eds) (n 29).

[141] *Pandey v India* (n 90), Originating Application filed by the plaintiff, 2, para [III] <http://blogs2.law.columbia.edu/climate-change-litigation/wp-content/uploads/sites/16/non-us-case-documents/2017/20170325_Original-Application-No.-___-of-2017_petition-1.pdf> accessed 14 July 2023.

[142] Jacqueline Peel and Jolene Lin, 'Climate Change Adaptation Litigation: A View from Southeast Asia' in Lin and Kysar (n 29).

[143] Satish Chand and Ron Duncan, 'Resolving Property Issues as a Precondition for Growth: Access to Land in the Pacific Islands' in Peter Larmour (ed), *The Governance of Comon Property in the Pacific Region* (ANU 2013) (arguing that security of access to land currently under communal ownership must be enhanced so as to encourage efficient use of this resource and enhance the inflow of technology and capital in such countries).

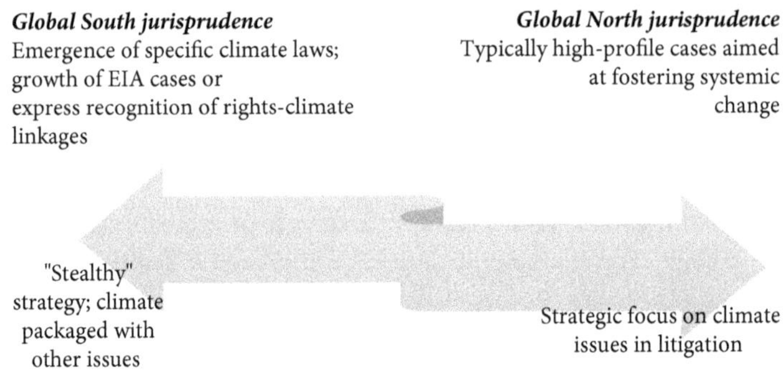


**Figure 3.4** Spectrum of climate litigation strategies

Achieving climate goals in litigation through other means is not a trend limited to the Global South. In strategic litigation, there will always be choices made by claimants about what legal avenues best serve the case and have the highest chances of achieving success.[144] Nonetheless, factors such as a conservative judiciary, higher prospects for success with tried-and-tested arguments, and a lower policy salience of climate issues compared with other environmental or public health concerns—some or all of which characterize most Global South countries—increase the chances that a more indirect route will be pursued in these climate cases. The pursuit of 'stealthy' climate litigation and the filing of 'core' cases which put issues of climate change law and policy at the front and centre are not incompatible trends and neither are they mutually exclusive. Across the board, we observe that many cases in the Global South docket adopt a 'stealthy' approach. At the same time, we also observe that in some jurisdictions such as Brazil, 'core' cases increasingly dominate the national docket. This points to the importance of noting that, in the process of drawing out trends, we are not making the claim that climate case-law developments across the Global South will embody all the characteristics we discuss here. Indeed, we will emphasize that the individual trajectory of climate litigation in each Global South jurisdiction will differ according to its particular social context, political settings, and legal culture.

In embedding climate change mitigation and adaptation concerns in wider issues of environmental protection, land use, disaster management, and natural resource conservation, Global South climate litigation is supporting the trend of mainstreaming climate change in post-Paris Agreement governance arrangements. This approach can be seen as an efficient use of limited public policy resources given the potential to achieve climate co-benefits through measures that address a range of environmental concerns. The Paris Agreement, as well as other

[144] Jacqueline Peel and Hari M Osofsky, 'Litigation as a Climate Regulatory Tool' in Christina Voigt (ed), *International Judicial Practice on the Environment: Questions of Legitimacy* (CUP 2019).

international sustainable development instruments including the United Nations Sustainable Development Goals (SDGs) and the Sendai Framework for Disaster Risk Reduction, endorse this broader implementation approach. For instance, Article 2(1) of the Paris Agreement declares that one of its aims is 'to strengthen the global response to the threat of climate change, in the context of sustainable development and efforts to eradicate poverty'. By framing climate change as one issue in a panoply of factors that must be addressed to ensure sustainable development and environmental protection, Global South climate jurisprudence reinforces the need for climate governance to be part of, rather than separate from, broader global environmental governance.

## 4. TRANSNATIONAL CLIMATE LITIGATION: SEEKING CLIMATE ACCOUNTABILITY IN THE GLOBAL NORTH AND SUPRANATIONAL TRIBUNALS

The discussion in this chapter has thus far focused on cases that have been decided, are pending decision, or have recently been filed in Global South legal systems. There is, however, a growing body of cases that are being filed by Global South plaintiffs (either on their own or alongside Global North NGOs and individuals) in Global North courts as well as international and regional tribunals. There are also cases that may not be filed by plaintiffs who live in the Global South but seek to hold governments and corporations accountable for investments and project activities in the Global South which are contrary to the Paris Agreement and increase GHG emissions. We take the view that while these cases do not constitute Global South climate litigation, they are noteworthy for creating a unique transnational pathway for Global South inhabitants seeking climate justice. Such cases can also be the basis for creating transnational coalitions of actors seeking accountability from Global North governments and corporations for their failure to reduce GHG emissions and decarbonize their economic activities. As mentioned earlier, a summary of these 'transnational' cases can be found in the accompanying online case law resources supporting this book. At the time of writing, there are 35 'transnational' cases.

### 4.1 Proceedings in Supranational Tribunals

Thirteen cases have been filed in international and regional tribunals. Of these cases, it should be noted that four of them are, strictly speaking, not cases but advisory opinions. An advisory opinion is legal advice given by an international or regional court at the request of a specialized agency or a government.[145]

[145] See eg art 96 of the Charter of the United Nations and art 191 of the UN Law of the Sea Convention.

A Request for an Advisory Opinion from the Inter-American Court of Human Rights Concerning the Interpretation of Article 1(1), 4(1) and 5(1) of the American Convention on Human Rights (OC-23/17) was made by Colombia in 2016 to seek legal clarification on whether the right to a healthy environment is a human right.[146] On 9 January 2023, the Foreign Ministers of Chile and Colombia requested an advisory opinion from the Inter-American Court of Human Rights on the scope of state obligations for responding to the climate emergency under the frame of international human rights law and, specifically, under the American Convention on Human Rights.[147] This request builds on the earlier advisory opinion in which the Court recognized the justiciable right to a healthy environment.

In December 2022, the Co-Chairs of the Commission of Small Island States on Climate Change and International Law, Prime Minister Gaston Browne of Antigua and Barbuda, and Prime Minister Kausea Natano of Tuvalu, submitted a request for an advisory opinion to the International Tribunal for the Law of the Sea (ITLOS) on the obligation of states under the 1982 United Nations Convention on the Law of the Sea (UNCLOS) to protect and preserve the marine environment in relation to climate change impacts, including ocean warming, sea level rise, and ocean acidification.[148] States and international organizations provided written statements on the questions submitted to ITLOS for the advisory opinion over the course of 2023 with oral proceedings taking place in September 2023.[149] In another set of proceedings that is complementary to this request to ITLOS, the UN General Assembly adopted a resolution on 29 March 2023 requesting an advisory opinion from the International Court of Justice (ICJ) on the obligations of states with respect to climate change.[150] These complementary efforts to clarify the legal

[146] For discussion, see Christopher Campbell-Duruflé and Sumudu Anopama Atapattu, 'The Inter-American Court's Environment and Human Rights Advisory Opinion: Implications for International Climate Law' (2018) 8(3–4) Climate Law 321–37.

[147] An unofficial translation of the request can be found here: <http://climatecasechart.com/non-us-case/request-for-an-advisory-opinion-on-the-scope-of-the-state-obligations-for-responding-to-the-climate-emergency/>. For discussion, see Maria Antonia Tigre, Natalia Urzola, and Juan Sebastián Castellanos, 'A Request for an Advisory Opinion at the Inter-American Court of Human Rights: Initial Reactions' (*Climate Law: A Sabin Center blog*, 17 Februrary 2023) <https://blogs.law.columbia.edu/climatechange/2023/02/17/a-request-for-an-advisory-opinion-at-the-inter-american-court-of-human-rights-initial-reactions/> accessed 17 July 2023.

[148] Commission of Small Island States on Climate Change and International Law, 'Request for Advisory Opinion' (ITLOS, 12 December 2022) <www.itlos.org/fileadmin/itlos/documents/cases/31/Request_for_Advisory_Opinion_COSIS_12.12.22.pdf> accessed 15 July 2023.

[149] Request for an Advisory Opinion submitted by the Commission of Small Island States on Climate Change and International Law (List of Cases No 31, Order 2023/1, 15 February 2023) ITLOS <www.itlos.org/fileadmin/itlos/documents/cases/31/C31_Order_2023-1_15.02.2023_Readable.pdf> accessed 15 July 2023.

[150] Res A/77/L/58 United Nations General Assembly, Request for an Advisory Opinion of the International Court of Justice on the Obligations of States in Respect of Climate Change, Doc A/77/L.58 (1 March 2023). For discussion, see Jacqueline Peel and Zoe Nay, 'The UN is Asking the International Court of Justice for its Opinion on States' Climate Obligations. What Does this Mean?' (*The Conversation*, 3 April 2023) <https://theconversation.com/the-un-is-asking-the-international-court-of-justice-for-its-opinion-on-states-climate-obligations-what-does-this-mean-202943> accessed 15 July 2023.

obligations of states to prevent and redress the adverse effects of climate change represent a significant step forward in the development of standards and parameters to assess state compliance with international law concerning climate change, as well as to provide domestic courts greater clarity on the obligations of states under international law.

## 4.2 France's Corporate Duty of Vigilance Law (*Loi de Vigilance*)

In 2017, France was the first country in the world to adopt a law that requires large companies to carry out human rights and environmental due diligence (HREDD) and publish a 'vigilance plan' annually.[151] The vigilance plan must identify risks and establish measures to prevent severe impacts on human rights and the environment resulting from the company's own activities, activities of companies it controls directly or indirectly, and its subcontractors and suppliers. The law also provides for civil liability. Harmed individuals can bring a civil lawsuit (based on French tort law) to seek damages for harm resulting from a company's failure to comply with its vigilance obligations, where compliance would have prevented the harm.[152] French multinationals, including nuclear company EDF, water supplier Suez, bank BNP Paribas, and agri-food firm Danone, have been sued under this law.[153] The claims have been diverse, ranging from plastics pollution to loans to companies that contribute to deforestation. For current purposes, three cases that have been filed under this law are of interest: the *Total Uganda* case, the *EDF Mexico* case, and the *Casino* case. The *Total Uganda* case is the first duty of vigilance decision by the French courts and therefore particularly noteworthy.

In June 2019, Friends of the Earth France, Survie, and four Ugandan NGOs served a formal notice to Total Energies, pointing out their failure to comply with the corporate duty of vigilance law including taking measures to address the forced displacement of thousands of people by the Tilenga and the East African Crude

[151] Loi n° 2017-399 du 27 mars 2017. An English translation of the Duty of Vigilance Law, inserted in art L 225-102-4 of the French Commercial Code, can be found here: European Coalition for Corporate Justice, 'French Duty of Vigilance law—English Translation' (*Business & Human Rights Resource Centre*, 14 December 2016) <www.business-humanrights.org/en/latest-news/french-duty-of-vigilance-law-english-translation/> accessed 15 July 2023.

[152] For discussion, see Elsa Savourey and Stéphane Brabant, 'The French on the Duty of Vigilance: Theoretical and Practical Challenges Since its Adoption' (2021) 6(1) Business and Human Rights Journal 141.

[153] Swann Bommier, Lucie Chatelain, and Camille Loyer, 'Follow Up on Current Cases: July 2021' (*Duty of vigilance radar*, July 2021) <https://vigilance-plan.org/wp-content/uploads/2021/07/A4-VF-FICHES-UK-060721-xxs.pdf> accessed 15 July 2023; Stéphane Mandard, 'Danone Taken to French Court by Environmental Campaigners over Plastic Use' *Le Monde* (9 January 2023) <www.lemonde.fr/en/environment/article/2023/01/09/danone-taken-to-french-court-by-environmental-campaigners-over-plastic-use_6010918_114.html> accessed 15 July 2023; Peggy Hollinger 'Danone Lawsuit Could be Test Case for New War on Plastics' *Financial Times* (18 January 2023) <www.ft.com/content/5a575e20-2d49-475e-9936-47c4bf641678> accessed 16 July 2023.

Oil Pipeline (EACOP) projects.[154] Subsequently, after the lapse of the mandatory three-month period during which Total Energies had the opportunity to respond to the notice and address inadequacies in implementing its vigilance plan, the NGOs proceeded to file a lawsuit against Total Energies in September 2019 seeking interim measures to halt the projects.[155] The question of which courts had jurisdiction over the matter led to years of delay. On 28 February 2023, the Judicial Court of Paris ruled that the case was inadmissible, namely that the evolution of the plaintiff's demands—which initially focused on human rights and then extended to climate and environmental impact—which were not reflected in the June 2019 formal notice to Total Energies rendered the notice procedurally flawed.[156] The court ruled that the formal notice had to be sufficiently precise to identify the shortcomings in the company's vigilance plan to allow for amicable settlement. Further, the aim of the formal notice procedure is to enable the company to achieve compliance with its vigilance plan. As such, the plaintiff's requests and complaints must be clearly stated in the formal notice. The court's order is currently on appeal.

At the heart of the *EDF Mexico* case was a proposed wind farm to be developed by EDF (a French multinational electric utility company that is majority-owned by the French state) in Mexico on land that belonged to the Zapotec and other communities of Union Hidalgo, a small municipality in the Mexican state of Oaxaca.[157] The Indigenous communities opposed the project on several grounds, principally the failure by ECF to obtain their Free and Prior Informed Consent (FPIC), and sought to challenge the project through domestic litigation, a complaint to the French National Contact Point for alleged violation of the Organisation for Economic Cooperation and Development (OECD) Guidelines for Multinational

[154] Les Amis de la Terre France and Survie, 'Total, See You in Court' (*Total, See You in Court*) <www.totalincourt.org/> accessed 16 July 2023 is a comprehensive bilingual website about this case.

[155] Tribunal judiciaire de Nanterre, 30 janvier 2020, N°R.G 19/02833—N° Portalis DB3R-W-B7D-VIPX.

[156] Eleonore Hannezo, Jean-Charles Jaïs, and Claudia Cavicchioli, 'First Decision by French Courts on Duty of Vigilance Law: Dismissal of Claims Lodged by Six NGOs against TotalEnergies' (*Linklaters*, 1 March 2023) <https://sustainablefutures.linklaters.com/post/102i8vw/first-decision-by-french-courts-on-duty-of-vigilance-law-dismissal-of-claims-lod> accessed 16 July 2023; Sarah White and Camilla Hodgson, 'French Court Pushed Back on Total Case in Tide of Climate Litigation' *Financial Times* (1 March 2023) <www.ft.com/content/10bae5b1-c19f-4b4e-b298-cfd88ed1ec89?shareType=nongift> accessed 16 July 2023; Emilie Vasseur and others, 'Business and Human Rights: First French Case-law on the Duty of Vigilance—Judges Adopt a Cautious Approach to Avoid Judicial Interference in Corporate Management' (*Mayer Brown*, 14 March 2023) <www.eyeonesg.com/2023/03/business-and-human-rights-first-french-case-law-on-the-duty-of-vigilance-judges-adopt-a-cautious-approach-to-avoid-judicial-interference-in-corporate-management/> accessed 16 July 2023; Zérah Brémond, 'Corporate Duty of Vigilance and Environment: Some Lessons Drawn from the EDF and the TotalEnergies Cases' (*Verfassungsblog*, 6 April 2023) <https://verfassungsblog.de/corporate-duty-of-vigilance-and-environment/> accessed 16 July 2023.

[157] European Center for Constitutional and Human Rights (ECCHR), ProSESC, and Terre Solidaire, 'Case Report—Wind Farm in Mexico: French Energy Firm EDF Disregards Indigenous Rights' (*ECCHR*, October 2020) <www.ecchr.eu/fileadmin/Fallbeschreibungen/20201013_Case_report_EDF_EN.pdf> accessed 16 July 2023.

Enterprises, and a suit in the French civil courts alleging that EDF's failure to respect the right to FPIC violated the French Duty of Vigilance Law.[158] The case is no longer live as the Mexican government has cancelled the contract with EDF to build this wind farm.[159] What is noteworthy is how the Indigenous communities and NGOs used the Duty of Vigilance Law alongside other accountability mechanisms to enforce corporate obligations to respect Indigenous and environmental rights.[160]

Finally, the *Casino* case represents a novel attempt to hold supply chain actors responsible for contributing to environmental and climate harm in the Global South. In March 2021, several NGOs brought this claim against the Casino corporation for the environmental and human rights harms attributable to its operations in the cattle industry in Brazil and Colombia.[161] Casino is the largest supermarket chain in both Brazil and Colombia where it operates through subsidiaries. The plaintiff NGOs allege that, despite Casino's stated commitment to eliminate deforestation and comply with Brazilian law (under which all actors in a production chain are deemed responsible for environmental damage with their knowledge), it continues to buy beef from slaughterhouses whose suppliers are known to act with impunity and have engaged in illegal land grabbing, slavery practices, and deforestation.[162] This case marks the first time that a supermarket chain has been taken to court for deforestation and human rights violations in its supply chain. It has been noted that this case demonstrates how the 'French duty of vigilance law is unprecedented in that it allows plaintiffs to circumvent the limited liability of corporations, increasing the likelihood that litigation will be brought under this law'.[163]

[158] ECCHR Mexico Wind Farm Case Report (n 157); OECD Watch, 'Union Hidalgo vs EDF Group' (*OECD Watch*, October 2020) <www.oecdwatch.org/complaint/union-hidalgo-vs-edf-group/> accessed 16 July 2023.

[159] ProDESC, 'Set Back to EDF in Mexico! The Zapotec Community of Unión Hildalgo and ProDESC Achieve the Definitive Cancellation of the Wind Park Megaproject "Gunna Sicarú"' (*ProDESC*, 2 June 2022) <https://prodesc.org.mx/en/set-back-to-edf-in-mexico-the-zapotec-community-of-union-hidalgo-and-prodesc-achieve-the-definitive-cancellation-of-the-wind-park-megaproject-gunaa-sicaru/> accessed 16 July 2023.

[160] For discussion, see Christopher Ewell, 'Reimagining the Renewable Energy Transition: The Potential for Mandatory Corporate Due Diligence to Ensure Respect for the Right to Free, Prior and Informed Consent' (2022) 47(2) Yale Journal of International Law 1 <https://openyls.law.yale.edu/bitstream/handle/20.500.13051/18188/Ewell%20Christopher%203.pdf?sequence=1&isAllowed=y>.

[161] Climate Case Chart, 'Envol Vert et al. v. Casino' (*Sabin Center for Climate Change Law*, 2 March 2023) <http://climatecasechart.com/wp-content/uploads/sites/16/non-us-case-documents/2021/20210302_13435_complaint.pdf> accessed 16 July 2023.

[162] ibid; Center for Climate Crime Analysis, 'Casino Case' (*Center for Climate Crime Analysis*) <https://climatecrimeanalysis.org/wp-content/uploads/2022/06/casino_case_-_english.pdf>.

[163] CLX: Climate Litigation Accelerator, 'Envol Vert v Casino—Tracing the Chain: Holding Supply Chain Actors Responsible' (*CLX*) <https://clxtoolkit.com/casebook/envol-vert-et-al-v-casino/> accessed 15 July 2023.

### 4.3 Seeking Damages

While there are numerous lawsuits pending against states in the European Union regarding their human rights obligations to protect the climate and against European companies for their role in climate change, *Lliuya v RWE* and *Asmania and Others v Holcim* stand out as the only cases that have been filed by affected parties in the Global South. While both cases are at relatively early stages of the adjudicatory process, they have forged a transnational litigation pathway for Global South claimants who seek financial contribution towards the costs of climate change adaptation from corporate actors that operate across the globe but typically have their headquarters in Global North jurisdictions and often are owned (at least partially) by Global North states.[164] These corporate actors are also major emitters of GHGs. Due to the novelty of their case theory, the two cases may have potentially far-reaching ramifications for holding corporate actors legally accountable for climate change-related damage or threats of damage.

Richard Heede's 2013 study broke new ground by being the first to map and quantify the cumulative emissions of the 90 largest carbon producers from the year 1854 to 2010, which roughly corresponds to the period since the Industrial Revolution to the present time.[165] This study calculated a percentage figure for the individual contribution of each 'carbon major' entity of two-thirds of all global anthropogenic carbon emissions.[166] This pivotal study formed the basis of the unprecedented lawsuit filed in November 2015 by Saul Luciano Lliuya, a farmer from the Andean region of Peru, against the German utility company RWE.[167] In this lawsuit, Lliuya seeks from RWE damages of US$21,000 to contribute to the costs of building defences against glacial lake flooding, landslides, and destruction of his property. The sum of US$21,000 amounts to 0.47% of the estimated costs of the adaptation works described above, which corresponds to the finding in Heede's *Carbon Majors* study that 0.47% of carbon dioxide that has been emitted into the atmosphere between 1854 and 2010 can be traced back to RWE. The lawsuit is based on the claim for removal and injunction pursuant to section 1004 of the German Civil Code and, at its core, is concerned with the impending violation of an individual legal interest (property).[168] At the time of writing, *Lliuya v RWE* is pending

[164] See Climate Accountability Institute, 'Carbon Majors 2018 Data Set Released December 2020' <https://climateaccountability.org/carbonmajors_dataset2020.html> accessed 17 July 2023.

[165] Richard Heede, 'Tracing Anthropogenic Carbon Dioxide and Methane Emissions to Fossil Fuel and Cement Producers, 1854–2010' (2014) 122 Climatic Change 229.

[166] ibid.

[167] LG Essen, 23 November 2015, 2 O 285/15, 14/0354Z/R/rv.

[168] For discussion, see eg Philipp Semmelmayer, 'Climate Change and the German Law of Torts' (2021) 22 German Law Journal 1569; Vedantha Kumar and Will Frank, 'Holding Private Emitters to Account for the Effects of Climate Change: Could a Case Like Lliuya Succeed under English Nuisance Laws?' (2018) 12(2) Carbon & Climate Law Review 110–23.

decision after judges of the Higher Regional Court of Hamm (OLG Hamm), legal advisors, and experts travelled to Peru to gather evidence in 2022.

In a lawsuit that mirrors *Lliuya v RWE*, Asmania and three other residents of the Indonesian island of Pari are suing Swiss cement giant Holcim for causing climate change and are seeking compensation for 'climate damages'.[169] The plaintiffs argue that the 1,500 inhabitants of Pari are at significant risk of losing their livelihoods due to sea level rise and flooding, even though they have contributed little to global GHG emissions. Citing a 2022 study by Richard Heede, the plaintiffs demand that Holcim pay 0.42% of the cost of damage that has already occurred and is impending, and of new flood protection measures, which corresponds to the company's contribution of 0.42% of global fossil fuel and cement emissions since 1751.[170] The plaintiffs in this case are also asking Holcim to commit to more rapidly reducing its carbon emissions. The case is supported by aid group Swiss Church Aid HEKS/EPER, the European Centre for Constitutional and Human Rights, and the Indonesian environmental organization WALHI.[171] These three organizations are supporting the lawsuit with a wider 'Call for Climate Justice' advocacy campaign, which has already helped generate significant media coverage of the lawsuit in established news outlets including *The Jakarta Post*, *Financial Times*, and SWI swissinfo.ch.[172] The lawsuit was filed in February 2023 after the conciliation process failed and is pending at the time of writing.

## 5. CONCLUSION

This chapter provided an overview of the 'Global South docket' which, at the time of writing, comprises of 128 cases. The cases show that a variety of legal arguments are raised by plaintiffs, including infringement of constitutional rights, flaws in environmental impact assessments for proposed projects, and inadequacies in the state's implementation of climate change laws and policies. The chapter then proceeded to distil some characteristics of climate cases in the Global South

[169] HEKSEPER, 'An Island Demands Justice' <https://en.heks.ch/medien/island-demands-justice> accessed 17 July 2023.

[170] Richard Heede, 'Carbon History of Holcim Ltd: Carbon Dioxide Emissions from 1950–2021' (*Call for Climate Justice*, 7 July 2022) <https://callforclimatejustice.org/wp-content/uploads/Heede-Report.pdf> accessed 17 July 2023.

[171] HEKSEPER (n 169).

[172] ibid; for examples of media coverage: AFP, 'Pulau Pari Islanders Sue Cement Giant Holcim over Climate Damage' *The Jakarta Post* (14 July 2022) <www.thejakartapost.com/business/2022/07/14/pulau-pari-islanders-sue-cement-giant-holcim-over-climate-damage.html> accessed 16 July 2023; Camilla Hodgson, 'Indonesia Islanders File Climate Lawsuit against Swiss Cement Group Holcim' *Financial Times* (1 February 2023) <www.ft.com/content/e1fd3eba-64a0-476a-85cb-fb8fa7084fb0> accessed 16 July 2023; Emma Farge, 'Indonesian Residents File Climate Case against Holcim in Swiss Court' (*swissinfo*, 1 February 2023) <www.swissinfo.ch/eng/reuters/indonesian-residents-file-climate-case-against-holcim-in-swiss-court/48249618> accessed 16 July 2023.

as a basis for reframing and broadening our collective knowledge of the phenomenon of transnational climate litigation. Finally, the chapter provided insights into cases that are not strictly part of the Global South docket but have been filed by affected parties in the Global South seeking legal accountability and climate justice in Global North courts as well as international and regional tribunals.

Following this survey of climate litigation across the Global South, Chapter 4 undertakes a closer examination of the jurisprudence of the 'frontrunner jurisdictions' of India, Brazil, and South Africa where most of this litigation is occurring. With a focus on the experience of these 'frontrunner' jurisdictions, Chapter 4 also seeks to shed light on the factors and conditions that support the emergence of climate change litigation in the Global South more broadly.

# 4

# The Emergence of Climate Change Litigation in the Global South

## 1. INTRODUCTION

The emergence of climate change litigation in the Global South is happening in a small group of countries that fall within the Middle-Income Countries (MICs) category. As of 2023, there are 108 MICs, mostly in Asia and Latin America, and they occupy the socio-economic middle ground between the lowest-income countries of the Global South and the high-income countries of the Global North.[1] Middle-income countries are a varied group socially, culturally, politically, and economically but tend to confront a common set of challenges. They include maintaining sustainable economic growth while curbing rising income inequality and environmental degradation; providing a safety net for their citizens who are adversely affected by globalization and drastic economic transformation; tackling corruption; meeting the Sustainable Development Goals (SDGs) on, for example, healthcare and access to education; and building effective governance institutions and maintaining social and political stability in the face of citizens' growing expectations of the state and quality of life.[2]

Although MICs generally have adequate institutional capacity, institutions are typically weaker than their counterparts in developed countries. In some MICs, '[t]heir institutional fields are hybrid, in that they exhibit elements of capacity and incapacity, and are marked by contrast—for example, between an authoritarian

[1] Nada Hamadeh and others, 'World Bank Group Country Classifications by Income Level for FY24 (July 1, 2023 – June 30, 2024)' (*World Bank Blogs*, 30 June 2023) <https://blogs.worldbank.org/opendata/new-world-bank-group-country-classifications-income-level-fy24> accessed 2 August 2023.

[2] Randall Peerenboom, 'Law and Development in Middle-Income Countries: Introduction' in Randall Peerenboom and Tom Ginsburg (eds), *Law and Development of Middle-Income Countries: Avoiding the Middle-Income Trap* (CUP 2014) 1. There is a vast body of literature on MICs. See eg Shekhar Aiyar and others, 'Growth Slowdowns and the Middle-Income Trap' (2013) International Monetary Fund Working Papers WP/13/71 <www.imf.org/external/pubs/ft/wp/2013/wp1371.pdf> accessed 24 July 2023; José Antonio Alonso and José Antonio Ocampo, *Trapped in the Middle?: Developmental Challenges for Middle-Income Countries* (OUP 2020); Indermit Gill and Homi Kharas, *An East Asian Renaissance: Ideas for Economic Growth* (The World Bank 2007) <https://openknowledge.worldbank.org/handle/10986/6798> accessed 24 July 2023; Geoffrey Garrett, 'Globalisation's Missing Middle' (2004) 83(6) Foreign Affairs 84.

*Litigating Climate Change in the Global South.* Jolene Lin and Jacqueline Peel, Oxford University Press.
 DOI: 10.1093/oso/9780191926525.003.0004

executive captured by private interests on the one hand, and an independent judiciary that protects rights on the other'.[3] Such conditions tend to create 'the perfect storm for strategic litigation: a legal system that provides an ideal institutional framework to litigate, and conditions of life for the majority of the people that generate disputes and demands that become major cases'.[4] It has been noted that among the economic variables that favour the judicialization of conflicts regarding socio-economic rights, 'income level and distribution stand out. By definition, middle-income countries have more resources than low-income countries, but have a distribution of wealth just as or more unequal than the latter.'[5]

In some Global South jurisdictions, climate change litigation is emerging because the courts are the only available institution through which civil society (and citizens, more broadly) is able to engage in and influence climate change governance.[6] Civil society groups, youth, and other potential litigants have also been inspired by lawsuits in the Global North, which have yielded climate-protective rulings and mobilized climate consciousness in society.[7] Climate change lawsuits are also often filed—despite hostile conditions—as part of broader advocacy strategies.[8] In such cases, winning in the courtroom is not a major consideration. Other outcomes, such as publicity and the opportunity to highlight the perceived inadequacies in governmental and/or corporate climate change policies, are deemed sufficiently valuable to pursue litigation.[9] In Global South countries with an established tradition of public interest environmental litigation, court action is a matter

[3] Alonso and Ocampo, Introduction (n 2) 43.

[4] Jason Brickhill, 'Strategic Litigation in a Perfect Storm—South Africa' (*OpenGlobalRights*, 2 April 2019) <www.openglobalrights.org/strategic-litigation-in-a-perfect-storm-south-africa/> accessed 24 July 2023.

[5] César Rodríguez-Garavito and Diana Rodríguez-Franco, *Radical Deprivation on Trial: The Impact of Judicial Activism on Socioeconomic Rights in the Global South* (CUP 2015) 13.

[6] See, for discussion, Louis J Kotzé and Anél du Plessis, 'Putting Africa on the Stand: A Bird's Eye View of Climate Litigation on the Continent' (2020) 50 Environmental Law 615. For broader discussion about the turn to the courts because of democratic deficit, such as where the legislature is captured by powerful elite, see Torbjörn Vallinder, 'The Judicialization of Politics. A World-Wide Phenomenon: Introduction' (1994) 15 International Political Science Review 91; Ran Hirschl, 'The Judicialization of Politics' in Robert E Goodin (ed), *The Oxford Handbook of Political Science* (OUP 2011); Mary L Volcansek, 'Judicialization of Politics or Politicization of the Courts in New Democracies?' in Christine Landfried (ed), *Judicial Power: How Constitutional Courts Affect Political Transformations* (CUP 2019).

[7] eg *Urgenda Foundation v The State of the Netherlands (Ministry of Infrastructure and the Environment)* [2018] HAZA C/09/00456689 (The Hague Court of Appeal, 9 October 2018); *Friends of the Irish Environment v The Government of Ireland* [2020] IESC 49 (Supreme Court of Ireland); *Neubauer v Germany* (BvR 2656/18/1 BvR 78/20/1 BvR 96/20/1 BvR 288/20) (German Federal Constitutional Court, 29 April 2021).

[8] Interviews 1 and 2. See also Lisa Vanhala, 'Legal Opportunity Structures and the Paradox of Legal Mobilization by the Environmental Movement in the UK' (2012) 46 Law and Society Review 523, in which the author explores the phenomenon in the United Kingdom that environmental NGOs continue to pursue legal action despite facing a hostile legal opportunity structure (LOS).

[9] 'Even a negative ruling on the merits can expose where failures occurred, vindicate salient facts, and place the onus for future action on another branch of government.' Ketan Jha, 'Networked Public Interest Litigation: A Novel Framework for Climate Claims?' in Jolene Lin and Douglas A Kysar (eds), *Climate Change Litigation in the Asia Pacific* (CUP 2020) 47.

of course. In India and Pakistan, for instance, 'almost every major environmental issue that has made the news ends up before the High Court or the Supreme Court through a petition or is taken up by the Supreme Court of its own volition under its *suo motu* jurisdiction'.[10] In these jurisdictions, litigants have filed core climate change lawsuits in the form of public interest litigation, as well as periphery cases involving statutory interpretation and application of environmental and natural resources laws.

Climate change litigation in the Global South nonetheless is in its infancy and is by no means a universal phenomenon. As one eminent jurist has put it, 'judicial approaches [to climate change litigation] in different countries vary according to the constitutional and legal frameworks, responsiveness of the courts and robustness of the civil society. Some countries, therefore, play a more significant role in climate change litigation compared to others.'[11] Our survey of the Global South docket has allowed us to identify three jurisdictions—Brazil, India, and South Africa—as the climate litigation 'frontrunners' with the highest number of cases in Latin America, Asia and the Pacific, and Africa respectively. With a focus on the experience of these 'frontrunner' jurisdictions, we have sought to gain a better understanding of the general conditions that support the emergence of climate change litigation in the Global South more broadly. [12]

Through an analytical survey of the climate change case law, perusal of secondary literature on the political economy, legal system, and environmental politics in these three jurisdictions, and interviews, we have identified five conditions as being key explanatory factors for the emergence of climate change litigation:

1. access to justice, which includes broad standing rules and the removal of barriers such as hefty court fees and security of costs orders;

[10] Waqqas Ahmad Mir, 'From Shehla Zir to Asghar Leghari: Pronouncing Unwritten Rights Is More Complex than a Celebratory Tale' in Lin and Kysar (eds) (n 9) 286.

[11] Syed Mansoor Ali Shah, 'Foreword' in Lin and Kysar (eds) (n 9) xii.

[12] The body of scholarship on LOS has been instructive in this regard. LOS is a theoretical framework that has been used to explain why social movements employ litigation, as opposed to lobbying or protesting, as a strategy. Hilson has applied LOS theory to the women's, environmental, animal welfare, and lesbian and gay movements: Chris Hilson 'New Social Movements: The Role of Legal Opportunity' (2002) 9 Journal of European Public Policy 238. Other early adopters include Anderson, and Wilson and Rodríguez-Cordero who applied LOS theory to the gay rights movements in the United States and Costa Rica respectively, and Case and Givens who applied LOS theory to the study of racial equality in the European Community: see Ellen Ann Andersen, *Out of the closets and into the courts: Legal opportunity structure and gay rights litigation* (University of Michigan Press 2005); Bruce M Wilson and Juan Carlos Rodríguez-Cordero, 'Legal Opportunity Structures and Social Movements: The Effects of Institutional Change on Costa Rican Politics' (2006) 39 Comparative Political Studies 325; Rhonda Evans Case and Terri E Givens, 'Re-engineering Legal Opportunity Structures in the European Union? The Starting Line Group and the Politics of the Racial Equality Directive' (2010) 48 Journal of Common Market Studies 221. Setzer and Benjamin draw upon LOS approaches in their survey of climate litigation in the Global South: Joana Setzer and Lisa Benjamin, 'Climate Litigation in the Global South: Constraints and Innovations' (2019) 9 Transnational Environmental Law 77.

2. a climate-conscious judiciary by which we mean, inter alia, judges who are receptive to creative arguments that use existing environmental and natural resources laws as the basis for climate change litigation. In some cases, a climate-conscious judge goes beyond mere receptivity and is willing to use judicial power to address the climate emergency;
3. progressive environmental jurisprudence and/or the existence of climate change laws, policies, and institutions;
4. constitutional protection for the environment including the right to a clean environment; and
5. a well-developed civil society.

Not all five conditions always exist in every Global South jurisdiction with climate cases but, by and large, they complement each other to create an environment that supports the emergence of climate change litigation. What our analysis shows is that while climate change litigation generally takes a different form in the Global South when compared to the Global North—for example, there are more periphery cases and more frequent use of rights-based arguments—the conditions that support climate change litigation in the Global South are not significantly different from those in the Global North. This suggests that if litigation is to be promoted as a climate change governance tool in the Global South, policy attention and resources must continue to be directed at establishing and strengthening the fundamental pillars of effective legal institutions and access to justice.[13]

Following this Introduction, Section 2 provides a descriptive account of three jurisdictions—India, South Africa, and Brazil—that have been identified as the climate litigation 'lead' jurisdictions in Asia and the Pacific, Africa, and Latin America respectively. Section 3 discusses the five conditions: (1) access to justice; (2) a climate-conscious judiciary; (3) progressive environmental jurisprudence and/or the existence of climate change laws, policies, and institutions; (4) constitutional protection for the environment; and (5) robust civil society. Section 4 concludes with a summary of the ideas advanced in this chapter.

[13] This coheres with the work the United Nations Environment Programme (UNEP) is doing to support the environmental rule of law as part of Paris implementation. For example, UNEP published the first ever global assessment of environmental rule of law in 2019, which found that weak enforcement of the rule of law was a global trend that is exacerbating environmental threats. See 'Environmental Rule of Law: First Global Report' (*United Nations Environment Programme*, 24 January 2019) <www.unep.org/resources/assessment/environmental-rule-law-first-global-report> accessed 25 July 2023. Its second global environmental rule of law report was published in November 2023: see 'Environmental Rule of Law: Tracking Progress and Charting Future Directions' (*United Nations Environment Programme*, 22 November 2023), available at https://www.unep.org/resources/publication/environmental-rule-law-tracking-progress-and-charting-future-directions, accessed 22 January 2024.

## 2. THE CLIMATE CHANGE LITIGATION 'FRONTRUNNER' JURISDICTIONS

Across Asia and the Pacific, Africa, and Latin America, climate change lawsuits have been filed mostly against governments to compel them to implement domestic climate change policies as well as to address climate change through existing environmental law statutes and regulatory frameworks. Our analytical survey of the case law has led us to identify three jurisdictions as the 'frontrunners' with the highest number of climate change lawsuits in their respective geographical regions. They are India, South Africa, and Brazil.

This section provides a descriptive account of the rise of climate change litigation in these 'frontrunner' jurisdictions. It is beyond the scope of this chapter to provide detailed case studies, and this is not what is intended. The primary aim of this section is to give readers a more contextual understanding of how climate change litigation has come to be in these 'frontrunner' jurisdictions. It also provides the analytical fodder for identifying the common factors or conditions that are key indicators for the emergence of climate change litigation in the Global South, which is the focus of Section 3 of this chapter.

### 2.1 India

With a population of 1.4 billion people, India is the most populous country in the world.[14] The country also has one of the world's largest economies.[15] While it is one of the fastest growing economies in the world, it is also one of the most unequal. The top 10% of India's population hold nearly 75% of the total national wealth.[16] Industrialization, rapid population growth, and growing consumerism have placed great pressure on the environment. In a list of the 15 most polluted cities in

[14] 'UN DESA Policy Brief No. 153: India Overtakes China as the World's Most Populous Country' (*United Nations Department of Economic and Social Affairs*, 24 April 2023) <www.un.org/development/desa/dpad/publication/un-desa-policy-brief-no-153-india-overtakes-china-as-the-worlds-most-populous-country/> accessed 25 July 2023.

[15] India was the fifth largest economy in the world in 2019 but slipped one place to sixth largest in 2020 because strict lockdowns in response to the Covid pandemic adversely affected its economy; Yen Nee Lee, 'Here Are the 10 Biggest Economies in the World—Before the Pandemic vs. Now' *CNBC* (20 April 2021) <www.cnbc.com/2021/04/21/coronavirus-worlds-10-biggest-economies-before-covid-pandemic-vs-now.html> accessed 25 July 2023.

[16] 'India: Extreme Inequality in Numbers' (*Oxfam International*) <www.oxfam.org/en/india-extreme-inequality-numbers> accessed 25 July 2023: India has 119 billionaires, while 63 million Indians are pushed into poverty because of healthcare costs every year. The Covid-19 pandemic has worsened economic inequality and will hurt certain groups, especially women and young people, more than others: Delphine Strauss, 'Pandemic Hits Poor Hit Hardest as Inequality Rises' *Financial Times* (3 November 2020) <www.ft.com/content/750eb552-639e-4fa0-941b-4f3f57f1a8d4> accessed 25 July 2023; Oliver Tonby and Anu Madgavkar, 'Countering the Gender-Regressive Shock of COVID-19 in Asia' *Financial Times* (24 August 2020) <www.ft.com/partnercontent/mckinsey/countering-the-gender-regressive-shock-of-covid-19-in-asia.html> accessed 25 July 2023.

the world in 2021, all but two were Indian cities.[17] With nearly 30% of its population living in poverty without access to electricity, India is the world's third largest emitter of carbon and it is likely that it will soon overtake China and the United States with its increasing reliance on fossil fuel for energy.[18] India therefore has difficult choices to make in the face of competing priorities of economic growth, energy security, and climate change. In terms of its institutions, the Indian judiciary is globally recognized for its progressive jurisprudence on environmental matters and, at least until the Modi administration came to power in 2014, has played an important role in filling the governance gap created by divided and weak executive and legislative branches of government.[19]

In our analysis at the time of writing, the Indian climate change litigation docket comprises of 18 cases—two core cases and 16 periphery cases. The majority of cases—11 out of 18 cases—were adjudicated in the National Green Tribunal (NGT) and the remaining seven were adjudicated in the Supreme Court and High Courts. As the following discussion shows, India has emerged as the climate litigation 'frontrunner' in the Asia and Pacific region because it has a proactive judiciary with a track record of being progressive in environmental matters, and Indian citizens have greater access to justice compared to many other jurisdictions. Standing rules are broad and some of the procedural hurdles that typically are major obstacles to litigation elsewhere have been removed to promote public interest litigation (PIL). Non-governmental organizations (NGOs) and concerned citizens are also using their experience in environmental litigation to bring climate change lawsuits.[20] India does not have a dedicated climate change law, but several environmental laws serve as 'hooks' for climate change litigation, including laws regulating air pollution, forestry, and environmental impact assessment.[21]

[17] Owen Mulhern, 'Mapping the Most Polluted Cities in the World' (*Earth.Org*, 30 August 2021) <https://earth.org/data_visualization/mapping-the-most-polluted-cities-in-the-world/> accessed 25 July 2023. The Ganges, which is the third largest river in the world, is also one of the most polluted because of, inter alia, the dumping of raw sewage and plastic waste; Donald G McNeil Jr, 'The Ganges Brims With Dangerous Bacteria' *The New York Times* (23 December 2019) <www.nytimes.com/2019/12/23/health/ganges-drug-resistant-bacteria.html> accessed 25 July 2023.

[18] Jocelyn Timperley, 'The Carbon Brief Profile: India' (*CarbonBrief*, 14 March 2019) <www.carbonbrief.org/the-carbon-brief-profile-india> accessed 25 July 2023.

[19] Throughout the 1980s till the sixteenth general elections in 2014 which ushered in the Modi administration, the Indian political space was defined by unstable coalitional politics which impeded decisive and firm policymaking. With governmental weakness and Parliament riddled with corruption and criminality, the Supreme Court became a prominent institution of agenda-setting and policymaking in environmental governance and other policy arenas: Arun K Thiruvengadam, *The Constitution of India: A Contextual Analysis* (Bloomsbury 2017) 129–34.

[20] In contrast, it has been argued that there has not been widespread engagement with climate change by civil society actors at the local level, even if they are dealing with its social and environmental impacts, because of sympathy with the government's international position in the climate change negotiations that by and large '[conserves] a (carbon intensive) development space and ensures that the obligations imposed on developing countries like itself are kept at a minimum': Birsha Ohdedar, 'Climate Change Litigation in India and Pakistan: Analyzing Opportunities and Challenges' in Ivano Alogna, Christine Bakker, and Jean-Pierre Gauci (eds), *Climate Change Litigation: Global Perspectives* (Brill 2021) 106.

[21] Chandra Bhushan and Tarun Gopalakrishnan, 'Environmental Laws and Climate Action: A Case for Enacting a Framework Climate Legislation in India' (*Shakti Sustainable Energy Foundation*

It is noteworthy that contemporary politics, namely an authoritarian government which has sought to shrink the space for environmental advocacy and clamp down on environmental activism, has hindered the development of climate change litigation in India. Tighter restrictions on NGOs receiving financial contributions from foreign sources,[22] and the arrest of a climate change youth activist charged with sedition (which carries a life imprisonment term) for her alleged role in creating an online toolkit that incited violence during protests against new agricultural laws, contributing to fear of adverse consequences that might result from challenging the government's performance on climate action.[23] Instead, petitioners

*and International Forum for Environment, Sustainability and Technology*, 2021) <https://shaktifoundation.in/wp-content/uploads/2021/12/Environmental-Laws-and-Climate-Action.pdf> accessed 25 July 2023.

[22] The Foreign Contribution (Regulation) Act 2010 (FCRA) (India) aims to (a) regulate the acceptance and use of foreign contribution or hospitality by certain individuals and entities (including NGOs), and (b) prohibit the acceptance and use of foreign contribution or hospitality for any activities which are 'detrimental to the national interest'. The Act has an expansive definition of 'organisations of a political nature' including an organisation of 'farmers, workers, students, youth based on caste, community, religion, language or otherwise, which is not directly aligned to any political party, but whose objectives . . . include steps towards advancement of political interests of such groups', as well as any organization which 'habitually engages itself in or employs common methods of political action' such as strikes in support of political causes; Foreign Contribution (Regulation) Rules 2011 (India), r 3. Amendments to the FCRA introduced in 2020 brought in further restrictions, eg on transferring foreign contributions. *The Hindu* reported in September 2020 that the licences of 20,674 NGOs or associations registered under the FCRA had been cancelled; Vijaita Singh, 'Government Suspends FCRA Clearance of 4 Christian Groups' (*The Hindu* (6 September 2020) <www.thehindu.com/news/national/government-suspends-fcra-clearance-of-4-christian-groups/article32535766.ece> accessed 25 July 2023. Later that month, Amnesty International halted its India operations: K. C. Deepika, 'Amnesty International Halts India Operations' (*The Hindu*, 29 September 2020) <www.thehindu.com/news/national/amnesty-international-halts-india-operations/article32721447.ece> accessed 25 July 2023. The International Commission of Jurists has asserted that the FCRA is 'not compliant with international law and standards', and that it 'imposes undue restrictions and burdens on the legitimate activities of human rights defenders and activists while placing them at great risk': 'India's Foreign Contribution Regulation Act: A Tool to Silence Indian Civil Society Organizations' (Briefing Paper, International Commission of Jurists 2020) <www.icj.org/wp-content/uploads/2020/12/India-FCRA-Briefing-Paper-2020-ENG.pdf> accessed 25 July 2023. Michelle Bachelet, United Nations High Commissioner for Human Rights, has likewise expressed concerns that the FCRA is 'open to abuse, and that it is indeed actually being used to deter or punish NGOs for human rights reporting and advocacy': 'Bachelet Dismayed at Restrictions on Human Rights NGOs and Arrests of Activists in India' (*United Nations Human Rights Office of the High Commissioner*, 20 October 2020) <www.ohchr.org/en/press-releases/2020/10/bachelet-dismayed-restrictions-human-rights-ngos-and-arrests-activists-india> accessed 25 July 2023. At the time of writing, the Indian Supreme Court is hearing a batch of petitions challenging the constitutional validity of the 2020 Amendment: 'FCRA Amendments Aim to Prevent Foreign Powers Interfering with Internal Polity, Diversion of Funds by NGO: Centre Tells Supreme Court' (*Live Law*, 22 October 2021) <www.livelaw.in/top-stories/fcra-amendments-2020-supreme-court-ngos-foreign-powers-centre-diversion-of-funds-184100> accessed 25 July 2023. One of the affidavits, *Noel Harper v Union of India*, WP (C), Petition No 566/2021, Supreme Court of India is available at <www.livelaw.in/pdf_upload/fcra-amendment-2020-centres-counter-affidavit-2-402735.pdf> accessed 25 July 2023.

[23] Suchitra Mohanty and Zeba Siddiqui, 'Indian Judge Sees No Reason to Detain Climate Activist Charged with Sedition' *Reuters* (23 February 2021) <www.reuters.com/article/us-india-farms-protests-idUSKBN2AN13D> accessed 25 July 2023; Sameer Yasir, 'Climate Activist Jailed in India as Government Clamps Down' *New York Times* (15 February 2021) <www.nytimes.com/2021/02/15/world/asia/climate-activist-jailed-india.html> accessed 25 July 2023; Amy Kazmin, 'Detention Stokes Fear Among India's Young Climate Change Activists' *Financial Times* (20 February 2021) <www.ft.com/content/29d82d57-882e-4aaa-b58c-941310b0d2e5> accessed 25 July 2023; Interviews 3 and 4.

have generally used legal arguments that rely on human and constitutional rights jurisprudence and enforcement of existing environmental laws; climate change concerns are peripheral and 'under the radar'. Gitanjali Gill and Gopichandran Ramachandran argue that 'the Indian judiciary through a step-change or a small win process underpinned by the environmental rule of law contributes towards an evolving climate change transformation'.[24] It is a process of incremental steps which can be taken because they 'do not rock the boat' or cause 'paralysing schisms', unlike proposals for more drastic change.[25]

The following discussion introduces the NGT as a significant venue for climate change litigation in India. The discussion highlights some of the cases in the Indian climate change litigation docket that have been decided by the NGT. It then proceeds to briefly introduce PIL which has not been a significant legal route for climate change litigation but is a hallmark of the Indian environmental legal framework and should be included in any discussion about climate change litigation in India as a potential driver for developing climate change jurisprudence. The Supreme Court, as India's apex court, also has a role to play in advancing sustainability and climate change governance through its exercise of appellate jurisdiction. Finally, the Indian PIL tradition has engendered an active environmental bar and civil society. These actors are using their courtroom experience and legal expertise to litigate climate change issues, demonstrating the importance of a robust civil society in advancing climate change litigation.

### The National Green Tribunal

The NGT has jurisdiction over all civil cases where 'a substantial question relating to the environment . . . is involved and such question arises out of the implementation of the enactments specified in Schedule 1' of the National Green Tribunal Act 2010 (NGT Act 2010).[26] The laws listed in Schedule 1 include the Forest (Conservation) Act 1980, the Air (Prevention and Control of Pollution) Act 1981, and the Biological Diversity Act 2002. As the government's climate change mitigation and adaptation obligations do not arise from any of the seven laws listed in Schedule 1 of the NGT Act 2010, it has been argued that the tribunal does not

[24] Gitanjali N Gill and Gopichandran Ramachandran, 'Sustainability Transformations, Environmental Rule of Law and the Indian Judiciary: Connecting the Dots Through Climate Change Litigation' (2021) 23 Environmental Law Review 228, 239.

[25] ibid 240.

[26] The National Green Tribunal Act 2010 (No 19 of 2010) (India) (NGT Act). The preamble states the overarching goal of this law, that is, 'to provide for the establishment of a National Green Tribunal for the effective and expeditious disposal of cases relating to environmental protection and conservation of forests and other natural resources including enforcement of any legal right relating to environment and giving relief and compensation for damages to persons and property and for matters connected therewith and incidental thereto'. The NGT is made up of legal and scientific experts who work together as joint decision-makers and adjudicators to dispense environmental justice: NGT Act, ss 4(1)–(2), 5(1)–(2).

have jurisdiction over climate change-based claims.[27] The NGT's decisions in two core cases—*Gaurav Kumar Bansal v Union of India*[28] and *Pandey v Union of India and Others*[29]—could be characterized as being cautious, arguably evidencing the tribunal's cognizance of its jurisdictional limit, especially when it is invited to intervene in the government's policy decisions. In *Gaurav Kumar Bansal v Union of India*, the petitioner petitioned the NGT to hold the central government and state governments accountable for implementing the National Action Plan on Climate Change. In a short decision of two pages, the NGT directed the state governments to comply with the directions issued by the Ministry of Environment, Forests and Climate Change (MoEFCC).

In *Pandey v Union of India*, the petitioner was a nine-year-old girl represented by her legal guardian. She argued that she had standing on the basis that children today face greater risks of being adversely affected by climate change.[30] In this first case of its kind in India, Pandey argued that the non-implementation of various environmental laws has led to adverse climate change impacts across the country, that the state was not taking effective science-based climate mitigation measures, and was therefore in breach of its obligations pursuant to the public trust doctrine under which the state holds the country's resources on trust for future generations. The applicant also raised the argument that the state's inaction meant that it was not fulfilling its Paris Agreement obligations. In a submission spanning nearly 50 pages, the applicant provided a comprehensive overview of the impacts of climate change that current and future generations of Indians have to experience including glacial melt, extreme weather events, and impacts on human health.[31] The applicant also put forward a cogent argument that many of the climate impacts 'can be averted, minimised and mitigated if effective, science-based measures are taken under the existing environmental legal framework in India' including the laws listed in Schedule 1 of the NGT Act 2010.[32] In a short decision issued two years after the case was filed in 2017, the tribunal disposed of the matter by concluding that it was not necessary to make any orders because '[t]here is no reason to presume that [the] Paris Agreement and other international protocols are not

[27] eg in *Gaurav Kumar Bansel v Union of India* (Order of the Tribunal) (National Green Tribunal Principal Bench, New Delhi, 23 July 2015), the Ministry of Environment, Forests and Climate Change (MoEFCC) filed an affidavit raising objections to the NGT exercising jurisdiction over this case as the issue of implementing the National Action Plan on Climate Change and the State Action Plans on Climate Change does not fall squarely under any of the laws listed in sch 1 of the NGT Act 2010.

[28] *Gaurav Kumar Bansal v Union of India*, Original Application, No 498 of 2014, National Green Tribunal, 23 July 2015.

[29] *Pandey v Union of India*, Original Application No 187/2017, National Green Tribunal, 15 January 2019.

[30] On this and the subsequent details, see *Pandey v Union of India*, Original Application No 187/2017, National Green Tribunal, 25 March 2017.

[31] ibid.

[32] ibid 24.

reflected in the policies of the Government of India or are not taken into consideration in granting environmental clearances'.[33]

In a third core case, *Indian Council for Enviro-legal Action (ICELA) v Ministry of Environment, Forest and Climate Change*, the NGT did not hesitate to take a purposive reading of the Environmental Protection Act 1986 and the NGT Act 2010 to conclude that it had jurisdiction to decide on the petition seeking the tribunal's intervention to stop the manufacturing of a hydrofluorocarbon (HFC-23) which is a greenhouse gas.[34] However, the tribunal concluded that, as there was no domestic legal regulation of HFC-23 in place, the issue was solely one to be dealt with under international treaty law and 'a part of policy of Government of India at the global level'.[35] While the tribunal recognized that there was a public policy gap to fill as HFC-23 is a greenhouse gas that contributes to the problem of global warming, it held that it was not in the position to make policy when the issue had environmental as well as other socio-economic implications.[36] The tribunal concluded by directing the relevant authorities to examine the HFC-23 regulatory regime.

The NGT has also adjudicated several periphery cases where climate change regulation is being advanced via the enforcement of the forestry and environmental laws listed in Schedule 1 of the NGT Act 2010. By addressing the fundamental drivers behind climate change, for example deforestation that causes loss of carbon sinks that can only be poorly substituted by replanting trees, these periphery cases can go a long way to advancing climate protection. *Sher Singh v State of Himachal Pradesh* is an interesting case in point.[37] This is a case where the NGT exercised *suo moto* jurisdiction and, as such, brought this case 'on its own motion' to push for better tourism and traffic management in the Rohtang Pass of the Himalayan range.[38] The Rohtang Pass, at a height of about 13,000 feet above sea level, is widely known as the 'Crown Jewel' of Himachal Pradesh and attracts a large number of tourists.[39] In its decision, the NGT described in detail the adverse ecological and environmental impacts of unbridled tourism, including glacial retreat that is attributable to high levels of black carbon emissions from tourist vehicles. The tribunal noted that black carbon 'is believed to be the biggest contributor to global warming

[33] *Pandey v Union of India* (n 29) 2.

[34] Original Application No 170/2014, National Green Tribunal, 10 December 2015. Hydrofluorocarbons are listed as a greenhouse gas (GHG) in Annex A of the Kyoto Protocol to the United Nations Framework Convention on Climate Change.

[35] *Indian Council for Enviro-legal Action* (n 34) 23.

[36] ibid.

[37] Also cited as *Court on its own Motion v State of Himachal Pradesh*, Application No 237/2013, CWPIL No 15 of 2010, National Green Tribunal, 6 February 2014.

[38] In *suo moto* proceedings, a superior court commences proceedings in the public interest and acts of its own volition in the absence of parties. Galanter argues that these *suo moto* 'initiatives derive from judicial reaction to a newspaper article or a letter (the so-called "epistolary jurisdiction"), which have become a familiar feature of the Indian judicial scene, rare but highly visible': Marc Galanter, 'Snakes and Ladders: Suo Motu Intervention and the Indian Judiciary' (2014) 10 Florida International University Law Review 69, 74.

[39] *Sher Singh v State of Himachal Pradesh* (n 37) [3].

after carbon dioxide.'[40] Guided by the aim of preventing further environmental degradation by reducing vehicular pollution in particular, the tribunal made an order setting out detailed directions to various agencies. These directions included: (1) ordering the state government to 'install computerized weigh-in-motion systems to check over-loading of the vehicles' at the entry point to the Rohtang Pass; (2) ordering that government and private vehicles entering the Rohtang Pass be subjected to random pollution checks to ensure that they are adhering to prescribed emissions standards; and (3) finding that 'vehicles which are more than 10 years old . . . shall be phased out and should not be permitted to operate or ply on the route to Rohtang Pass from Vashishta.'[41] To deal with deforestation, the NGT ordered the state government to 'take immediate and effective measures for reforestation' with specific recommendations for the tree species that can be planted at various altitudes.[42] The NGT also ordered that a monitoring committee consisting of various ministry officials including the Director of Tourism (Government of Himachal Pradesh) and the Secretary of Environment (Government of Himachal Pradesh) and an environmental sciences expert be established. The committee would be responsible for monitoring compliance, and cases of non-compliance would be reported to the NGT for further action: 'The Monitoring Committee shall submit quarterly reports to the NGT, clearly stating non-compliance with the directions, if any, the persons responsible . . . and also suggestions . . . as it may consider appropriate in order to make further improvements.'[43]

The NGT is well known for its broad standing rules which are seen to afford greater access to environmental justice compared to the courts of general jurisdiction (ie the High Courts and the Supreme Court). Section 18(2) of the NGT Act 2010 states that an application to the NGT for grant of relief or compensation or dispute settlement may be made by '(a) the person, who has sustained the injury; or (b) the owner of the property to which the damage has been caused; or . . . (e) *any person aggrieved, including any representative body or organization*' (emphasis added). Section 16 of the Act states that '[a]ny person aggrieved by' an order or decision made pursuant to the laws listed in Schedule 1 may file an appeal with the NGT. As held in *Vimbal Bhai v Ministry of Environment and Forests*, '[t]he person aggrieved in environmental matters must be given a liberal construction and needs to be flexible.'[44]

In *Vimbal Bhai*, the appellants were social activists and researchers challenging the validity of the forest clearance permit that had been issued to clear forests to make way for the proposed Vishnugad-Pipalkoti hydropower project. The respondent Ministry of Environment and Forests sought a dismissal of the appeal

[40] ibid [4].
[41] ibid [23].
[42] ibid [30].
[43] ibid [38(ix)].
[44] Appeal No 5 of 2011, National Green Tribunal, 14 December 2011, 7.

by arguing that the appellants were not aggrieved or injured persons pursuant to the NGT Act 2010 and therefore did not have standing to approach the tribunal. The NGT was not convinced by this argument and granted standing to the appellants on the grounds that the terms 'aggrieved person' and 'injured persons' should be flexibly interpreted to enable the law to advance environmental protection effectively.[45] Further, the tribunal noted that Article 51A of the Indian Constitution states that it was the duty of every citizen to protect and improve the natural environment. In the tribunal's view, '[o]nce the [constitutional] mandate is of every citizen, any person can approach this Tribunal complaining [of] environmental threat in the activities of the State or any organization or individual'.[46]

Observers have commented that '[t]he very fact that a villager can approach the NGT is itself a victory in terms of access to justice'.[47] It is noteworthy that apart from broad standing rules, the NGT's procedures are designed to enable citizens to approach the tribunal relatively easily. Citizens can approach the NGT without formal legal representation and it is not uncommon for petitioners to speak in their own languages and dialects.[48] About 20% of cases adjudicated by the NGT annually are argued in person by the petitioners.[49] Court fees have been deliberately kept modest so that the financial costs of litigation do not deter citizens from approaching the NGT to resolve environmental disputes or challenge a governmental decision that is deemed to be environmentally harmful.[50] The NGT is also not bound by formal rules of evidence and civil procedure rules, unlike the courts of general jurisdiction.[51] Section 19(2) of the NGT Act 2010 confers power on the NGT to determine its own rules of procedure. The NGT is therefore able to develop 'made for purpose' rules that enhance access to justice and promote innovative problem-solving approaches.[52] Finally, to become more accessible, especially in remote parts of the country, the NGT has adopted the circuit procedure of courts going to the people. For example, Shimla has received circuit benches from New Delhi where the NGT's principal bench is located.[53]

The NGT enjoys a high degree of credibility and is widely perceived by ordinary citizens as a driving force that promotes and supports environmental justice. It

[45] ibid 6.
[46] ibid 7.
[47] 'Environmental Courts and Tribunals—2021: A Guide for Policy Makers' (*United Nations Environment Programme*, 5 July 2022) 8 <www.unep.org/resources/publication/environmental-courts-and-tribunals-2021-guide-policy-makers> accessed on 20 August 2023.
[48] ibid.
[49] ibid.
[50] The National Green Tribunal (Practices and Procedure) Rules 2011 (India), r 12(2) provides for an application or appeal where no compensation is sought to be accompanied by a fee of a thousand Indian rupees (approximately US$13).
[51] NGT Act, ss 19(1), 19(3).
[52] NGT Act, s 20 states: 'The Tribunal shall, while passing any order or decision or award, apply the principles of sustainable development, the precautionary principle and the polluter pays principle.'
[53] NGT/PB/157/2013/331, 20 December 2013 (office order).

has been argued that the NGT has been 'stubbornly determined and successful' in enforcing environmental laws.[54] Its case law is evidence of 'its ability and willingness to require transparency and accountability that ultimately constitutes the building blocks of good governance and environmental democracy'.[55] It is therefore in no small measure due to the NGT that climate change litigation has emerged and is developing apace in India.

### Public interest litigation

In the late 1970s and early 1980s, the Indian Supreme Court started to develop PIL to address issues concerning the poor and marginalized sections of Indian society.[56] This was a period after the Emergency which has been regarded as a 'cataclysmic event in India's political and constitutional history'.[57] The court's reputation was badly affected as it was seen to have capitulated to the will of an authoritarian regime and was not doing enough to protect the rule of law.[58] Embarking 'on a series of unprecedented and electrifying initiatives' including expanding *locus standi* and the appointment of investigative commissions, the court sought to repair its reputation through promoting PIL.[59] In terms of constitutional competence, PIL cases come within the Supreme Court's writ jurisdiction under Article 32 of the Constitution for the enforcement of fundamental rights and under Article 226 for the High Courts.[60]

From the 1990s onwards, environmental issues dominated the Supreme Court's PIL docket.[61] Article 21 of the Indian Constitution (right to life and liberty) was

[54] Gitanjali Nain Gill, 'Mapping the Power Struggles of the National Green Tribunal of India: The Rise and Fall?' (2020) 7 Asian Journal of Law and Society 85, 113.

[55] ibid.

[56] Thiruvengadam (n 19) 128. Upendra Baxi prefers the term 'social action litigation': see Upendra Baxi, 'Taking Suffering Seriously: Social Action Litigation in the Supreme Court of India' (1985) 4 Third World Legal Studies 107. He argues that the term PIL is used in the United States, but PIL in India differs from American PIL in significant ways. First and foremost, American PIL tends to be funded by government and private foundations and its focus is less on state repression and governmental lawlessness than it is on public participation in governmental decision-making: ibid 109–10.

[57] Thiruvengadam (n 19) 126. During the Emergency, which officially lasted from 25 June 1975 to 21 March 1977, civil liberties were suspended and all governmental actions were immune from judicial scrutiny, amongst other things.

[58] ibid 127.

[59] Jayanth K Krishnan, 'Legitimacy of Courts and the Dilemma of their Proliferation: The Significance of Judicial Power in India' in Jiunn-rong Yeh and Wen-Chen Chang (eds), *Asian Courts in Context* (CUP 2014) 297. Upendra Baxi has argued that the Supreme Court's activism for the poor and marginalized sectors of Indian society in the aftermath of the Emergency was an attempt to 'bury its emergency past': Upendra Baxi, *The Indian Supreme Court and Politics* (Eastern Book Company 1980).

[60] While PIL in the Supreme Court is extensively documented, much less attention is paid to PIL or PIL-like cases in the High Courts. For a fascinating account of PIL in the Indian High Courts, see Jill Cottrell, 'Courts and Accountability: Public Interest Litigation in the Indian High Courts' (1992) 11 Third World Legal Studies 199.

[61] There is a large body of scholarship on environmental PIL in India. See eg Lavanya Rajamani, 'Public Interest Environmental Litigation in India: Exploring Issues of Access, Participation, Equity, Effectiveness and Sustainability' (2007) 19 Journal of Environmental Law 293; Michael F Faure and AV Raja, 'Effectiveness of Environmental Public Interest Litigation in India: Determining the Key Variables' (2010) 21 Fordham Environmental Law Review 239; Gitanjali Nain Gill, 'Human Rights and

judicially interpreted to include implicitly the right to a clean environment in 1991, and in a matter of a few years, the court was faced with so many environmental PIL cases that it noted '[a]t this point in time, the effect of the quality of the environment on the life of the inhabitants is much too obvious to require any emphasis or elaboration'.[62] Through PIL, the Supreme Court has sought to address several environmental issues including cleaning the Ganges river,[63] relocating hazardous industries in Delhi,[64] requiring compulsory environmental education,[65] and re-directing an illegally diverted river.[66]

There are a number of key features of the PIL process. First, the 'foremost change' that has enabled the flourishing of PIL was an expansion of the standing rules for initiating legal proceedings.[67] Given that the *raison d'être* for PIL is to provide legal redress to those who are too poor to approach the courts or are unaware of their legal rights, the court granted standing to social activists and lawyers to bring cases in the public interest.[68] Public interest litigation cases do not even need to be filed in the traditional sense. In many instances, the court has exercised *suo moto* jurisdiction to address issues that have come to a judge's attention, for example through newspaper reports. Legal formalities, such as the need to submit a formal writ petition to the courts, have also been done away with. *Sunil Batra v*

the Environment in India: Access through Public Interest Litigation' (2012) 14 Environmental Law Review 200.

[62] *MC Mehta v Union of India* (1998) (9) SCC 589, 590.

[63] *MC Mehta v Union of India* (1998) (6) SCC 63.

[64] *MC Mehta v Union of India* (1996) (5) SCC 281.

[65] *MC Mehta v Union of India* (1998) (n 63).

[66] *MC Mehta v Kamal Nath* (2000) 6 SCC 213. An overview of all the cases brought by MC Mehta concerning environmental protection can be found at 'Welcome to M.C. Mehta Environmental Foundation' (*MC Mehta Environmental Foundation*, 2021) <https://mcmef.org/> accessed on 1 July 2023. Perhaps one of the most extraordinary examples of Indian environmental PIL is the 'Omnibus Forest Protection case' that commenced with Writ Petition (Civil) Number 202 of 1995. In 1995, TN Godavarman Thirumulpad filed a writ petition with the Supreme Court to protect the Nilgiris forest land from deforestation by illegal timber operations. This led to the court making a far-reaching order to examine 'all aspects relating to the National Forest Policy' (*TN Godavarman Thirumulkpad v Union of India* (1997) (2) SCC 267). The court, amongst other things, issued a complete ban on the felling of trees in certain areas as well as the movement of timber from any of the seven northeastern states to any other part of the country. Regional and state-level committees were created to monitor the implementation of the court's sustainable forestry directives. Through its continuing mandamus, the court maintained oversight of what has effectively evolved into a new forestry governance regime over the 20 years that this case remains open. For a useful summary of this landmark case, see 'T.N. Godavarman Thirumulkpad vs Union of India & Ors' (12 December 1996, *United Nations Environmental Programme Law and Environment Assistance Platform*) <https://leap.unep.org/countries/in/national-case-law/tn-godavarman-thirumulkpad-vs-union-india-ors> accessed 31 July 2023.

[67] KG Balakrishnan CJ, 'Annual Lecture 2008: Growth of Public Interest Litigation in India' (*Singapore Academy of Law*, 8 October 2008) <www.sal.org.sg/Newsroom/Speeches/Speech-Details/id/450> accessed 31 July 2023.

[68] SP Sathe, 'Judicial Activism: The Indian Experience' (2001) 6 Washington University Journal of Law and Policy 29, 71–72. This has arguably given rise to the problem of a 'clique' of public interest litigators and petitioners dominating the PIL agenda and therefore serving the interests of a distinctive group, typically middle-class, rather than the poor and marginalized in society: Rajamani (n 61) 305.

*Delhi Administration* was initiated by a letter written by a prisoner to a Supreme Court judge alleging brutal prisoner abuse.[69] The court accepted the letter as a writ petition, stating that 'technicalities and legal niceties are no impediment to the court entertaining even an informal communication as a proceeding for habeas corpus if the basic facts are found'.[70]

The remedies that have been created by the judiciary to achieve social change are a third distinctive feature of PIL in India. The traditional constitutional remedies that can be invoked by Indian judges are those of *habeas corpus*, *quo warranto*, *mandamus*, *prohibition*, and *certiorari*. However, Indian judges have pushed the boundaries by developing the 'continuing mandamus' which involves the passing of regular judicial directions and court monitoring of their implementation.[71] Finally, the nature of PIL proceedings is distinctly different from the common law framework of adversarial adjudication. In PIL proceedings, the judge takes on a more active role in posing questions to the parties and exploring solutions. The process is more akin to collective problem solving, and the courts have developed the practice of setting up 'fact-finding commissions' to provide expert insights.[72] Senior counsel may also be appointed as *amicus curiae* on a case-by-case basis.[73]

In *Association For Protection of Democratic Rights v the State of West Bengal*, the issue before the Supreme Court was whether the government of West Bengal should be allowed to fell trees for road construction. Some of the trees to be felled are more than 150 years old.[74] The court noted that 'it is essential to strike the right balance between environmental conservation and protection on one hand, and the right to development on the other, while articulating the doctrine of sustainable development'.[75] Further, the issue of felling trees for infrastructure development was said to '[assume] significance from the perspective of climate change as a growing national and international concern'. The court referred to the National Action Plan on Climate Change (specifically its commitment to increase tree cover from 23% to 33%) and India's Paris Agreement Nationally Determined Contribution which included a commitment to create additional carbon sequestration of 2.5–3 billion tonnes of carbon dioxide equivalent through more forest and tree cover by 2030. In its order issued on 25 March 2021, the court directed the constitution of an expert committee to develop a set of guidelines to govern decision-making with respect to tree felling for development projects and further specified that the guidelines were

[69] 1980 SCR (2) 557.

[70] ibid. This letter-based petition system has been formalized as 'epistolary jurisdiction': Balakrishnan CJ (n 67).

[71] The term 'continuing mandamus' was used for the first time in *Vineet Narain v Union of India* (1998) (1) SCC 226 and has been described as an innovative procedure that enables the courts to ensure proper implementation to redress the wrongs suffered by rights-bearers (*Lok Prahari through its General Secretary SN Shukla Ias (Retd) v Union of India* (Writ Petition No 1236 of 2019) [50]).

[72] Balakrishnan CJ (n 67).

[73] ibid.

[74] Special Leave Petition (Civil) No 25047 of 2018 (Supreme Court of India, 25 March 2021).

[75] ibid 2.

to, inter alia, provide a valuation mechanism for the value of ecosystem services provided.[76] A lawyer was appointed as *amicus curiae* to assist the committee in the legal aspects of developing the guidelines.[77]

The discussion about PIL in India may seem unnecessary as it can be argued that there is little likelihood of a climate change PIL case being filed at the Indian Supreme Court for a number of reasons. First, since the creation of the NGT, the Supreme Court has transferred all environmental cases to the tribunal for disposal to avoid the likelihood of conflicting decisions between the NGT and the courts of general jurisdiction.[78] Secondly, activists and other potential litigants are deterred from filing a PIL case because it can take years for a case to be heard and even longer for the Supreme Court to render its decision. As described by Gill, this is 'an ossified litigious legal system that challenges and possibly surpasses the court delays described in Jardine v Jardine in Dicken's *Bleak House*. Delay is not a recent phenomenon and can be traced back to the time of the Raj. It is a result of court clogging, adjournments, missing papers, absent witnesses, and conscious delaying tactics by both lawyers and the parties.'[79] Thirdly, PIL has a complex legacy and there is a large body of scholarly critique of the judiciary's approach in the past two decades.[80] It would not be possible to examine the critiques in detail here, but briefly, there is criticism of judicial overreach and over-extension into powers that are normally reserved for democratically elected government.[81] There are also questions of effectiveness (how effective were the PIL lawsuits in achieving the stated environmental goals?) and sustainability.[82] Kodiveri also argues that litigating climate change in the courts ought to be pursued cautiously because research has shown that many PIL cases have resulted in the violation of rights of those who are meant to be protected, especially marginalized communities.[83] Nonetheless, this chapter includes a discussion of PIL for the sake of completeness. A sketch of the Indian climate change litigation landscape which does not include Supreme Court PIL

[76] ibid 5.

[77] ibid 6.

[78] The Supreme Court in *Bhopal Gas Peedith Mahila Udyog Sangathan v Union of India* (2012) (8) SCC 326, 347 transferred all active and pending environmental cases to the NGT, stating that this would 'help in rendering expeditious and specialized justice in the field of environment'.

[79] Gitanjali Nain Gill, 'Mapping the Power Struggles of the National Green Tribunal of India: The Rise and Fall?' (2020) 7 Asian Journal of Law and Society 85, 94.

[80] See eg Ashok H Desai and S Muralidhar, 'Public Interest Litigation: Potential and Problems' in BN Kirpal and others (eds), *Supreme but not Infallible: Essays in Honour of the Supreme Court of India* (OUP 2000); Surya Deva, 'Public Interest Litigation in India: A Critical Review' (2009) 28 Civil Justice Quarterly 19; Varun Gauri 'Public Interest Litigation in India: Overreaching or Underachieving?' (2009) The World Bank Policy Research Working Paper Series 5109 <https://ideas.repec.org/p/wbk/wbrwps/5109.html> accessed 1 August 2023

[81] See eg Anuj Bhuwania, *Courting the People: Public Interest Litigation in Post-Emergency India* (CUP 2017).

[82] Rajamani (n 61).

[83] Arpitha Kodiveri, 'The Promises and Challenges of Climate Change Litigation in India' (*OpenGlobalRights*, 28 June 2020) <https://www.openglobalrights.org/promises-and-challenges-of-climate-change-litigation-in-india/> accessed 1 August 2023

would be incomplete because of the potential for a rights-based environmental PIL lawsuit that has climate change at its core to be filed at the Supreme Court. Rights-based climate change litigation has become increasingly prominent in recent years, creating strategies that can be replicated and transplanted in various jurisdictions including India.[84] Public interest litigation is a potential vehicle for future strategic climate change litigation in India, particularly test cases in which climate change arguments are premised on rights-based environmental jurisprudence.

## 2.2 South Africa

While there has been a steady rise in climate change litigation globally, there have been few cases related to climate change in Africa.[85] This is because the ability to litigate in most African countries is unfortunately limited by significant capacity constraints including weak legislative and regulatory frameworks, poor governance and law enforcement, limited access to financial and legal resources, and the absence of robust human rights protection mechanisms.[86] South Africa is 'firmly out of the starting blocks' with the highest number of decided and pending climate change cases amongst the African jurisdictions even though it can be argued that South Africa's climate change litigation potential is yet to be fully realized.[87] In our analysis at the time of writing, the South African climate change litigation docket comprises of 16 cases—six core cases and 10 periphery cases.

South Africa is a relatively young constitutional democracy. In 1994, the country made its transition to democratic rule after hundreds of years of colonial rule followed by decades of apartheid rule.[88] The legacies of colonialism and

[84] For discussion, see eg Kumaravadivel Guruparan and Harriet Moynihan, 'Climate Change and Human Rights-Based Strategic Litigation' (*Chatham House*, 11 November 2021) <www.chathamhouse.org/2021/11/climate-change-and-human-rights-based-strategic-litigation> accessed 2 August 2023; César Rodríguez-Garavito (ed), *Litigating the Climate Emergency: How Human Rights, Courts and Legal Mobilization Can Bolster Climate Action* (CUP 2022); Jacqueline Peel and Hari M Osofsky, 'A Rights Turn in Climate Change Litigation?' (2017) 7 Transnational Environmental Law 37; Chapter 3 of this book.

[85] Our analysis shows that, as of mid-2023, there are 16 decided and pending cases in Africa (see Chapter 3 of this book). For an excellent analysis of climate change litigation on the African continent, see the special issue in (2021) 15 Carbon and Climate Law Review.

[86] Sam Adelman, 'Climate Change Litigation in Africa: A Multi-Level Perspective' in Alogna, Bakker, and Gauci (eds) (n 20) 275. One of the legacies of colonialism is that the legal systems in many sub-Saharan states are underdeveloped and there is relatively weak civil society. See Makau Mutua, 'Africa and the Rule of Law' (2016) 13(23) Sur International Journal on Human Rights 159.

[87] Tracy-Lynn Field, 'Climate Change Litigation in South Africa: Firmly Out of the Starting Blocks' in Alogna, Bakker, and Gauci (eds) (n 20) 173.

[88] See Robert Ross, *A Concise History of South Africa* (2nd edn, CUP 2014): a more recent book on the history of South Africa is a noteworthy and important contribution to the scholarship because it 'puts black people at the centre, for a change'; Thula Simpson, *History of South Africa: From 1902 to 2021* (OUP 2022): the book covers 12 decades, from the surrender of Boer guerillas in the Second Anglo-Boer War in 1902 to the looting spree in July 2021 in two of South Africa's provinces; see Kenneth S Broun and others, *Black Lawyers, White Courts: The Soul of South African Law* (Ohio University Press 1999) on the experience of black lawyers during the Apartheid era.

apartheid—patterns of deep inequality, racism, and poverty—continue to dominate the daily lives of South Africans and produce serious tensions and violent confrontations at times.[89] The Covid-19 pandemic worsened living conditions and will continue to have long-term economic repercussions as South Africa has experienced its steepest economic recession in over a century.[90] Due to its great reliance on fossil fuels in its primary energy mix, South Africa is the thirteenth largest GHG emitter in the world.[91] The country is already experiencing significant climate change-related impacts, particularly water scarcity.[92] From 2015 to 2018, Cape Town experienced the worst drought in almost a century.[93] While Cape Town managed to narrowly escape the dire prospect of zero water supply due to draconian water rationing measures and good rains in 2018, rising demand and drought conditions are escalating a national water crisis.[94]

South Africa has emerged as the climate litigation 'frontrunner' in Africa because of its functional legal system and active civil society. Its constitutional democracy and enforceable Bill of Rights are a rare beacon on the African continent.[95] There is an independent judiciary and a plethora of environmental laws that can be used in conjunction with the Promotion of Access to Information Act[96] and the

[89] In discussing how the South African Constitutional Court has identified dignity, equality, and freedom as the core human values underpinning the Bill of Rights, Klug goes on to point out that '[t]his placing of rights at the centre of the post-apartheid project is challenged daily by the contextual realities of the new South Africa, where exposure to violence, poverty and government incapacity continue to shape the lives of the majority of the country's inhabitants': Heinz Klug, *The Constitution of South Africa: A Contextual Analysis* (Hart Publishing 2010) 5.

[90] Roxanne Henderson, 'South Africa Tallies Virus Blow, Warns on Economic Scarring' *Bloomberg* (27 May 2021) <www.bloomberg.com/news/articles/2021-05-27/south-africa-tallies-up-virus-damage-warns-on-economic-scarring> accessed 1 August 2023; Emmanuel Sekyere and others, 'The Impact of COVID-19 in South Africa' (*Wilson Center*, April 2020) <www.wilsoncenter.org/publication/impact-covid-19-south-africa> accessed 1 August 2023; Channing Arndt and Sherman Robinson, 'Who Has Been Hit Hardest by South Africa's Lockdown? We Found Some Answers' *The Conversation* (11 June 2020) <https://theconversation.com/who-has-been-hit-hardest-by-south-africas-lockdown-we-found-some-answers-138481> accessed 1 August 2023.

[91] Coal meets around 70% of installed power generation capacity in South Africa: 'South Africa' (*International Energy Agency*) <www.iea.org/countries/south-africa> accessed 1 August 2023; 'Each Country's Share of CO2 Emissions' (*Union of Concerned Scientists*, 12 July 2023) <www.ucsusa.org/resources/each-countrys-share-co2-emissions> accessed 12 July 2023.

[92] World Meteorological Organisation, 'State of the Climate in Africa 2019' (2020) WMO-No 1253 <https://library.wmo.int/index.php?lvl=notice_display&id=21778#.YZxzAFMRVTa> accessed 1 August 2023

[93] 'South Africa Grapples with Worst Drought in 30 Years' *BBC* (30 November 2015). <www.bbc.com/news/world-africa-34884135> accessed 1 August 2023; Danielle Torrent Tucker, 'In a Warming World, Cape Town's "Day Zero" Drought Won't Be an Anomaly, Stanford Researcher Says' *Stanford News* (9 November 2020) <https://news.stanford.edu/2020/11/09/cape-towns-day-zero-drought-sign-things-come/> accessed 1 August 2023; Gina Ziervogel, 'What Cape Town's Drought Can Teach Other Cities About Climate Adaptation' *The Conversation* (3 April 2019) <https://theconversation.com/what-cape-towns-drought-can-teach-other-cities-about-climate-adaptation-114379> accessed 1 August 2023.

[94] Jon Heggie, 'Day Zero: Where Next?' *National Geographic* (10 February 2021) <www.nationalgeographic.com/science/article/partner-content-south-africa-danger-of-running-out-of-water> accessed 1 August 2023.

[95] Constitution of the Republic of South Africa, Act 108 of 1996, ch 2.

[96] Act 2 of 2000 (South Africa).

Promotion of Administrative Justice Act[97] to launch climate change litigation, and fairly broad standing rules. Prior to the adoption of the Constitution and the end of apartheid in 1994, judicial review did not exist.[98] There was no written or unwritten constitution that enshrined fundamental principles against which executive action could be assessed. However, with the adoption of a written constitution (and Bill of Rights) and building on the experience of anti-apartheid lawyers who devoted their practice to protecting the rights of the oppressed majority, South Africa has built a tradition of social justice litigation and a culture of seeking rights protection through the courts. There exists in South Africa an ecosystem of activist lawyers, NGOs, and foundations that have gathered experience in the use of litigation as part of a multi-pronged strategy to achieve social progress on issues ranging from housing, the right to basic education, and, more recently, climate change.[99] Advocacy groups and NGOs are supporting and launching climate change lawsuits as part of their wider advocacy campaigns to phase out coal and embark on a just energy transition. Like India, South Africa does not have a dedicated climate change law, but several environmental laws serve as 'hooks' for climate change litigation, including laws relating to water, protected areas, and environmental impact assessment (EIA).

The following section will provide a brief analysis of decided and pending cases in the South African docket. Following the success of *Earthlife Africa v Minister of Environmental Affairs*, similar lawsuits challenging the Environment Impact Assessment (EIA) process were launched to halt proposed coal-fired power plant projects.[100] In November 2021, a landmark constitutional challenge was filed in the North Gauteng High Court against the South African government by a coalition of advocacy groups and activist lawyers. The applicants argued that the government's plans to build more coal-fired power plants pose significant unjustifiable threats to constitutional rights. This case heralded the arrival of rights-based climate change litigation in South Africa, following in its rich tradition of constitutional litigation to protect rights that are enshrined in the Bill of Rights.

### Focus on project-based litigation

*Earthlife Africa v Minister of Environmental Affairs* (the *Thabemetsi* case), decided in 2017, was South Africa's first core climate change case and has since served as a 'model' or 'template' for subsequent litigation that focuses on challenging projects

97 Act 3 of 2000 (South Africa).

98 Kenneth S Broun, 'Black Lawyers under Apartheid: The Soul of South African Law' (2001) 27(2) MILLENNIUM 33, 37.

99 On social justice litigation/PIL in Africa, see eg Steven Budlender, Gilbert Marcus, and Nick Ferreira, *Public Interest Litigation and Social Change in South Africa: Strategies, Tactics and Lessons* (The Atlantic Philanthropies 2014); Jason Brickhill, *Public Interest Litigation in South Africa* (Juta 2018); Marius Pieterse, *Rights-Based Litigation, Urban Governance and Social Justice in South Africa: The Right to Joburg* (Routledge 2019).

100 [2017] 2 All SA 519 (GP).

which threaten to lock South Africa in its current GHG-intensive mode of energy production instead of fostering the urgently needed energy transition.[101] The case proved to be a victory for the plaintiff environmental justice groups after three years of court proceedings, community mobilization, advocacy, and public engagement. In 2015, while South Africa ratified the Paris Agreement, as many as 13 new coal power plants were being planned under the country's independent power producer (IPP) coal baseload procurement programme.[102] Thabametsi Power Company (Pty) Ltd won the bid for the planned Lephalale facility in Limpopo province. Thabametsi needed to obtain an environmental authorization in order to construct the power plant and proceeded to do so.[103] At that time, neither the National Environmental Management Act (NEMA) nor the EIA Regulations required the submission of a climate change impact assessment report or even the express consideration of climate change. In March 2015, the Chief Director of the Department of Environmental Affairs granted the authorization.[104] The EIA report for the project did not include a quantification of the projected GHG emissions from the project. There was also no consideration of the impact of the proposed project on water scarcity in the region. Following an unsuccessful appeal to the Minister of Environmental Affairs, environmental NGO Earthlife Africa challenged Thabametsi's environmental authorization in court in 2017.

In a landmark judgment issued in March 2017, the Pretoria High Court set aside the environmental authorization, holding that the Minister was obliged to consider climate change impacts but had failed to do so in reaching her decision to grant the environmental authorization.[105] The court ordered the Minister to constitute the appeal afresh, now taking climate change impact assessment reports into account.[106] In January 2018, the Minister again upheld Thabametsi's environmental authorization on the basis that extensive climate change mitigation measures had been added and even though Thabametsi's own climate change impact assessment showed that it would be one of the most GHG-intensive plants in the world, she held that this did not constitute a 'fatal flaw'.[107] This time, Earthlife Africa and

[101] ibid.

[102] IPP Coal, 'About the Coal Baseload IPP Procurement Programme' (*IPP Coal*) <www.ipp-coal.co.za/About> accessed 1 August 2023. South Africa ratified the Paris Agreement on 1 November 2016: see United Nations Climate Change, 'South Africa' (*United Nations Climate Change*) <https://unfccc.int/node/61201> accessed 1 August 2023.

[103] *Earthlife* (n 100) [2].

[104] ibid.

[105] For detailed analysis of this groundbreaking judgment, see Tracy-Lynn Humby, 'The Thabametsi Case: Case No 65662/16 Earthlife Africa Johannesburg v Minister of Environmental Affairs' (2018) 30(1) Journal of Environmental Law 145; Marjoné van der Bank and Jaco Karsten, 'Climate Change and South Africa: A Critical Analysis of the *Earthlife Africa Johannesburg v Minister of Energy 65662/16 (2017)* Case and the Drive for Concrete Climate Practices' (2020) 13 Air Soil and Water Research 1.

[106] *Earthlife* (n 100) [119]–[122].

[107] Minister of Environmental Affairs, Republic of South Africa, 'Appeal decision' (LSA 142346, Minister of Environmental Affairs, Republic of South Africa, 2018) [4.5] <https://cer.org.za/wp-content/uploads/2018/01/Thabametsi-Appeal-Decision-30-January-2018-2.pdf> accessed 1 August 2023.

another NGO, groundWork, collaborated to launch new court proceedings.[108] In this second judicial review, the applicants argued that the Minister had to exercise her own discretion and consider the project-specific impacts of the Thabametsi project instead of treating the national electricity plan or any other policy document as being determinative of her decision.[109]

The second judicial review was complemented by a suite of legal challenges of other regulatory approvals given to Thabametsi (and other coal power plants) including the water use licences, air pollution emission licences, and energy regulatory permits.[110] This multi-pronged litigation strategy was effective in putting pressure on the project developer and its financiers.[111] In addition, there was significant community mobilization as well as shareholder activism to put pressure on the banks and public financial institutions (such as the Development Bank of South Africa and the Public Investment Corporation) that eventually agreed to withdraw from funding the Thabametsi project.[112] By November 2020, Thabametsi agreed to settle the litigation with Earthlife Africa and groundWork and set aside its

[108] The court documents pertaining to this second judicial review (Thabametsi II) can be found online at 'The Proposed Thabametsi IPP: Earthlife Africa Johannesburg v Department of Environmental Affairs, Thabametsi Power Project (Pty) Ltd and Others' (*Centre for Environmental Rights*) <https://cer.org.za/programmes/pollution-climate-change/litigation/the-proposed-thabametsi-ipp-earthlife-africa-johannesburg-v-department-of-environmental-affairs-thabametsi-power-project-pty-ltd-and-others> accessed 1 August 2023.

[109] *Earthlife Africa Johannesburg v Minister of Environmental Affairs*, Case Number 65662/16, 'Notice of Motion' (High Court of South Africa, Gauteng Division, Pretoria, 23 August 2016) [3] <https://cer.org.za/wp-content/uploads/2014/06/Thabametsi-Notice-of-Motion.pdf> accessed 1 August 2023.

[110] Nicole Loser, 'Climate Change Litigation in Africa: A Focus on Project-Based Litigation' (*Climate Change Litigation Initiative*, 14 September 2021) <www.c2li.org/climate-change-litigation-in-africa-a-focus-on-project-based-litigation/> accessed 1 August 2023.

[111] ibid.

[112] Reuters Staff, 'Marubeni to Pull Out of South African Coal-Fired Power Plant Project' *Reuters* (12 November 2020) <www.reuters.com/article/us-climatechange-safrica-coal-idUSKBN27R2OQ> accessed 1 August 2023; 'No More Overseas Coal Power Projects, Says South Korea's Kepco' *Global Construction Review* (16 October 2020) <www.globalconstructionreview.com/no-more-overseas-coal-power-projects-says-south-ko/>; BankTrack, 'Thabametsi Coal Power Plant' (*BankTrack*, 27 November 2020) <www.banktrack.org/project/thabametsi_coal_fired_power_plant> accessed 1 August 2023; Chris Yelland and Roger Lilley, 'Nedbank Withdraws Funding for New Coal IPPs' *Daily Maverick* (30 January 2019) <www.dailymaverick.co.za/article/2019-01-30-nedbank-withdraws-funding-for-new-coal-ipps/> accessed 1 August 2023. In South Africa, there are a growing number of environment-focused investor advocacy groups which, amongst other things, file climate risk-related resolutions within certain banks and have driven an agenda on the fiduciary duties of pension fund trustees, particularly as they relate to fossil fuel investments. Just Share characterizes itself as 'a non-profit shareholder activism and responsible investment organisation' which focuses its advocacy on three issues: inequality, climate change, and diversity: see Just Share, 'New Organisation Launched to Promote the Use of Investor Power for a Fairer South Africa' (*Just Share*) <https://justshare.org.za/op-eds/new-organisation-launched-to-promote-the-use-of-investor-power-for-a-fairer-south-africa/> accessed 1 August 2023. The Chartered Governance Institute of Southern Africa (CGISA) issued a Best Practice Guide on 'Shareholder Activism and the Role of the Company Secretary' in May 2020, in which it identified that '[c]limate change has become a critical issue for many shareholders, who are demanding greater disclosure by companies on efforts undertaken to combat the potential risks to climate and the environment as a whole': see Sabrina Paxton, 'CGSIA Best Practice Guide: Shareholder Activism and the Role of the Company Secretary' (*Chartered Governance Institute of Southern Africa*, May 2020) s 3.2 <https://chartsec.co.za/documents/members/CGISA%20Best%20Practice%20Guide%20Shareholder%20Activism.pdf> accessed 1 August 2023.

environmental authorization.[113] This setting aside means that, should Thabametsi wish to revive the project in the future, it would have to seek a new authorization.

After the court held in *Thabametsi* that section 240 of NEMA must be interpreted to include climate change impacts as part of the potentially significant environmental impacts that must be assessed in an EIA, civil society organizations have challenged the EIAs of a further five proposed coal-fired power plants under the IPP coal baseload procurement programme.[114] In the case of the proposed KiPower plant, for example, the environmental authorization was granted in October 2015 and the internal appeal to the Minister against the department's decision to grant the authorization was made in December 2015.[115] One of the grounds of appeal was that the Department of Environmental Affairs failed to apply the cautious approach that should be adopted in decision-making pursuant to NEMA.[116] In November 2016, the Minister issued her decision to dismiss the appeal. In relation to the ground of appeal that a climate change impact assessment was required, the Minister held that there was no legal basis to require an assessment of climate change impacts within the EIA framework.[117] After the *Thabametsi* judgment, groundWork applied to court for an order (after another unsuccessful appeal to the Minister) to review and set aside the environmental authorization as well as the Minister of Environmental Affairs' dismissal of the appeal against the Department's decision, namely on the ground that the authorization and appeal decision were wrongly made in the absence of a climate change impact assessment (applying *Thabametsi*). Kuyasa Mining (Pty) Ltd initially opposed the application but, on 1 November 2017, informed the Centre for Environmental Rights (the lawyers acting for groundWork) that it would conduct a climate change assessment and amend its environmental authorization.[118]

[113] *Earthlife Africa NPC v Minister of Environmental Affairs*, Case No 21559/2018, 'Draft Order' (High Court of South Africa, Gauteng Division, Pretoria, 19 November 2020) <https://cer.org.za/wp-content/uploads/2020/11/21559-2018-19-November-2020.pdf> accessed 1 August 2023.

[114] The list of cases and sets of documents pertaining to each litigation are available at Centre for Environmental Rights, 'Pollution & Climate Change—Litigation' (*Centre for Environmental Rights*) <https://cer.org.za/programmes/pollution-climate-change/litigation> accessed 2 August 2023.

[115] Centre for Environmental Rights, 'Challenge of the Decision of the Department of Environmental Affairs to Grant Environmental Authorisation' (*Centre for Environmental Rights*) <https://cer.org.za/programmes/pollution-climate-change/litigation/groundwork_kipower/environmental-authorisation> accessed 2 August 2023.

[116] '[T]he Authorization was granted without adequate information about the full implications of the Project for health and for its contribution to climate change and adaptation to a changed climate': see 'Appeal pursuant to Section 43(2) of the National Environmental Management Act, 1998 against environmental authorisation granted to Kuyasa Mining (Pty) Ltd on behalf of KiPower (Pty) Ltd on 21 October 2015' (10 December 2015) [98] <https://cer.org.za/wp-content/uploads/2014/06/CER-KiPower-Appeal-Submission-10-December-2015.pdf> accessed 2 August 2023.

[117] Minister of Environmental Affairs, Republic of South Africa, 'Appeal decision' (LSA 149588, Minister of Environmental Affairs, Republic of South Africa, 8 November 2016) <https://cer.org.za/wp-content/uploads/2014/06/KI-Power-Appeal-Decision.pdf> accessed 2 August 2023, 18.

[118] Centre for Environmental Rights, 'Challenge of the Decision' (n 115).

Several of the lawsuits, with a focus on preventing more new coal-fired power plants from being built, have adopted multi-pronged strategies in the sense that the proposed project is faced with legal challenges on multiple fronts including EIA, water licensing, air pollution and minimum emission standards, and constitutionality. While it is resource-intensive and requires deep legal expertise to implement such an approach, it can be argued that the strategy is highly effective in demonstrating that climate change litigation represents a real business risk for carbon-intensive industries and the urgent need for transformational change in its business practices towards decarbonization. Secondly, the strategy provides ample opportunity to keep the proposed project under public scrutiny and in the media spotlight. This helps generate the attention needed to ensure that the issues of energy transition and tackling climate change continue to be on the public agenda and that the government and corporations are held accountable.

### Going beyond coal

While the majority of climate change litigation in South Africa is likely to continue to involve administrative law challenges to halt the construction of new coal-fired power plants, increasing levels of shareholder activism and civil society mobilization are likely to lead to more complex and novel causes of action being advanced in the courtroom to compel corporates and government agencies to act on climate change. The *#CancelCoal* case heralds the arrival of constitutional rights-based climate change litigation in South Africa and there is great potential for transnational borrowing to inform litigation premised on, for example, the French Corporate Duty of Vigilance Law.[119]

It is noteworthy that voluntary associations have brought cases to resist the movement away from fossil fuels. For instance, *Coal Transporters Forum v Eskom Holdings Ltd* was initiated by the Coal Transporters Forum, a voluntary association representing companies that transport coal.[120] The case was an attempt to stop Eskom Holdings from signing any further power purchase agreements with independent renewable energy power producers and to declare existing agreements invalid. The application was dismissed in March 2019 largely on the grounds that the Forum's claims had no merit or evidential basis.[121] It can be argued that such litigation, motivated by social resistance to change that threatens livelihoods and jobs, highlights the importance of a just transition that provides alternative employment and retraining to take advantage of new employment opportunities. These lawsuits are uncommon, but more can be anticipated as governments and corporations face escalating pressure to make the transition away from fossil fuels.

[119] See ch 3, s 3.4.2 on the French Duty of Vigilance Law.
[120] [2019] ZAGPPHC 76.
[121] ibid.

Subnational climate change litigation that involves local and federal governments in legal tussles over jurisdictional powers to implement climate change and energy policies is less frequently seen in the Global South compared to the Global North (particularly in the United States). The case of *The City of Cape Town v National Energy Regulator of South Africa and Minister of Energy* is therefore noteworthy.[122] In this case, Cape Town applied to the court to declare that the Minister of Energy does not have to approve the purchase of renewable energy from independent power producers.[123] The city had tried to diversify its energy portfolio but was not able to do so because the Minister of Energy did not give the requisite approval. The case was eventually set aside by the court as the court deemed it to be an intergovernmental dispute that should be settled out of court.[124] It can be argued that this was a lost opportunity to clarify the powers of local authorities and empower subnational governments to take concrete action to reduce GHG emissions.

## 2.3 Brazil

Climate change litigation has been emerging in Latin America, with significant and high-profile lawsuits filed, for example, in Columbia and Peru.[125] Up until 2020, the two global climate change litigation databases managed by the Sabin Center for Climate Change Law (Columbia University) and the Grantham Research Institute on Climate Change and the Environment (London School of Economics and Political Science)[126] identified only six cases of climate change litigation in Brazil. However, there has been a sharp rise in the Brazilian docket with nine major

[122] Case no 51765/2017; Notice of Motion is available at <https://cer.org.za/wp-content/uploads/2017/08/1-Notice-of-motion-and-founding-affidavit.pdf> accessed 2 August 2023; Antony Sguazzin, 'Cape Town in Landmark Suit to Open South Africa's Power System' *Bloomberg* (12 May 2020) <www.bloomberg.com/news/articles/2020-05-12/cape-town-seeking-energy-independence-awaits-court-ruling> accessed 2 August 2023.

[123] *The City of Cape Town v National Energy Regulator of South Africa*, Case Number 51765/2017, 'Notice of Motion' (High Court of South Africa, Gauteng Division, Pretoria, 26 July 2017) [1] <https://cer.org.za/wp-content/uploads/2017/08/1-Notice-of-motion-and-founding-affidavit.pdf> accessed 2 August 2023.

[124] 'The City of Cape Town v. National Energy Regulator of South Africa and Minister of Energy' (*Grantham Research Institute on Climate Change and the Environment*) <https://climate-laws.org/geographies/south-africa/litigation_cases/the-city-of-cape-town-v-national-energy-regulator-of-south-africa-and-minister-of-energy> accessed 2 August 2023.

[125] eg *Future Generations v Ministry of the Environment* (2018) STC4360-2018, Number 11001-22-03-000-2018-00319-01. English translations of key excerpts from the Supreme Court's decision can be found at Dejusticia, 'Climate Change and Future Generations Lawsuit in Colombia: Key Excerpts from the Supreme Court's Decision' (*Dejusticia*, 13 April 2018) <www.dejusticia.org/en/climate-change-and-future-generations-lawsuit-in-colombia-key-excerpts-from-the-supreme-courts-decision/> accessed 2 August 2023.

[126] The Sabin Center for Climate Change Law database can be found at <http://climatecasechart.com/> accessed 2 August 2023; the Grantham Research Institute on Climate Change and the Environment database can be found at <https://climate-laws.org> accessed 2 August 2023.

lawsuits filed in 2020, four in 2021, and three in 2022.[127] These lawsuits marked the beginning of a new era of climate change litigation in Brazil and the jurisdiction is now the 'frontrunner' in the Latin American region. Brazil is fertile ground for climate change litigation because it is home to the Amazonian rainforest—the 'lungs of planet Earth'[128]—which are under grave threat, and it has a robust legal framework to tackle environmental and climate change issues. Public prosecutors (the *Ministerio Publico*) play a distinctive role in enforcing environmental laws in Brazil and have brought many of the lawsuits discussed in this section. Brazil has a well-developed civil society which, during the Bolsanaro administration, encountered threats by a government that was determined to 'end all activism in Brazil'.[129] It is noteworthy that climate change litigation and environmental advocacy occur despite Brazil being one of the most dangerous places in the world for environmental defenders.[130]

Brazil is the seventh largest emitter of GHGs globally.[131] Unlike other major emitters such as India, South Africa, and China which have high levels of GHG emissions due to their reliance on fossil fuels to produce electricity, the majority of Brazil's carbon emissions are produced by agriculture and land use change (including the conversion of forests into farmland for soy production and cattle ranching).[132] More than half of Brazil's emissions are attributable to deforestation,

[127] The five 2021 cases do not include *Envol Vert v Casino* (filed in 2021), a lawsuit brought by an international coalition of NGOs against the French supermarket chain Casino for its involvement in the cattle industry in Brazil and Colombia, which is known to be a key driver of deforestation. See Chapter 3, Section 3.4 for discussion of *Envol v Casino* and the potential for the duty of vigilance to be a vehicle for transnational climate change litigation that seeks to hold corporations accountable for their acts and omissions that contribute to the main drivers of climate change, eg deforestation and financing fossil fuel extraction activities.

[128] Aylin Woodward, 'The "Lungs of the Planet" are on fire' (*World Economic Forum*, 22 August 2019) <www.weforum.org/agenda/2019/08/amazon-burning-unseen-rate/> accessed 2 August 2023.

[129] Adriana Ramos, 'CLIMATE CHANGE: 'There Is No Respect for the Role of Civil Society' (*CIVICUS*, 27 September 2019) <www.civicus.org/index.php/media-resources/news/interviews/4080-climate-change-there-is-no-respect-for-the-role-of-civil-society> accessed 2 August 2023.

[130] Stuart Braun, '5 Deadly Countries for Environmental Defenders' *Deutsche Welle* (29 July 2020) <www.dw.com/en/5-deadly-countries-for-environmental-defenders/a-54298499> accessed 2 August 2023.

[131] Johannes Friedrich and others, 'This Interactive Chart Shows Changes in the World's Top 10 Emitters' (*World Resources Institute*, 10 December 2020) <www.wri.org/insights/interactive-chart-shows-changes-worlds-top-10-emitters> accessed 2 August 2023.

[132] Cattle ranching generates about 44% of Brazil's GHGs, and Brazil is the second largest exporter of beef and soybeans globally: Larissa Basso, 'Brazilian Energy-Related Climate (In)action and the Challenge of Deep Decarbonization' (2019) 62 Revista Brasileira de Estudos Políticos <www.scielo.br/j/rbpi/a/dh7YhQb77PVksb6w7RmkHZk/?lang=en> accessed 2 August 2023. Between 1990 and 2008, Brazilian soybean production increased nearly 200%, and land conversion to soy crops currently accounts for up to 18% of deforestation in Brazil: Britaldo Soares-Filho and others, 'Challenges for Low-Carbon Agriculture and Forest Conservation in Brazil' (2012) Environmental Safeguards Unit Technical Notes No IDB-TN-385, Inter-American Development Bank <https://publications.iadb.org/publications/english/viewer/Challenges-for-Low-Carbon-Agriculture-and-Forest-Conservation-in-Brazil.pdf> accessed 2 August 2023.

particularly in the Amazon region.[133] Safeguarding the Amazon biome is a critical priority, not just for the eight Amazon basin states.[134] It is also a global priority because of the Amazon's planetary importance. Spanning across 1,600 million acres, the Amazon contains the world's single largest tropical rainforest and largest river basin system.[135] The Amazon is home to more than 10% of the world's known wildlife species and is therefore a biological diversity hotspot.[136] Due to its sheer size, the Amazon is a globally important carbon sink and therefore a priority for global climate change policy.[137] It is estimated that the ecosystem services provided by the Amazon contribute as much as 8 billion dollars (USD) to Brazil's economy alone.[138] These environmental services include storing carbon (climate change mitigation), recycling water that is important for rainfall in the Amazon as well as surrounding regions, and maintaining biodiversity. It has therefore been argued that the protection of the Amazon should be considered a common concern of humankind.[139] It is also important to note that the Amazon is home to about 400 tribes and the Indigenous communities of the Brazilian Amazon alone have legal title to around 1 million square kilometres of these forests.[140]

On 1 January 2019, Jair Bolsonaro took office as the thirty-eighth president of Brazil. His response to the massive Amazon rainforest fires of 2019 and his anti-environmentalist rhetoric drew international attention to the serious environmental crisis during his presidency.[141] Through policies that dismantled environmental regulations and severely curtailed the financial budgets of environmental regulatory agencies, 'the administrative and institutional frameworks of climate change and environmental protection in Brazil [were] on the verge of collapse'

133 Datta Aparajita and Krishnamoorti Ramanan, 'Understanding the Greenhouse Gas Impact of Deforestation Fires in Indonesia and Brazil in 2019 and 2020' (2022) 4 Frontiers in Climate <www.frontiersin.org/articles/10.3389/fclim.2022.799632/full> accessed 2 August 2023.

134 They are Bolivia, Brazil, Colombia, Ecuador, Guyana, Peru, Suriname, and Venezuela.

135 'Why Protect the Amazon' (*Amazon Conservation*) <www.amazonconservation.org/the-challenge/why-the-amazon/> accessed 23 August 2023.

136 'About the Amazon' (*World Wide Fund for Nature*) <https://wwf.panda.org/discover/knowledge_hub/where_we_work/amazon/about_the_amazon/> accessed 2 August 2023.

137 William Boyd, 'Ways of Seeing in Environmental Law: How Deforestation Became an Object of Climate Governance' (2010) 37 Ecology Law Quarterly 843.

138 Jon Strand and others, 'Spatially Explicit valuation of the Brazilian Amazon Forest's Ecosystem Services' (2018) 1 Nature Sustainability 657.

139 Beatriz Garcia, *The Amazon from an International Law Perspective* (CUP 2011) 304.

140 Daniel Nepstad and others, 'Inhibition of Amazon Deforestation and Fire by Parks and Indigenous Lands' (2006) 20(1) Conservation Biology 65.

141 Brian Clark Howard, 'Brazil Leads World in Reducing Carbon Emissions by Slashing Deforestation' *National Geographic* (6 June 2014) <www.nationalgeographic.com/science/article/140605-brazil-deforestation-carbon-emissions-environment> accessed 2 August 2023; Herton Escobar, '"A hostile environment." Brazilian Scientists Face Rising Attacks from Bolsonaro's Regime' *Science* (7 April 2021) <www.science.org/content/article/hostile-environment-brazilian-scientists-face-rising-attacks-bolsonaro-s-regime> accessed 2 August 2023; Roberto Goulart Menezes and Ricardo Barbosa Jr, 'Environmental Governance under Bolsonaro: Dismantling Institutions, Curtailing Participation, Delegitimising Opposition' (2021) 15 Zeitschrift für Vergleichende Politikwissenschaft 229.

during his administration.[142] In addition, by promoting 'economic development' in the Amazon, the Bolsonaro administration escalated the environmental crisis to the point that scientists were concerned that climate change, deforestation, and fires could cause the Amazon to reach a tipping point and dry out.[143] It is against this backdrop that NGOs, political parties, and the *Ministerio Publico* started to file lawsuits in 2020 that became the vanguard of climate change litigation in Brazil.[144]

The following section first briefly describes the applicable environmental legal framework in Brazil. An analysis of some of the key cases in the Brazilian docket follows. As a matter of practice, most environmental public interest litigation in Brazil is typically brought by the prosecutors, rather than NGOs as would be common in other countries including India and South Africa.[145] That recent cases have been filed by NGOs and political parties as well marks a notable departure from common practice and arguably highlights how various sectors of Brazilian society felt compelled to put a stop to the environmental damage and regulatory rollback by the Bolsonaro administration.[146] It has also been pointed out that the prosecutors are overwhelmed by heavy workloads and therefore are playing a less prominent role in bringing climate lawsuits as one might expect.[147] Indirect, 'stealthy' litigation has been the dominant approach in Brazil, with climate change embedded in litigation strategies focused on the enforcement of existing forestry and environmental laws. However, it is clear that the more recent cases filed since 2020 are building the foundations for more direct legal challenges which compel the Brazilian courts to address climate change in a substantial way.

[142] Caio Borges, 'Will Climate Change Litigation Save the Brazilian Amazon?' (*OpenGlobalRights*, 23 September 2020) <www.openglobalrights.org/will-climate-change-litigation-save-the-Brazilian-Amazon/> accessed 2 August 2023. See Kathryn Hochstetler, 'Climate Institutions in Brazil: Three Decades of Building and Dismantling Climate Capacity' (2021) 30 Environmental Politics 49 for a detailed and insightful analysis about the evolution of climate institutions in Brazil across different presidencies.

[143] Lisa Viscidi and Nate Graham, 'Brazil Was a Global Leader on Climate Change. Now It's a Threat' *Foreign Policy* (4 January 2019) <https://foreignpolicy.com/2019/01/04/brazil-was-a-global-leader-on-climate-change-now-its-a-threat/> accessed 3 August 2023; Ignacio Amigo, 'When Will the Amazon Hit a Tipping Point?' *Nature* (25 February 2020) <www.nature.com/articles/d41586-020-00508-4> accessed 3 August 2023; Lucas Ferrante and Philip M Fearnside, 'Brazil's New President and "Ruralists" Threaten Amazonia's Environment, Traditional Peoples and the Global Climate' (2019) 46(4) Environmental Conservation 261.

[144] Alessandra Lehmen describes this wave of litigation starting from 2020 onwards as 'the renaissance of climate litigation in Brazil': Alessandra Lehmen, 'Advancing Strategic Climate Litigation in Brazil' (2022) 22 German Law Journal 1471, 1472.

[145] Interview 2; Antonio Herman Benjamin and Nicholas Bryner, 'Brazil' in Emma Lees and Jorge E Viñuales (eds), *The Oxford Handbook of Comparative Environmental Law* (OUP 2019) 17; see Lesley K McAllister, *Making Law Matter: Environmental Protection and Legal Institutions in Brazil* (Stanford Law Books 2008) 152–53, explaining that environmental groups usually bring their complaints to the Member of Parliament and stating that about 97% of environmental public civil actions have been filed by prosecutors. See Chapter 5 of this book for more discussion about the role of prosecutors in climate change litigation.

[146] Brazil is indeed unique in this regard as we have not seen political parties file climate change lawsuits in other jurisdictions either in the Global South or the Global North.

[147] Discussion during by-invitation only conference, 'Climate Change Litigation in the Global South' (National University of Singapore, 22–23 May 2023).

## Environmental legal framework

The Federal Constitution of 1988 provides the institutional framework for three levels of government—federal, state, and municipal—in Brazil. It grants federal (and state) ownership of and power to regulate certain natural resources. For example, mineral resources are exclusively owned by the federal government.[148] Enforcement and lawmaking powers on environmental matters are distributed to both federal and state governments though federal law tends to dominate because it is superior to conflicting state or municipal law.[149] Environmental law in Brazil was transformed by the 1988 Constitution when environmental rights, responsibilities, and principles were explicitly included. Article 225 of the Constitution states that '[a]ll have the right to an ecologically balanced environment, which is an asset of common use and essential to a healthy quality of life, and both the Government and the community shall have the duty to defend and preserve it for present and future generations'. The incorporation of other values such as transparency, judicial independence, and prosecutorial independence also strengthened the legal system's environmental law enforcement capabilities.[150]

The public civil action in Brazil is a vehicle for the protection of diffuse and collective rights and can be used to seek compensation for environmental harm and to prevent further harm. The 1985 Public Civil Action Act[151] confers standing on certain entities to file a class action. They are the public prosecutor's office, the public defenders, the federal government, the states, the federal district and the municipal districts, government agencies, semi-public corporations, and civil associations. Although the Act does not focus on environmental concerns only, it increases the ability of the *Ministerio Publico* to compel compliance with environmental laws and seek damages for harm done.[152] In practice, the public civil action has proven to be a highly effective tool at securing compliance because many of those being prosecuted find it better to negotiate and comply than to be dragged through years of litigation.[153]

[148] Constituição Federal de 1988, art 20(ix).

[149] Antonio Herman Benjamin and Nicholas Bryner, 'Brazil' in Lees and Viñuales (eds) (n 145) 4.

[150] For discussion, see eg 'Justice and Socio-environmental Protection in the Brasilian Amazonia' (*Conselho Nacional de Justiça*, 2020) <www.cnj.jus.br/wp-content/uploads/2020/11/JUSTICE-AND-SOCIO-ENVIRONMENTAL-PROTECTION-IN-THE-BRASILIAN-AMAZONIA_V-6_2020-12-16.pdf> accessed 3 August 2023; Geanluca Lorenzon, 'Corruption and the Rule of Law: How Brazil Strengthened Its Legal System' (*CATO Institute*, 20 November 2017) <www.cato.org/policy-analysis/corruption-rule-law-how-brazil-strengthened-its-legal-system> accessed 3 August 2023; Manuellita Hermes, 'The Brazilian Federal Supreme Court as Guardian of the Ecosystem: An Emerging Framework for Ecological Justice' (*IACL-AIDC Blog*, 23 February 2021) <https://blog-iacl-aidc.org/2021-posts/2021/2/23/the-brazilian-federal-supreme-court-as-guardian-of-the-ecosystem-an-emerging-framework-for-ecological-justice> accessed 3 August 2023.

[151] Lei da Ação Civil Pública, Lei n 7,347 de 24 de julho de 1985.

[152] Nicholas S Bryner, 'Brazil's Green Court: Environmental Law in the Superior Tribunal de Justiça (High Court of Brazil)' (2012) 29 Pace Environmental Law Review 470, 480.

[153] It has been estimated that about 90% of such suits are settled: Bernardo Mueller, 'The Fiscal Imperative and the Role of Public Prosecutors in Brazilian Environmental Policy' (2010) 32 Law and Policy 104, 107, citing Maria Tereza Sadek and others, *Justiça e cidadania no Brasil* (SciELO—Centro Edelstein 2009).

As soon as the Fifteenth Conference of the Parties (COP-15) of the UN Framework Convention on Climate Change (UNFCCC) in Copenhagen concluded, then Brazilian President Luiz Inacio Lula da Silva signed Law 12,187/2009 on the National Policy on Climate Change and Brazil's international commitments to reduce GHG emissions.[154] The law enshrined Brazil's Copenhagen Accord commitment of reducing between 36.1% and 38.9% of its projected GHG emissions by 2020.[155] Article 5 states that Brazil must promptly comply with all its commitments pursuant to the UNFCCC and other international legal instruments on climate change that the country eventually signs. Eighteen instruments are listed in Article 6 to implement the goals of the law, including action plans to control deforestation, the establishment of GHG emissions inventories and standards for emissions reduction, a Climate Change Fund, and fiscal measures including tax incentives to encourage GHG emissions reduction. The Climate Change Fund was created to support research and activities related to mitigation and adaptation. A committee made up of 10 governmental representatives and five civil society representatives is responsible for managing the Fund.[156] Article 3 of the Law establishing the Climate Change Fund (Law 12,114 of 2009) states that revenue for the Fund will be drawn from contributions from national and international organizations, allocations in the annual federal budget, and up to 60% of the royalties received by the Ministry of Environment from oil exploitation would be earmarked for this fund. The dormancy of this Climate Change Fund during the Bolsonaro administration led to the first climate change lawsuit in the Supreme Court (see discussion in Chapter 1 and below on the *Climate Fund Case*).

On 1 January 2023, former president Luiz Inácio Lula da Silva was inaugurated as the thirty-ninth President of Brazil. President Lula had campaigned on a platform of, amongst other things, bringing the country back to the global stage as an environmentally responsible major developing power and 'to make the

[154] 'Brazil Signs Law Cutting CO2 Emissions' *The Sydney Morning Herald* (30 December 2009) <https://amp.smh.com.au/world/brazil-signs-law-cutting-co2-emissions-20091230-ljc2.html> accessed 3 August 2023. For analysis relating to the National Policy on Climate Change, see Karen Alvarenga Oliveira, 'Brazil' in Kevin R Gray, Richard Tarasofsky, and Cinnamon Carlarne (eds), *The Oxford Handbook of International Climate Change Law* (OUP 2016); Natascha Trennepohl, 'Brazil's Policy on Climate Change: Recent Legislation and Challenges to Implementation' (2010) 4(3) Carbon and Climate Law Review 271.

[155] Given the nature of Brazil's political structure, state and municipal governments play an important role in implementing laws and policies to achieve the targets set at the federal level. Some state governments have set their own targets. For example, the State Parliament of Sao Paulo in 2009 approved a Climate Change Bill of Brazil (Law No 12.798 of 2009) which set a GHG emissions reduction target of 30% of 2005 levels by 2012 through several measures focused on transportation, energy efficiency, waste management, and land use. Sao Paulo is the first city in Brazil to pass such legislation: 'Sao Paulo Climate Change Policy (Sao Paulo City)' (*United Nations Sustainable Development Goals*) <https://sdgs.un.org/partnerships/sao-paulo-climate-change-policy-sao-paulo-city> accessed 3 August 2023.

[156] A useful diagram setting out the Fund's governance structure can be found at 'Climate Finance Architecture in Brazil' (*Vivideconomics*, March 2012) <https://assets.publishing.service.gov.uk/media/57a08a6e40f0b652dd000718/DEWPoint_A0510-A0513_Feb2012_Climate_Finance_Architecture_in_Brazil.pdf> accessed 3 August 2023, 5.

preservation of the Amazon a top priority'.[157] However, just six months later, a controversial proposal to drill for oil in the sea off the mouth of the Amazon river emerged as a major challenge for Lula as he seeks to balance his campaign pledges of environmental protection and economic developments in Brazil.[158]

### Overview of the Brazilian docket

The pre-2020 Brazilian docket consisted of a handful of cases where climate change (as a factual or legal issue) was peripheral to the central issue of forest protection and land use.[159] For example, *Public Prosecutor's Office v Oliveira and Others* is a consolidation of cases by Sao Paulo's state prosecutor against farmers to stop them from setting fires to clear land as part of sugar cane harvesting.[160] The arguments against the practice of burning included air pollution and the emission of GHGs.[161] Filed in 2004, *Public Prosecutor's Office v H Carlos Schneider S/A Comercio e Industria and Others* was a case whereby the Federal Public Prosecutor's Office argued that the defendants had illegally removed mangrove forests.[162] In the High Court of Justice's 2007 decision that upheld the original judgment finding the defendants guilty and requiring them to restore the mangroves, it was noted that mangroves serve a number of important ecological functions including coastal protection in the face of rising sea levels due to climate change.

As mentioned earlier, the year 2020 marked a milestone in the emergence of climate change litigation in Brazil with prominent NGOs and political parties turning to litigation as a last resort as they believed that civil society 'can't change the mindset of the government'.[163] These post-2020 cases are core cases which, for example, seek judicial clarification of the government's legal duties that derive from national and international climate change policies to halt deforestation, the largest driver of climate change in Brazil. The following discussion is broadly divided into two sections, one on cases that have been brought against private actors and the other on cases that have been brought against the government.

[157] Roula Khalaf, Michael Stott, and Michael Pooler, 'FT Interview: Brazil's Lula on the Prospects of an Extraordinary Comeback' *The Financial Times* (11 July 2022) <www.ft.com/content/228e0615-0dae-4ba6-afdb-c3db0fcdd15a> accessed 3 August 2023.

[158] Michael Pooler, 'Brazil Divided Over Oil Drilling Proposal off Amazon Delta' *The Financial Times* (8 June 2023) <www.ft.com/content/800693a0-fc52-408e-83bd-0b8e63f38b3a> accessed 3 August 2023.

[159] For discussion of these cases, see Gabriel Wedy, 'Climate Litigation in the Brazilian Superior Courts' (2018) Innovations in the Ecological Rule of Law 208 <http://papers.ssrn.com/abstract=3818602> accessed 3 August 2023.

[160] AgRg nos EDcl no RECURSO ESPECIAL No 1.094.873- SP (2008/0215494-3).

[161] ibid.

[162] RECURSO ESPECIAL No 650.728-SC (2003/0221786-0).

[163] Sam Cowie, 'Brazilian Government Taken to Court for Assault on Environment, Climate' (*Mongabay*, 10 June 2020) <https://news.mongabay.com/2020/06/brazilian-government-taken-to-court-for-assault-on-environment-climate/> accessed 3 August 2023.

### Climate change litigation against private actors

#### *Sao Paulo Public Prosecutor's Office v KLM and Others* (the *Airlines* case)

In 2010, the State of Sao Paulo's Public Prosecutor Office filed 35 cases to compel airlines that make use of the Sao Paulo international airport to offset their emissions. The prosecutors argued that, under the National Environmental Policy Act of 1981 (NEPA) civil liability regime, there was a direct link between the airlines' activities and the GHG emissions causing damage to the climatic system. It was argued that compensation for the harm caused could take the form of reforestation in the municipality of Guarulhos or in the same watershed or payments into the State Fund for Reparation of Diffuse Injured Interests which would be responsible for disbursing the funds for ecological restoration activities including reforestation.[164]

The court rejected several of the cases on the basis that it lacked jurisdiction. The judges examined the merits of the case in three lawsuits. One lawsuit against KLM made it all the way to the Superior Tribunal de Justice (High Court of Brazil) which rendered its decision on 14 December 2020.[165] It was noted that the court of first instance dismissed the case against KLM because the court was of the view that the act of emitting carbon dioxide and other GHGs does not constitute an illicit act. If this were the case, then drivers of all motor vehicles would be committing environmental offences. Further, the court held that if the claim against KLM succeeded, it would be akin to imposing new environmental regulations after the approval licences to operate a hydroelectric plant or mine have been obtained by the project developers. This would violate the principle of maintaining legal stability, which, in the court's view, would be unacceptable. The appeal to the Regional Federal Court of Appeal (3rd Regional) was also unsuccessful. The court affirmed that the federal aviation authority (ANAC) was the rightful authority to regulate airline activities. Even though it is undeniable that air transport emits polluting gases, it would be regulatory overreach for the courts to impose obligations on KLM that are additional to those imposed by ANAC. The court also highlighted that KLM was merely carrying out its economic activity and therefore was not committing illegal acts. In fact, the services provided by airlines were of high social relevance and any imposition of obligations and additional costs on airlines should be carefully considered from the perspective of air travel costs and consumer welfare.

The prosecutor's office made a final appeal to the High Court which was dismissed.[166] The High Court reiterated the fundamental misgiving that the judiciary had with this case: ie that the prosecutor's office was seeking compensation from a company that is not carrying out any illegal activities and is, in reality, seeking to impose unprecedented environmental regulations on the airlines'

[164] RECURSO ESPECIAL No. 1856031-SP (2020/0001750-7).

[165] ibid. The information contained in the rest of this paragraph is drawn from page 4 of the High Court's judgment.

[166] ibid 6.

commercial activity.[167] The court held that it could not impose such regulations without violating the doctrine of separation of powers and potentially destabilizing the rule of law. The final ground for dismissing the appeal lay in the court's recognition that the ANAC has already put in place a regulatory framework to monitor GHG emissions from international air transport as part of its implementation of the Carbon Offsetting and Reduction Scheme for International Aviation (CORSIA). The court's implicit view was that this created a sound and predictable regulatory response that takes the political and technical complexities into account, a fine balance that the court was reluctant to disturb.

***IBAMA (Federal Environmental Agency) v Siderugica Sao Luiz Ltd and Geraldo Magela Martins (Steel Company** case)*

In July 2019, the Brazilian Attorney-General's office acting on behalf of the federal environmental agency (the Brazilian Institute of the Environment and Natural Resources or IBAMA) filed a public civil action against a steel company (Siderugica Sao Luiz Ltd) and its administrator or managing director (Mr Martins) for the fraudulent use of illegal charcoal in its plants in the state of Minas Gerais.[168] In the petition commencing the lawsuit, the government plaintiff alleged that the certificates of origin for the charcoal being acquired for the plants were forged and falsified to hide their place of origin. The real place of origin for the illegal charcoal could be the Atlantic Forest, the Cerrado, or even the Amazon.[169] The plaintiff emphasized that the use of charcoal from unregulated sources and the falsifying of certificates amounted to an illegal act that directly contributes to deforestation. IBAMA also claimed that the company led a scheme of illegal charcoal production, from which they gained significant profits. It is noteworthy that IBAMA found that the defendant company had used 44,636,000 cubic metres of illegal charcoal, which is a staggering amount when one considers that the average capacity of one coal truck is 60 cubic metres.[170]

The petition identified that the use and production of illegal charcoal caused various types of environmental harm including the loss of vegetation, the impairment of the hydrological cycle due to the loss of vegetation, soil erosion, biodiversity loss, and climate change via the emissions of GHGs from burning the charcoal, as well as loss of carbon sinks.[171] Highlighting the company's key role as consumer of the charcoal, IBAMA sought to hold the defendant company and its managing director strictly, jointly, and severally liable for the environmental damage caused

[167] ibid 7. The information contained in the rest of this paragraph is drawn from page 7 of the High Court's judgment.

[168] *Federal Environmental Agency (IBAMA) v Siderúrgica São Luiz Ltd and Martins*, Numero: 1010603-35.2019.4.01.3800.

[169] ibid [15].

[170] ibid [18]–[19].

[171] ibid [21].

by deforestation due to production of the illegal charcoal and contributing to climate change (as production and burning of the charcoal releases GHGs).[172]

To calculate the quantum of damages sought from the defendants, IBAMA's spatial modelling and calculations allowed it to measure the amount of GHGs released into the atmosphere as a result of burning the illegal charcoal. The quantum of damages would be the amount of GHGs multiplied by the social cost of carbon (ie the cost of the damage created by one extra tonne of carbon dioxide emissions). In the petition, IBAMA pointed out that many jurisdictions and international organizations including the United States and the Organisation for European Economic Co-operation (OECD) use the social cost of carbon as a policy tool to quantify and compare the costs and benefits of specific regulations and policies.[173] Adopting OECD's midpoint estimate of the cost of carbon in 2020, IBAMA argued that the social cost of carbon that the defendants ought to pay was 60 euros per tonne of carbon dioxide equivalent.[174]

IBAMA sought compensatory damages as well as environmental remediation.[175] In its decision issued on 3 February 2021, the court denied the plaintiff's claim. The presiding judge was of the view that an environmental compliance programme is an internal environmental management tool and there is no legal basis to compel companies to implement such programmes.[176] While it is not impossible for the court to judicially compel the defendant company to implement such a programme in defence of the fundamental right to an ecologically balanced environment, the judge did not think the facts of the case merited such an outcome.[177] The judge noted that IBAMA's formal notifications to the company about its allegedly illegal conduct were draw up in November 2012, which meant—at the time

[172] According to the plaintiff, the managing director could be held liable on the basis of criminal law, civil law, and administrative law. Art 2 of Law 9,605 of 1998 extends criminal liability to the director, administrator, auditor, manager, agent, or representative of a legal entity if it can be shown that they knew of the criminal conduct but omitted to prevent it when they could do so. On the civil law basis, strict liability applies alongside the theory of integral risk. Acting as the director of the company, Mr Martins had administrative and managerial powers to direct the company's activities. As such, his liability ought to be joint and several with the company: ibid [38]–[50].

[173] ibid [175]–[180]. On the social cost of carbon, see Isabella Backman, 'Stanford Explainer: Social Cost of Carbon' *Stanford News* (7 June 2021) <https://news.stanford.edu/2021/06/07/professors-explain-social-cost-carbon/> accessed 3 August 2023; 'Social Cost of Carbon' (*Energy Policy Institute at the University of Chicago*) <https://epic.uchicago.edu/area-of-focus/social-cost-of-carbon/> accessed 3 August 2023; Paul Watkiss and Thomas E Downing, 'The Social Cost of Carbon: Valuation Estimates and Their Use in UK Policy' (2008) 8(1) The Integrated Assessment Journal 85.

[174] *Ibama v Siderúgica São Luiz* Petition (n 168) [181].

[175] More specifically, the agency sought: (1) reforestation of an equivalent land area; (2) compensation for environmental damage; (3) compensation for causing climate change harm based on the social cost of carbon, to be invested in the creation of carbon sinks; (4) disgorgement of profits gained from the use and production of illegal charcoal; (5) an order compelling the company to put in place an environmental compliance programme; (6) an order withholding tax benefits and access to lines of credit (amounting in total to 16 million reais R$16,000,000) at least until the environmental compliance programme is implemented.

[176] *IBAMA v Siderúrgica São Luiz* 03/02/2021 Decision (n 168) 3.

[177] ibid.

of the judgment—nearly nine years have elapsed since the offences were allegedly committed. While there were good indications in the evidence that the defendant company was indeed committing the said offences, the court ruled it could not now order the company to implement the programme when it was not clear that it was still using and producing illegal charcoal. More damning was the realization that the company might have ceased operations three years ago and was no longer in existence. The final order by the judge sought clarification on these factual issues and denied the plaintiff's request for an injunction. At the time of writing, it is not clear that there will be an appeal.

### Climate change litigation against the government

#### *Laboratorio de Observatorio do Clima v Minister of Environment and Brazil*

On 26 October 2021, just about a week before the 2021 United Nations Climate Change Conference (COP-26), 70 NGOs led by Laboratorio de Observatorio do Clima filed a class action at the federal court of Amazonas against the Ministry of Environment and the Brazilian government. The plaintiffs contended that the National Policy on Climate Change was inadequate and needed to be updated according to the best available science and the latest Intergovernmental Panel on Climate Change (IPCC) Sixth Assessment report to ensure that the reduction of GHG emissions by Brazil is consistent with a 1.5-degree Celsius global warming scenario.

The plaintiff NGOs grounded their claim in the constitutionally recognized right to a healthy environment, as well as fundamental rights such as the right to life, dignity, health, food, and housing. They argued that the state has legal obligations arising from the Paris Agreement as well as national climate change law to reduce GHG emissions, effectively implement (and update) domestic climate change policies, and strive for the 'highest possible ambition'[178] in reducing GHGs. In its first Nationally Determined Contribution (NDC) under the Paris Agreement, Brazil committed to reducing GHG emissions by 37% by 2025 and by 43% by 2030 (using 2005 emissions levels as the baseline). The government submitted its updated NDC in December 2020 in which the same percentages of emissions reductions were stated but the baseline had been amended, which effectively rendered the updated NDC less ambitious than the first. In addition, the commitment to reduce deforestation by 80% by 2030 was in the first NDC but removed from the updated one. The plaintiff NGOs argued that Brazil's updated NDC amounted to a violation of the principle of non-regression in international environmental law

[178] This echoes art 4(3) of the Paris Agreement which states: 'Each Party's successive nationally determined contribution will represent a progression beyond the Party's then current nationally determined contribution and reflect its highest possible ambition, reflecting its common but differentiated responsibilities and respective capabilities, in the light of different national circumstances.'

and, more specifically, Article 4(3) of the Paris Agreement. The case is still pending decision at the time of writing.

*Six Youths v Minister of Environment and Others*

This lawsuit was also initiated to challenge the government's move to decrease the ambition of Brazil's NDC. On 13 April 2021, six youth activists filed a lawsuit alleging that the government of Brazil violated the Paris Agreement, Article 225 of Brazil's Constitution which enshrines the right to an ecologically balanced environment, and the National Policy on Climate Change by updating its NDC in such a way that it was less ambitious than the previous one. The updated NDC maintained the goal of a 43% emissions cut by 2030 but increased the estimated baseline emissions from 2005, effectively allowing Brazil to emit 400 million more tonnes of carbon dioxide equivalent by 2030.

On 28 May 2021, the court decided in favour of the defendant government. It held that the NDC cannot be considered less ambitious since it included a carbon neutrality goal. At the time of writing, an appeal is pending.

*PSB and Others v Brazil* (*Climate Fund* case)

A historic milestone, this is the first climate change case to be decided by the Brazilian Supreme Court. As briefly discussed in Chapter 1, Case ADPF708 centres on Brazil's Climate Fund which was established in 2009 (by Law No 12.114/09) as a financial mechanism of the National Policy on Climate Change. The Climate Fund is mandated to direct its annual budget towards funding projects, enterprises, and research on climate change mitigation and adaptation.

On 5 June 2020, four political parties brought this case to compel the executive to revive the proper functioning of the Fund by reconvening its committee and releasing the budgets that had already been allocated.[179] The parties filed an Action for Non-compliance with Fundamental Precept, arguing that the governmental inaction regarding the Climate Fund amounted to a violation of its constitutional and international legal obligations. What was remarkable was the decision by the presiding judge, Judge Barroso, to hold a public hearing to obtain an 'objective and official account' of the state of Brazil's environmental policy.[180] The public hearing

[179] The four political parties are the Brazilian Socialist Party (PSB), the Socialism and Freedom Party (PSOL), the Workers' Party (PT), and the Sustainability Network. Political parties in Brazil have the constitutional privilege of pursuing judicial relief for alleged violations of constitutional and human rights directly before the Supreme Court; Constitution of the Federative Republic of Brazil (1988), art 103. For early discussion of the case before the Supreme Court's decision, see Alessandra Lehmen, 'Advancing Strategic Climate Litigation in Brazil' (2021) 22 German Law Journal 1471, 1476.

[180] Joana Setzer, 'First Climate Case Reaches Brazil's Supreme Court' (*Grantham Research Institute on Climate Change and the Environment*, 30 September 2020) <www.lse.ac.uk/granthaminstitute/news/first-climate-case-reaches-brazils-supreme-court/> accessed 3 August 2023. Fewer than 8% of 556 Actions for Non-compliance with Fundamental Precept filed since 2000 have been successful, but

provided an important platform for concerned citizens to share their expertise and to dispel misinformation about the causes and consequences of climate change.

At the end of the public hearing, Justice Barroso presented a list of what he described as 'uncontroversial' points that emerged from the debates, including the fact that deforestation and burning increased significantly in the years 2019 and 2020 in Brazil and that there had been a notable reduction in fines and enforcement action by IBAMA.[181] Barroso also stated that the country's inaction, if proven, is 'potentially harmful from any perspective: environmental, social, cultural or economic'.[182]

On 30 June 2022, the Supreme Court issued its decision. It ruled that, on the basis of the separation of powers, the executive branch of government cannot ignore the mandate of the legislative branch. The executive branch therefore had a constitutional duty, bolstered by the constitutional right to a healthy environment, to allocate funds to the Climate Fund and execute the Fund's legislative mandate.[183] The court further ruled that the judiciary is required to act to prevent the further regression of environmental protection in Brazil. The Supreme Court clarified that environmental law treaties constitute a particular type of human rights treaty which enjoys 'supranational status'.[184] This means that they have a higher status above domestic laws in the legal hierarchy. Accordingly, any Brazilian law or decree that contradicts the Paris Agreement, including the NDC, may be invalidated on the basis that it is a direct violation of the Constitution and human rights. It can be argued that this is a landmark judgment which has the hallmark of a court taking a firm stance to play its role in halting (and perhaps even reversing) the regressive and environmentally destructive policies of the government of the day.

On the same day that the four political parties filed this case, they filed a similar case about the Amazon Fund.[185] A public hearing took place on 23 and 26 October 2020, focusing on the lack of implementation of the Amazon's environmental policies and the allocation of funds from the Amazon Fund. Justice Barroso was particularly interested in the role of the Amazon rainforest in climate change mitigation and expressly indicated that, given the close connection between both cases,

Justice Barroso, the presiding judge in this case, is widely considered to be one of the most progressive of the 11 justices of the Supreme Court.

[181] ibid.

[182] ibid. The public hearing took place at the Supreme Court over two days, and 66 experts including parliamentarians, representatives from the business and financial sector, and researchers provided testimony. Joana Setzer notes that having the Supreme Court convene such a hearing is unusual, if not unprecedented, and that this approach arguably helps to expand public participation in climate policy.

[183] *PSB v Brazil (on Climate Fund)*, 07/01/2022 Decision (unofficial translation) (*Grantham Research Institute on Climate Change and the Environment*) <http://climatecasechart.com/non-us-case/psb-et-al-v-federal-union/?mc_cid=5f5cba8306&mc_eid=7d825ad4d5> accessed 3 August 2023, [16], [28], [36], and [37].

[184] ibid [17].

[185] *PSB v Brazil (on Amazon Fund)* (2020) ADO 59/DF.

they could eventually be joined.[186] However, at the time of writing, a decision on *PSB and Others v Brazil (on Amazon Fund)* is pending.

### The future of climate litigation in Brazil

It is clear from the preceding discussion that climate change litigation in Brazil is unfolding rapidly. There is much to learn from the jurisprudence, and the litigation strategies adopted by the *Ministerio Publico*, political parties, and civil society organizations. While the earlier cases in the Brazilian docket were few and far between, we have witnessed a new period in the emergence of climate change litigation in Brazil from 2020 onwards, with the filing of many lawsuits challenging the Bolsanaro government's regulatory rollback and dismantling of climate change institutions like the Climate Fund and the Amazon Fund. Climate change litigation is also being deployed to halt the destruction of the Amazon. As the Amazon has global planetary significance, it would not be surprising to see more climate change lawsuits directed towards halting its decline and that international sources of funding will help drive the litigation.

The Brazilian legal system has an excellent tool kit in terms of existing climate change and environmental legislation, strategic litigation expertise, and progressive environmental rights-based jurisprudence. It is, of course, important not to paint an overly optimistic picture as it is well known that Brazil's legal system, not alike many others in the Global South, suffers from case overload, long delays, and enforcement difficulties.[187] Nonetheless, the overall picture is promising for concerted legal action to advance climate change protection.

Finally, a unique dimension to climate change litigation in Brazil involves lawsuits brought by actors outside the country. The *Casino* case targeting the Brazilian livestock industry is a good example of how the Duty of Vigilance in French law is being used to drive improvements in environmental and social governance in the Global South.[188] The request to the International Criminal Court (ICC) filed by AllRise, an Austrian climate change activism group,[189] to investigate whether

[186] Maria Antonia Tigre, 'Brazil's First Climate Case to Reach the Supreme Court' *Opinio Juris* (13 October 2020) <http://opiniojuris.org/2020/10/13/brazils-first-climate-case-to-reach-the-supreme-court/> accessed 3 August 2023; for details about the Fund, see its official website <www.amazonfund.gov.br/en/home/> accessed 3 August 2023. See also 'In Brazil, States and Federal Government Point Out the Need for Changes in the Amazon Fund' (*Governors' Climate & Forests Task Force*, 11 November 2020) <www.gcftf.org/brazil-states-federal-government-amazon-fund/> accessed 3 August 2023; Reuters Staff, 'Norway Stops Amazon Fund Contribution in Dispute with Brazil' *Reuters* (16 August 2019) <www.reuters.com/article/us-brazil-environment-norway-idUSKCN1V52C9> accessed 3 August 2023.

[187] See eg Matthew M Taylor, 'Citizens Against the State: The Riddle of High Impact, Low Functionality Courts in Brazil' (2005) 25(4) Brazilian Journal of Political Economy 418; Raymond Colitt and Stuart Grudgings, 'Brazil's Creaking Justice System Slow to Reform' *Reuters* (22 May 2008) <www.reuters.com/article/idUKN2249474820080522> accessed 3 August 2023.

[188] See discussion in Chapter 3, Section 4.2 of this book.

[189] All Rise, 'Communication under Article 15 of the Rome Statute of the International Criminal Court regarding the Commission of Crimes Against Humanity against Environmental Dependents and Defenders in the Brazilian Legal Amazon from January 2019 to present, perpetrated by Brazilian

the environmental policies of former president Bolsonaro and his administration constituted crimes against humanity is a novel example of climate change litigation against governments (potentially) reaching the international courts.

## 3. CONDITIONS FAVOURING THE EMERGENCE OF CLIMATE CHANGE LITIGATION IN THE GLOBAL SOUTH

The preceding discussion about the state of climate change litigation in the 'frontrunner jurisdictions' shows that there are common features across their socio-legal systems which are fundamental to the promotion of environmental rule of law. These same features also explain why climate change litigation has emerged more successfully in these jurisdictions compared to others in the Global South. These features are: (1) access to justice which includes broad standing rules and the removal of barriers such as hefty court fees and security of costs orders; (2) judicial capacity and receptivity to climate change litigation; (3) the existence of environmental (and climate change) laws and progressive environmental jurisprudence; (4) the right to a healthy environment; and (5) an active civil society.

It is not the case that every jurisdiction must have all five features in place for climate change litigation to develop or that the presence of these five features guarantees that climate change litigation will flourish. What our analysis suggests is that when these five features exist (and it is often the case that they exist to varying degrees, eg limited judicial capacity but a highly active environmental advocacy community), they complement each other to create an environment that is supportive of climate change litigation. A further point we would make is that these conditions that support climate change litigation in the Global South are not significantly different from those in the Global North. The conditions that promote environmental and climate justice in the Global North would do the same for the Global South. It stands to reason that if litigation is to be promoted as a climate change governance tool and mechanism for implementing Paris Agreement commitments in the Global South, policy attention and resources must continue to be directed at establishing and strengthening the fundamental pillars of effective legal institutions and access to justice. What is distinctive and different about the Global South experience with climate change litigation is that it generally takes on different characteristics, thereby reflecting the different socio-economic conditions, political arrangements, and institutional capacities in the Global South. Thus, we see a greater resort to rights-based litigation in the Global South compared to the Global North because of the tradition of litigation to protect socio-economic

President Jair Messias Bolsonaro and certain former and current principal actors of his administration' <https://climatecasechart.com/wp-content/uploads/non-us-case-documents/2021/20211012_14633_na.pdf> accessed 3 August 2023.

rights in Global South jurisdictions like South Africa and Colombia and the fact that most national constitutions of Global North jurisdictions do not recognize the right to a healthy environment.

While most of the discussion below draws and elaborates upon the examples of Brazil, South Africa, and India as 'frontrunner jurisdictions', we also use examples from other Global South countries where appropriate. To avoid repetition, the discussion in this section is kept concise as some of these factors, such as constitutional protection of the environment, have already been discussed in Chapter 3.

## 3.1 Access to Justice

### Rules on standing

The rules relating to standing, or *locus standi*, in public law are said to exist for the purpose of resolving the conflict between 'the desirability of encouraging people to participate actively in the enforcement of law, and the undesirability of encouraging meddlesome interlopers invoking the jurisdiction of the courts in matters in which they are not concerned'.[190] Standing rules vary across jurisdictions and litigants depend on these rules to be granted the right of access to courts, whether as individuals, a collective, a third party, or as *amicus curiae*. Cases are doomed to failure when narrow standing rules are a procedural hurdle that litigants—typically NGOs and citizen groups—are unable to overcome because they do not satisfy the conventional requirement of having sufficient interest to seek a judicial remedy.

Many Global South jurisdictions have narrow standing requirements.[191] However, in the jurisdictions where climate change litigation is emerging, broad standing rules exist either enshrined in the national constitutions and environmental legislation or as a result of judicial interpretation. The relatively liberal standing rules in India, South Africa, and Brazil have already been discussed in Section 2. We therefore illustrate this point with examples from other Global South

[190] Harry Woolf and others, *De Smith's Judicial Review* (7th edn, Sweet & Maxwell 2013) 66–67 [2-002].

[191] Examples include Malaysia and China. In China, the Civil Procedure Law requires a plaintiff to have a direct interest in the lawsuit. Jiangfeng Li argues that, '[i]n a climate change lawsuit, it is even more difficult for the plaintiff to establish that it has been directly affected by a particular defendant's GHG emissions or a government agency's inaction or improper action' and is therefore of the view that 'the difficulties of establishing standing may hinder private climate change lawsuits in China': Jiangfeng Li, 'Climate Change Litigation: A Promising Pathway to Climate Justice in China?' in Lin and Kysar (eds) (n 9) 342. In Malaysia, the Supreme Court decided in *Government of Malaysia v Lim Kit Siang, United Engineers (M) Berhad v Lim Kit Siang* [1988] 1 MLJ 50 that the applicant lacked standing because he could not distinguish himself from other road users and did not have sufficient interest in the project. Experts including Andrew Harding view this decision as detrimental to public interest litigation as NGOs are unable to bring lawsuits on behalf of communities: Andrew Harding, 'Public Interest Groups, Public Interest Law, and Development in Malaysia' (1992) 11 Third World Legal Studies 231. See further, Jacqueline Peel and Jolene Lin, 'Climate Change Adaptation Litigation: A View from Southeast Asia' in Lin and Kysar (eds) (n 9) 317.

jurisdictions. In the Philippines, for example, the Supreme Court in the seminal case *Minors Oposa v Secretary of the Department of the Environment and Natural Resources* ruled that the plaintiffs had standing to represent unborn generations in the protection of environmental rights.[192]

In Kenya, the new environmental law enacted in 1999 significantly broadened the standing requirements. Potential litigants were no longer required to demonstrate personal interest or injury.[193] Furthermore, the Kenyan Constitution stipulates that, in applications relating to the right to a clean and healthy environment, 'an applicant does not have to demonstrate that any person has incurred loss or suffered injury'.[194] The Climate Change Act 2016 has incorporated these liberal standing requirements.[195] Omuko-Jung has argued that this provision, coupled with the relevant provisions in the Constitution and National Environmental Management and Coordination Act, 'allows access to courts to anyone to hold private entities and government liable for their action or inaction relating to climate change'.[196]

A final example is that of Uganda. Its Constitution and National Environment Act grants legal standing to every person, including NGOs, without having to prove any injury or legally protected interest.[197]

### Rules on costs

Litigation requires the engagement of legal counsel, expert witnesses, and the payment of various court fees amongst other things. The costliness of litigation hinders access to justice for environmental NGOs and public-spirited citizens in both the Global North and the Global South. The rules relating to costs have been identified as the single most significant barrier to access to environmental justice in the United Kingdom and the same can be said about most Global South jurisdictions with a common law legal system.[198] In this section, we take security for costs as an

[192] 33 ILM 173 (1994) (Supreme Court of the Philippines, 30 July 1993).

[193] The National Environmental Management and Coordination Act 1999 (Kenya). s 3(4) states that a person seeking to litigate the right to a clean and healthy environment would 'have the capacity to bring an action notwithstanding that such a person cannot show that the defendant's act or omission has caused or is likely to cause him any personal loss or injury'.

[194] The Constitution of Kenya 2010, art 70(3).

[195] s 23 of the Act states: 'any person may, pursuant to Article 70 of the Constitution, apply to the Environment and Land Court (ELC) alleging that a person has acted in a manner that has or is likely to adversely affect mitigation and adaptation efforts.'

[196] Lydia A Omuko-Jung, 'The Evolving Locus Standi and Causation Requirements in Kenya: A Precautionary Turn for Climate Change Litigation' (2021) 15 Carbon and Climate Law Review 171, 181.

[197] Constitution of Uganda 1995, arts 50 and 137; National Environment Act 2019 (Uganda), s 4(2). For discussion, see Ben Kiromba Twinomugisha, 'Some Reflections on Judicial Protection of the Right to a Clean and Healthy Environment in Uganda' (2007) 3 Law, Environment and Development Journal 244; Joseph Oloka-Onyango, *When Courts Do Politics: Public Interest Law and Litigation in East Africa* (Cambridge Scholars Publishing 2017).

[198] As Chakrabarti and others have noted, the risk of an adverse costs order and the difficulties of quantifying this risk constitutes the biggest hurdle for public interest litigants: see discussion in Shami Chakrabarti, Julia Stephens, and Caoilfhionn Gallagher, 'Whose Cost the Public Interest' [2003] Public Law 697; Working Group on Access to Environmental Justice, *Ensuring Access to Environmental Justice*

example. Security for costs is a common law concept and is rare in civil law systems.[199] It refers to the security which a defendant may require from the plaintiff for payment of the costs that may be awarded to the defendant.[200] The first purpose is to protect the defendant who has no choice but to litigate or settle to avoid default judgment by ensuring an available fund to defray costs incurred by defending a frivolous claim.[201] The second purpose is to discourage the filing of unmeritorious claims.[202] Security for costs order applications have been used by defendant corporations to deter public interest litigants. The chilling effect of judicial discretion to make security for costs orders has been acknowledged to the extent that court systems that wish to promote access to justice have enacted rules that give the court discretion not to order that an applicant give security for costs where it is satisfied that the proceedings have been brought in the public interest.[203]

In Uganda, the Civil Procedure Rules provide that the court may, if it deems fit, order a plaintiff in any suit to give security for the payment of all costs incurred by the defendant.[204] In the seminal case of *Greenwatch and ACODE v Golf Course Holdings Ltd*, the court's order to the NGO to pay 50 million Uganda shillings was a death knell for the case, which could not proceed because the NGO could not afford to pay this hefty sum.[205] The construction of the golf course went ahead in a wetland and greenbelt area, illustrating the dramatic consequences of security for costs orders for public interest environmental litigation.[206] Since then, the Ugandan courts have shifted towards a more favourable position for NGO litigants by exercising their discretion to decline to grant security for costs orders simply because a petitioner was perceived to be financially vulnerable.[207]

The South African Constitutional Court developed a rule against imposing costs on losing parties who engage in good faith constitutional litigation to avoid the chilling effect of adverse costs orders (the *Biowatch* principle).[208] Unfortunately, at

*in England and Wales* (Report, May 2008) <http://assets.wwf.org.uk/downloads/justice_report_08.pdf> accessed 3 August 2023

[199] Security for costs is a common law concept and is rare in civil law jurisdictions. See Elaine Wong and others, 'A Global Perspective on Availability of Security for Costs and Claim in International Arbitration: Mirage or Oasis?' (*Herbert Smith Freehills*, 1 February 2017) <www.herbertsmithfreehills.com/hsfpdf/latest-thinking/a-global-perspective-on-availability-of-security-for-costs-and-claim-in?pdf=1> accessed 3 August 2023.

[200] Jonathan Law and Elizabeth A Martin, *A Dictionary of Law* (7th edn, OUP 2009) 421.

[201] ibid.

[202] ibid.

[203] eg Land and Environment Court Rules 2007 (New South Wales), r 4.2.

[204] The Civil Procedure Rules (No 71-1 of 2014) (Uganda), O 26 r 1.

[205] High Court Civil Suit (HCCS) No 834 of 2000 (High Court of Uganda).

[206] Ben Kiromba Twinomugisha, 'Some Reflections on Judicial Protection of the Right to a Clean and Healthy Environment' (2007) 3 Law, Environment and Development Journal 244, 253.

[207] See discussion in Samantha Atukunda K Mwesigwa and Peter Davis Mutesasira, 'Climate Litigation as a Tool for Enforcing Rights of Nature and Environmental Rights by NGOs: Security for Costs and Costs Limitations in Uganda' (2021) 15(2) Carbon and Climate Law Review 139.

[208] For discussion, see Tracy Humby, 'The Biowatch Case: Major Advance in South African Law of Costs and Access to Environmental Justice' (2010) 22 Journal of Environmental Law 125.

the time of writing, civil society organizations in South Africa are facing a judicial deviation from the *Biowatch* principle and are finding ways to deal with the threats of adverse costs orders.[209]

## 3.2 Climate-Conscious Judiciary

In its major study setting out a blueprint for tackling today's major environmental crises, UNEP articulated the importance of system-wide transformation of society including world views, norms, and values to a sustainable future.[210] The judiciary has been identified as a crucial partner in achieving sustainability and the Sustainable Development Goals (SDGs). Sustainable Development Goal 16(3) identifies the judiciary as an enabler to promote the environmental rule of law and ensure access to justice. For Justice Brian Preston, Chief Judge of the Land and Environment Court of New South Wales in Australia, adopting a climate-conscious approach to lawyering 'involves identifying the climate change issues and consequences of different courses of action and incorporating these into the preferred solution to the legal problem or dispute'.

To illustrate, the Indian NGT in *Nuggehalli Jayasimha v Government of NCT of Delhi* examined the climate change impact of livestock agriculture which is a major source of methane emissions.[211] The NGT directed the central pollution control board to develop a national livestock methane inventory and establish sustainability indicators for the management of livestock farming. In its decision, '[b]y consciously framing a direct link between methane emissions and livestock, the NGT raised awareness by constructing a climate sustainability narrative to adopt safeguards that are necessary to protect the environment and public health'.[212]

It can be argued that judicial receptivity and capacity to respond to creative, frequently rights-based, arguments and to interpret existing jurisprudence in light of the climate emergency is an important factor. In *DG Khan Cement Company Ltd v Government of Punjab*, Judge Syed Mansoor Ali Shah of Pakistan's Supreme Court gave a powerful articulation of his vision of the role of courts in climate change governance.

209 Lee-Anne Bruce, 'Public Interest Litigation Threats: Cost Orders' (*University of the Witwatersrand, Johannesburg*, 6 August 2021) <www.wits.ac.za/news/sources/cals-news/2021/public-interest-litigation-threats-cost-orders.html> accessed 10 August 2023.

210 'Making Peace with Nature: A Scientific Blueprint to Tackle the Climate, Biodiversity and Pollution Emergencies' (*United Nations Environment Programme*, 2021) <https://wedocs.unep.org/xmlui/bitstream/handle/20.500.11822/34948/MPN.pdf> accessed 10 August 2023.

211 Original Application No 46/2018, National Green Tribunal, 20 May 2020; 'Global Methane Assessment: Benefits and Costs of Mitigating Methane Emissions' (*United Nations Environment Programme*, 2021) <www.unep.org/resources/report/global-methane-assessment-benefits-and-costs-mitigating-methane-emissions> accessed 10 August 2023.

212 Gitanjali N Gill and Gopichandran Ramachandran, 'Sustainability Transformations, Environmental Rule of Law and the Indian Judiciary: Connecting the Dots through Climate Change Litigation' (2021) 23 Environmental Law Review 228, 241.

> This Court should be mindful that its decisions also adjudicate upon the rights of the future generations of this country. It is important to question ourselves; how will the future generations look back on us and what legacy we leave for them? This Court and the Courts around the globe have a role to play in reducing the effects of climate change for our generation and for the generations to come. Through our pen and jurisprudential fiat, we need to decolonize our future generations from the wrath of climate change, by upholding climate justice at all times.[213]

In the *Leghari* decision highlighted in previous chapters, Judge Shah described climate change as 'a defining challenge of our time' and 'a clarion call for the protection of fundamental rights of the citizens of Pakistan, in particular, the vulnerable and weak segments of the society who are unable to approach this Court'.[214] His Honour then proceeded to mandate the creation of a Climate Change Commission to ensure that the government took urgent action to implement their policies to address climate change impacts in Pakistan. The court also set time frames for policy implementation and reporting back to the court, retaining for itself a continuing mandamus jurisdiction to monitor the progress of that work.[215]

Judge Shah has also used the term 'climate judges with a common goal' to refer to judges who adjudicate matters with an eye towards the urgency of climate change mitigation and adaptation needs.[216] There are, of course, other judges in the Global South who may be less vocal about their value preferences but are equally committed to promoting climate change litigation as a regulatory response and governance tool. The point, to put it succinctly, is that in the Global South jurisdictions where climate change litigation has emerged, the presence of judges who are receptive to arguments concerning climate change is a notable factor in determining the potential for climate change litigation to develop.

It has been suggested that judicial approaches and attitudes in the Global South share some notable commonalities because many of these courts operate under similar conditions and therefore are likely to respond in a similar way.[217] Emily Barritt and Boitumelo Sediti have argued that, '[i]n conditions of entrenched

[213] CP1290-L\2019 (Supreme Court of Pakistan, 15 April 2021) [19]; 'Mr. Justice Syed Mansoor Ali Shah' (*Supreme Court of Pakistan*) <www.supremecourt.gov.pk/judges/honorable-judges/mr-justice-syed-mansoor-ali-shah/> accessed 10 August 2023.

[214] *Asghar Leghari v Federation of Pakistan* 2018 PLD 364 [6].

[215] ibid [4]–[6], [13]. For discussion, see Waqqas Ahmad Mir, 'From Shehla Zia to Asghar Leghari: Pronouncing Unwritten Rights Is More Complex than a Celebratory Tale' in Lin and Kysar (eds) (n 9). On the role of judicial commissions in Pakistani public interest environmental litigation like the Climate Change Commission, see Parvez Hassan, *Resolving Environmental Disputes in Pakistan: The Role of Judicial Commissions and Climate Justice in Pakistan* (Pakistan Law House 2018). Dr Parvez Hassan was chairperson of many of the judicial commissions. An electronic copy of the book is available for download at <https://library.lums.edu.pk/dr-parvez-hassan-publications> accessed 10 August 2023.

[216] Syed Mansoor Ali Shah, 'Foreword' in Lin and Kysar (eds) (n 9) xiii.

[217] Daniel Bonilla Maldonado, 'Towards a Constitutionalism of the Global South' in Daniel Bonilla Maldonado, *Constitutionalism of the Global South* (CUP 2013) 21.

poverty, inactive government and with the long shadow of colonialist and apartheid infrastructures, courts in the Global South have taken a transformative approach to adjudication, particularly when it comes to judging questions of fundamental rights'.[218] It is therefore possible to identify a common thread of adjudicative practice which has a more activist character, which arguably translates to greater receptiveness to climate change lawsuits.

However, it should be noted that several Global South jurisdictions also embrace a pro-development mentality which perceives climate and environmental protection as anti-development. Judges are not immune to and, in fact, can play a critical role in supporting such pro-development agendas. An example is *Wijana, Sufarlan, Astawa and Greenpeace Indonesia v Governor of Bali*, an administrative law case challenging the validity of the permit issued by the Governor of Bali for the development of a coal-fired power plant in Celukan Bawang, Bali.[219] The facts of this case are set out at length in the previous chapter. It has been argued that the failure of climate change litigation in Indonesia cannot be separated from the political economy of development in the country and the role that the judiciary perceives itself as playing in the quest for economic growth.[220] As climate litigation against coal-fired power plants challenges the consumption of coal as part of the government's national energy strategic projects as well as the powerful coal industry which plays a key role in Indonesia's economy, judges who view their role (and source of legitimacy) as supporting the governmental economic and energy security agenda are not likely to adopt a climate-conscious approach in their decision-making.

## 3.3 Progressive Environmental Jurisprudence and Climate Change Laws

The enactment of climate change legislation has clear benefits, including the articulation of norms and emissions reduction targets to which government agencies can be held accountable.[221] However, our analysis shows that, across the Global South, a lack of climate change laws has not been a major impediment to the emergence

[218] Emily Barritt and Boitumelo Sediti, 'The Symbolic Value of *Leghari v Federation of Pakistan*: Climate Change Adjudication in the Global South' (2019) 30(2) King's Law Journal 203, 208.

[219] *Ketut Mangku Wijana v Governor of Bali* (Decision Number 2/G/LH/2018/PTUN.DPS) (Administrative Court of Denpasar, 2018).

[220] Agung Wardana, 'Governing through Courts?—a Gloomy Picture of Climate Litigation in Indonesia' (*Völkerrechtsblog*, 21 March 2022) <https://voelkerrechtsblog.org/governing-through-courts/> accessed 10 August 2023.

[221] The GLOBE Climate Legislation Study is the most comprehensive audit of climate change laws globally. The latest 2015 study which covered 98 countries showed that climate laws have doubled in number every five years since the ratification of the Kyoto Protocol. The report can be downloaded at <https://globelegislators.org/globe-climate-legislation-study/global-climate-legislation-study-2015>. The World Bank's *Reference Guide to Climate Change Framework Legislation* defines climate change framework legislation as 'an overarching, multisectoral regulatory instrument that lays down general

of climate change litigation.[222] In India and South Africa, for example, litigants rely on several existing environmental laws such as EIA and air pollution control laws to advance climate-related claims. In Brazil, the first climate change lawsuit to reach the Supreme Court was about seeking governmental compliance with the national climate change law. Nonetheless, it has been argued that the true potential for climate change litigation in Brazil lies in its progressive environmental jurisprudence. Given the preference for indirect, periphery litigation in most Global South jurisdictions, we argue that a functioning environmental legal framework to provide 'hooks' for climate change litigation is more important than a specific climate change law, though undoubtedly the presence of the latter would add additional 'hooks' and perhaps also more core litigation.

In South Africa, for example, there are many environmental laws that govern specific natural resources and activities including the EIA Regulations 2014,[223] the Air Quality Act,[224] the Protected Areas Act[225]—which creates the legal framework governing the country's impressive network of protected areas including the renowned Kruger National Park—the National Forests Act,[226] and the National Water Act.[227] It is widely recognized that South Africa has impressive environmental legislation, but poor enforcement and fragmented governance have led to

principles and defines the institutional framework for climate change policy and implementation'. It identifies 12 elements of effective climate legislation. The reference guide can be downloaded from <https://openknowledge.worldbank.org/handle/10986/34972>.

222 However, see eg Jiangfeng Li, who argues that the absence of a climate change law is a serious impediment to the development of climate change litigation in China because judges and litigants alike struggle with broadening the interpretation of existing environmental laws to include GHG emissions and climate change; Jiangfeng Li, 'Climate Change Litigation: A Promising Pathway to Climate Justice in China?' in Lin and Kysar (eds) (n 9) 353.

223 Regulations 892 of 2014 (South Africa). Section 2 of the Environmental Impact Assessment Regulations states that '[t]he purpose of these Regulations is to regulate the procedure and criteria as contemplated in Chapter 5 of the [National Environmental Management Act] relating to ... applications for environmental authorisations for the commencement of activities, subjected to environmental impact assessment in order to avoid or mitigate detrimental impacts on the environment, and to optimise positive environmental impacts'.

224 National Environment Management: Air Quality Act, 39 of 2004 (South Africa).

225 National Environmental Management: Protected Areas Act, 57 of 2003 (South Africa). For discussion of how the principle of non-regression can be applied to interpretation of the Protected Areas Act with a view to halting the regressive impact of mining activities in protected areas, see Alexander Paterson, 'Protected Areas Law, Mining and the Principle of Non-Regression—A South African Perspective' [2017] South African Journal of Environmental Law and Policy 142. In *Mining and Environmental Justice Community Network of South Africa v Uthaka Energy (Pty) Ltd* (11761/2021) [2021] ZAGPPHC 195, the court set aside the decision of the Minister of Environmental Affairs and the Minister of Mineral Resources to grant permission to Atha-Africa Ventures 16 (Pty) Ltd to mine in the Mabola Protected Area on the basis that, inter alia, there was no evidence that the Ministers applied the NEMA principles and there was insufficient transparency and public participation. The judgment is available online at <https://cer.org.za/virtual-library/judgments/high-courts/mining-and-environmental-justice-community-network-of-south-africa-and-others-v-minister-of-environmental-affairs-and-others> accessed 10 August 2023.

226 Act 84 of 1998 (South Africa).

227 Act 36 of 1998 (South Africa).

continuing environmental degradation.[228] The conditions are therefore ripe for periphery climate change lawsuits that push for the proper enforcement of existing environmental laws. Given that South Africa is a water-stressed country and climate change will exacerbate water scarcity (and deny access to water for millions of South Africans) there is particular promise in the National Water Act and the National Forests Act to halt deforestation.[229] An urgent priority for South Africa is mitigating GHG emissions in the energy sector and reducing its reliance on coal. In this regard, climate lawsuits premised on statutory interpretation of the EIA Regulations have been a mainstay of the South African climate change docket.

The Promotion of Access to Information Act (PAIA) gives effect to the constitutionally protected right of access to information and establishes a right to access the records of both public and private bodies.[230] As Tracy-Lynn Field has pointed out, the successful internal appeal by the Centre for Environmental Rights (CER)

[228] See, for discussion, Louis J Kotze, 'Improving Unsustainable Environmental Governance in South Africa: The Case for Holistic Governance' (2017) 9 Potchefstroom Electronic Law Journal 74 <www.researchgate.net/publication/318396517_Improving_Unsustainable_Environmental_Governance_in_South_Africa_the_Case_for_Holistic_Governance> accessed 10 August 2023; Alexander Paterson and Louis J Kotze, *Environmental Compliance and Enforcement in South Africa: Legal Perspectives* (Juta 2017).

[229] The National Water Act, 36 of 1998 (South Africa) should be read with the Water Services Act, 108 of 1997 (South Africa) which gives effect to the right of access to sufficient water guaranteed in s 27(1)(b) of the Constitution of the Republic of South Africa, 1996. The subtle role that socio-economic rights can play in environmental governance was highlighted in *Mazibuko v City of Johannesburg* 2010 (4) SA 1, where the Constitutional Court had to flesh out the scope of the right to water for the first time. For discussion, see Bronwen Qumbu, 'The Role of the Courts in Advancing Water Security in South Africa' (2021) 24 Potchefstroom Electronic Law Journal; Louis J Kotze, 'Phiri, the Plight of the Poor and the Perils of Climate Change: Time to Rethink Environmental and Socio-economic Rights in South Africa?' (2010) 1 Journal of Human Rights and the Environment 135; Peter Danchin, 'A Human Right to Water? The South African Constitutional Court's Decision in the Mazibuko Case' (*EJIL:Talk!*, 13 January 2010) <www.ejiltalk.org/a-human-right-to-water-the-south-african-constitutional-court%E2%80%99s-decision-in-the-mazibuko-case/> accessed 10 August 2023; Tracy Humby and Maryse Grandbois, 'The Human Right to Water in South Africa and the *Mazibuko* Decisions' (2011) 51 Les Cahiers de droit 521.

[230] Act 2 of 2000 (South Africa). In *Arcelormittal South Africa v Vaal Environmental Justice Alliance* [2015] (1) SA 515 (SCA), the Supreme Court of Appeal upheld the environmental NGO's right of access to the respondent mining company's pollution records, thereby illustrating the value of the right of access to information in environmental and climate change litigation. For discussion about this Act, see Colin Darch and Peter G Underwood, 'Freedom of Information Legislation, State Compliance and the Discourse of Knowledge: The South African Experience' (2005) 37 International Information & Library Review 77; Wilhelm Peekhaus. 'South Africa's Promotion of Access to Information Act: An Analysis of Relevant Jurisprudence' (2014) 4 Journal of Information Policy 570; Rolf Sorensen, 'Impact of South Africa's Promotion of Access to Information Act After Three and a Half Years: A Perspective' (2003) 22 ESARBICA Journal 50. For discussion about the salience of this legislation to public interest litigation including environmental and climate change, see 'South Africa' (*UCL The Constitution Unit*) <www.ucl.ac.uk/constitution-unit/research/research-archive/foi-archive/international-focus/south-africa> accessed 10 August 2023; Michelle Toxopeüs and Louis J Kotzé, 'Promoting Environmental Justice through Civil-Based Instruments in South Africa' (2017) 13(1) Law, Environment and Development Journal 47; 'The Right of Access to Environmental Related Information—The Dawn of a New Era?' (*Hogan Lovells Publications*, December 2014) <www.hoganlovells.com/en/publications/the-right-of-access-to-environmental-related-information-the-dawn-of-a-new-era>. Some civil society organizations use the PAIA as a strategy to request for environmental records: see 'PAIA Requests for Environmental Records' (*Centre for Environmental Rights*) <https://cer.org.za/paia>.

to the Department of Forestry, Fisheries and the Environment for the GHG emissions data of some of South Africa's biggest GHG emitters highlights the importance of the PAIA for climate change litigation.[231]

In December 2019, the CER submitted an appeal pursuant to section 74 of the PAIA to the Department of Environment, Forestry and Fisheries challenging the department's decision to redact and withhold certain GHG emissions data in the emissions data reports, pollution prevention plans, and annual progress reports for a number of significant GHG emitters in South Africa including Glencore plc and Eskom Holdings SOC Ltd.[232] The main grounds for the appeal were that the Department had failed to provide adequate reasons for its decision and that it had interpreted the PAIA incorrectly.[233] The Minister of Forestry, Fisheries and the Environment largely upheld the appeal. The Minister held that adequate reasons for the redactions had not been provided as required by section 25(3)(a) of the PAIA.[234] In deciding whether to uphold the ground of appeal that the Department had failed to interpret PAIA appropriately, the Minister evaluated the redactions in the GHG emissions data reports, pollution prevention plans, and annual progress reports separately.[235] The Minister confirmed that the redaction of activity information in the emissions data report was justified 'because it provides related information which may be related to competition issues such as company investment decisions, market share, competitive advantage and trade secrets'.[236] Regarding the pollution prevention plans and the annual progress reports, the Minister did not think that redacting the information about projected emissions and actual emissions was justified. She ordered that access was to be provided within seven working days from the date of the conclusion of the national lockdown period, reiterating that '[t]he full access to this information will foster and contribute towards transparency and accountability which is owed to the public more so in matters pertaining to the environment'.[237]

The Promotion of Administrative Justice Act (PAJA) gives effect to section 33 of the Constitution which provides that everyone has the right to administrative action that is lawful, reasonable, and procedurally fair and that everyone whose rights have been adversely affected by administrative action has the right to be given written reasons.[238] Tracy-Lynn Field has pointed out that the PAJA's major value for climate change litigation lies in the grounds of review set out in

[231] Tracy-Lynn Field (n 87) 183.
[232] The appeal decision is available at 'Successful Appeal to the Department of Environment, Forestry and Fisheries regarding Greenhouse Gas Emission Data' (*Centre for Environmental Rights*, 13 December 2019) <https://cer.org.za/paia/appeal-to-the-department-of-environment-forestry-and-fisheries-regarding-greenhouse-gas-emission-documents> accessed 10 August 2023.
[233] ibid [1.11].
[234] Appeal Decision (n 232) [2.1.4].
[235] ibid [2.2.3].
[236] ibid [2.2.5].
[237] ibid [2.5.3]–[2.5.4].
[238] Act 3 of 2000 (South Africa).

section 6 of the Act.[239] In addition to the traditional grounds of review recognized in the common law, PAJA allows for judicial review of an administrative action on the basis of irrelevant considerations being taken into account or relevant considerations not being taken into account, unreasonableness, unlawfulness, and unconstitutionality.

*Africa Climate Alliance v Minister of Mineral Resources and Energy* (the #CancelCoal case) is the first constitutional challenge in South Africa concerning the government's plans to procure 1,500 megawatts of new coal-fired power electricity capacity.[240] The applicants argued, inter alia, that the impugned decisions to procure new coal-fired power ought to be set aside pursuant to section 6(2)(e)(iii) of PAJA because relevant considerations were not considered and section 6(2)(i) because the failure to have proper regard for the impact of new coal-fired power on constitutional rights was unlawful and unconstitutional.[241] It should be noted that there are many statutory provisions that can be used in conjunction with PAIA and PAJA to launch climate change litigation that challenges administrative decisions on, for example, energy policy and specific infrastructure projects.

In Brazil, the National Environmental Policy Act of 1981 (NEPA) created a national framework for comprehensive environmental governance. The NEPA's broad definition of a polluter allows for direct and indirect polluters to be held accountable for environmental damage. Article 14 of NEPA also establishes the rule of strict liability for environmental damage. The High Court, widely regarded as Brazil's 'green court' for its progressive jurisprudence that has strengthened environmental protection,[242] has interpreted the Article 14 civil environmental liability regime as having the following four characteristics.[243] First, a comprehensive risk doctrine must inform the application of the strict liability rule. This means that typical exclusions of liability such as third-party intervening events and force majeure cannot serve as defences.[244] Secondly, civil liability for causing environmental damage is on a joint and several basis. This is derived from the broad definition of 'polluter' in the NEPA and the rule in the Civil Code that joint and several liability

239 Tracy-Lynn Field (n 87) 183.

240 'Youth-led #CancelCoal Climate Case Launched against Government's Plans for New Coal-Fired Power' (*Life After Coal*, 17 November 2021) <https://lifeaftercoal.org.za/media/youth-led-cancelcoal-climate-case-launched-against-governments-plans-for-new-coal-fired-power> accessed 10 August 2023.

241 *Africa Climate Alliance v Minister of Mineral Resources & Energy*, Founding Affidavit, [416].

242 Nicholas S Bryner, 'Brazil's Green Court: Environmental Law in the Superior Tribunal de Justiça (High Court of Brazil)' (2012) 29 Pace Environmental Law Review 470, 532.

243 We are grateful to Professor Danielle de Andrade Moreira for taking the time to explain this body of law to us in detail.

244 Interview No 2. As Moreira and Herschmann explain, one of the consolidated precedents published by the Brazilian Superior Court of Justice in 2015 (*Jurisprudencia em Teses*, STJ, ed 30) states that strict liability for environmental damage is informed by full risk liability. See Danielle de Andrade Moreira and Stela Luz Andreatta Herschmann, 'The Awakening of Climate Litigation in Brazil: Strategies Based on the Existing Legal Toolkit' (2021) 59 Revista Direito, Estado e Sociedade 172, fn 38.

should be applied when more than one person contributes to the same, indivisible damage.[245] Thirdly, the question of 'Who is an indirect polluter?' casts a wide net over many possible actors, including those who have carried out a positive act, those responsible for an omission, those who permitted the act, those who were negligent in the sense that they should have exercised care but chose not to, those who financed the act, and those who benefited from the commission of the act.[246] Fourthly, the definition of environmental damage is broad and includes ecological harm, personal injury, and social/collective injury. As Moreira and Herschmann have argued, these unique aspects of the environmental civil liability regime are favourable for climate change litigation 'and the successes of 40 years of development of environmental litigation can serve as a "shortcut" to climate litigation'.[247]

## 3.4 Right to Healthy Environment

In Chapter 3, the rapid development of rights-based climate change litigation in the Global South was extensively discussed and will not be revisited here. There will also be discussion about rights-based litigation in Chapter 5, with a focus on the youth plaintiffs who have brought these lawsuits on behalf of future generations. In this section, it suffices to note that the Global South's rights-based route to climate change litigation 'is not serendipitous, or the result of the absence of specialized climate change legislation that litigants would otherwise have used in framing their cases. Instead, it is a route whose tracks were firmly laid over the last three decades through public interest law practice, research and judicial activism regarding constitutional rights in general and socioeconomic rights (SERs) in particular.'[248] Further, '[g]iven that most Southern countries joined—and, on some occasions, led—the "environmental rights revolution" by incorporating a right to a healthy environment in their constitutions and legal frameworks', there is a body of environmental constitutional case law that litigants can rely on to craft climate change arguments.[249]

Indeed, our analysis of the Global South docket shows that environmental rights (including the right to life that is judicially interpreted to encompass the right to a healthy environment) provide important climate change litigation 'hooks'. It is

[245] Código Civil, art 942.

[246] Brazilian Superior Court of Justice on causation: 'to determine the cause of an urban environmental damage and hold co-defendants jointly and severally liable, the following persons must be treated equally: whoever acts; whoever fails to act when should have acted; whoever is indifferent to others' actions; whoever remains in silence when they should have denounced; whoever finances others' actions; and whoever enjoys benefits from others' actions' (STJ 2nd Panel REsp 1.071.741/São Paulo). Reporting Justice Herman Benjamin, 24 March 2009. Moreira and Herschmann (n 244) fn 40.

[247] Moreira and Herschmann (n 244) 175.

[248] César Rodríguez-Garavito, 'Human Rights: The Global South's Route to Climate Litigation' (2020) 114 American Journal of International Law Unbound 40, 41.

[249] ibid.

primarily in those Global South jurisdictions that have recognized environmental rights in their legal systems that climate change litigation has emerged and has the potential to develop as a meaningful governance tool. As one example, the South African Bill of Rights is expansive, protecting a comprehensive range of political and civil rights as well as 'subsequent generations' which include socio-economic, cultural, and environmental rights.[250] Section 24 establishes the right of everyone: '(a) to an environment that is not harmful to their health or well-being; and (b) to have the environment protected, for the benefit of present and future generations, through reasonable legislative and other measures that (i) prevent pollution and ecological degradation; (ii) promote conservation; and secure ecologically sustainable development and use of natural resources while promoting justifiable economic and social development'. In the *#CancelCoal* case, the applicants argue, inter alia, that the procurement of new coal-fired power represents a serious threat to the constitutional rights of the people of South Africa, including section 24's right to environment and the rights to life, dignity, and equality.[251]

## 3.5 Active Civil Society

NGOs play a crucial role in launching climate change lawsuits and supporting the use of litigation as part of a multi-pronged advocacy strategy to drive climate action. For instance, a study that analysed some 1,130 cases decided by the Indian NGT between July 2011 and September 2015 showed that the most frequent plaintiffs were NGOs, social activists, and public-spirited citizens.[252]

As mentioned in Chapter 3, the *Carbon Majors Inquiry* by the Human Rights Commission of the Philippines is a global significant precedent.[253] It marks the first time that a human rights body has articulated the responsibility of transnational corporations for human rights violations resulting from climate change impacts. The Inquiry would not have been possible if it had not been initiated by a campaign led by local and global NGOs.[254]

A robust civil society and a vibrant practice of public interest law has equally created a firm basis for the pursuit of climate change litigation in South Africa.

[250] Constitution of the Republic of South Africa, Act 1089 of 1996, ch 2.

[251] *Africa Climate Alliance*, Founding Affidavit (n 241) [15].

[252] Gitanjali Nain Gill, *Environmental Justice in India: The National Green Tribunal* (Routledge 2017) 194–96. They account for 533 plaintiffs (47.2%) of 1,130 cases. Affected individuals, communities, and residents brought 17.7% of all cases.

[253] Commission on Human Rights of the Philippines, *National Inquiry on Climate Change* (Report, December 2022).

[254] Other than Greenpeace Southeast Asia (Philippines), the petitioners included Philippine Rural Reconstruction Movement, Philippine Alliance of Human Rights Advocates, Philippine Human Rights Information Center, Mother Earth Foundation, Ecowaste Coalition, 350.ORG East Asia, Asian Peoples' Movement on Debt & Development, and the Philippine Movement for Climate Justice, among others.

NGOs and activist lawyers have been the main driving force launching all the climate change lawsuits in South Africa to date. The Centre for Environmental Rights (CER) acted for the environmental NGOs, Earthlife Africa, and groundWork, in the coal-fired power plant cases. The CER comprises of 'activist lawyers who defend the right of communities and civil society organisations [in South Africa] to an environment not harmful to health or wellbeing for present and future generations'.[255] The Centre was established in October 2009 by eight non-profit civil society organizations in South Africa's environmental justice sector to provide legal and related support.[256] Their lawyers provide legal representation, legal advice, and advocacy support.[257] The CER has over 15 funders including the European Climate Foundation, the Ford Foundation, and the Open Society Human Rights Initiative.[258] Strong connections with the local community and global advocacy networks, adequate financial resources, high levels of legal expertise, and deep experience with the use of litigation as part of environmental advocacy strategies are all factors that place organizations like the CER in a favourable position to advance climate change litigation in South Africa.

While environmental litigation is traditionally within the domain of public prosecutors, NGOs in Brazil are not inexperienced with the use of strategic litigation to advance rights protection.[259] Galvanizing their expertise in public advocacy and campaigning, working with an epistemic climate community[260] made up of, inter alia, academics, judges, subnational governments, and corporations, and tapping into their domestic and global networks, NGOs have been involved in most of the high-profile cases. For example, *Laboratorio de Observatorio do Clima v Minister of Environment and Brazil* was filed by Observatorio do Clima, a prominent coalition of environmental NGOs including The Nature Conservancy, Amigos da Terra Amazônia Brasileira, Instituto de Pesquisa Ambiental da Amazônia, and ProNatura.[261] The coalition also provides legal and technical support to other

[255] 'Centre for Environmental Rights' (*Centre for Environmental Rights*) <https://cer.org.za/> accessed 22 January 2024.

[256] 'Overview' (*Centre for Environmental Rights*) <https://cer.org.za/about/overview> accessed 10 August 2023.

[257] 'Legal Services and Advice' (*Centre for Environmental Rights*) <https://cer.org.za/about/accessing-cer-services>.

[258] 'Funders' (*Centre for Environmental Rights*) <https://cer.org.za/about/funders> accessed 10 August 2023

[259] For a discussion on strategic litigation in Brazil, see Oscar Vilhena Vieira, 'Public Interest Law: A Brazilian Perspective' (2008) 13(1) UCLA Journal of International Law and Foreign Affairs 219; Mariana Prandini Assis, 'Strategic Litigation in Brazil: Exploring the Translocalisation of a Legal Practice' (2021) 12 TLT 360; Ana Paula de Barcellos, 'Sanitation Rights, Public Law Litigation, and Inequality: A Case Study from Brazil' (2014) 16(2) Health and Human Rights.

[260] See generally Peter M Haas, *Epistemic Communities, Constructivism, and International Environmental Politics* (Routledge 2016).

[261] 'History' (*Observatorio do Clima*) <www.oc.eco.br/en/quem-somos/nossa-historia/> accessed 10 August 2023.

NGOs and political parties seeking to launch climate cases and has acted as *amicus curiae* in several cases.[262]

It is notable that, unlike South Africa and India where the NGOs filing climate change lawsuits are typically environmental NGOs, not all the NGOs active in the climate change litigation space in Brazil focus on environmental issues. Instituto Alana, for instance, is an NGO focused on promoting children's rights.[263] Similarly, Conectas Direitos Humanos is a human rights NGO which is involved in filing cases at the national and international levels, especially in the Supreme Federal Court and in the Inter-American (OAS) and UN human rights systems.[264] The protection of the environment is only a subset of Conectas Direitos Humanos' work.[265] The diversity of NGOs is testament to Brazil's vibrant civil society. It also shows that fruitful partnerships are emerging amongst civil society actors traditionally working on human rights, Indigenous rights, and protection of environmental resources such as forests. These partnerships influence the litigation strategies and legal arguments advanced in court, for example by highlighting the close links between deforestation and its consequent adverse impacts on human rights and the rights of Indigenous peoples and forest communities.

We conclude that an active civil society plays a vital role in enabling the emergence of climate change litigation. In Chapter 5, we will elaborate on how various civil society actors and NGOs facilitate different modes of emergence of climate change litigation. Through their participation in domestic and transnational networks, they foster North-South and South-South cooperation as well as amplifying the impact of climate change litigation overall, as successful strategies are replicated and lessons from failure are shared.

## 4. CONCLUSION

In this chapter, we discussed the state of climate change litigation in three 'frontrunner' Global South jurisdictions which, at the time of writing, have the highest number of decided and pending cases in their respective geographical regions. The experience of South Africa, India, and Brazil shows some common features. Climate change litigation has taken off in these jurisdictions because

[262] See 'Legal Issues' (*Observatorio do Clima*) <www.oc.eco.br/en/atuacao-juridica/> accessed 10 August 2023.

[263] 'Alana' (*Alana*) <https://alana.org.br/en/> accessed 10 August 2023. It has a programme called 'Prioridade Absoluta' (Absolute Priority) which develops advocacy initiatives and strategic litigation against government institutions, with the goal of safeguarding children's rights. One branch of its work explicitly tackles climate justice: 'Prioridade Absoluta' (*Prioridade Absoluta*) <https://alana.org.br/en/legal-actions-climate-crisis/> accessed 10 August 2023.

[264] 'Litigation' (*Conectas Direitos Humanos*) <www.conectas.org/en/litigation/> accessed 10 August 2023.

[265] 'Defense of Socioenvironmental Rights' (*Conectas Direitos Humanos*) <www.conectas.org/en/socioenvironmental-rights/> accessed 10 August 2023.

there is a deep frustration amongst their civil societies about the profound lack of policy action in the face of the climate emergency. Many of the NGOs in these jurisdictions have significant prior experience in the use of litigation to achieve public policy goals and advance socio-economic rights. Tapping into this experience, the NGOs have turned their attention to climate change. The constitutionally protected right to a healthy environment and several environmental laws provide bases for advancing peripheral as well as core climate change lawsuits.

Our study of these 'frontrunner' jurisdictions, as well as the other jurisdictions whose cases are recorded in our Global South docket, show that there are five common features that are key explanatory factors for the emergence of climate change litigation. Section 3 canvassed these factors in some detail, including the presence of a robust civil society to launch lawsuits and the level of access to justice that the local communities have.

This conclusion highlights a salient point that arises from the preceding discussion, which is that the conditions that support climate change litigation in the Global South are not significantly different from those in the Global North. Thus, although the types of climate change cases emerging in the Global South have some distinctive features compared to Global North jurisprudence (eg a greater prevalence of periphery cases and rights-based arguments), the conditions that support climate change litigation in the first place are broadly similar. It cannot be overemphasized that if litigation is to be promoted as a climate change governance tool in the Global South and/or as a way of achieving climate justice and implementing Paris Agreement commitments, policy attention and resources must continue to be directed at establishing and strengthening the fundamental pillars of effective legal institutions and access to justice.[266] Finally, this chapter captured some of the most exciting developments in the world of climate change litigation in the Global South. In Chapter 5, we turn our attention to the actors who are mobilizing to take forward climate change litigation in the Global South.

[266] This coheres with the work UNEP is doing to support the environmental rule of law as part of Paris implementation. For example, UNEP published the first ever global assessment of environmental rule of law in 2019, which found that weak enforcement of the rule of law was a global trend that is exacerbating environmental threats. See 'Environmental Rule of Law: First Global Report' (*United Nations Environment Programme*, 2019) <www.unep.org/resources/assessment/environmental-rule-law-first-global-report> accessed 10 August 2023. UNEP reached similar conclusions in its second global report on the Environmental Rule of Law published in November 2023 (see n 13).

# 5
# Actors in Global South Climate Change Litigation Mobilization Efforts

## 1. INTRODUCTION

Who are the people mobilizing the law in Global South jurisdictions to strengthen climate change policy responses and halt projects that run counter to 'net zero' targets? The existing literature holds few answers to this question, which is an important one because understanding who is bringing lawsuits, who is funding litigation, as well as who is supporting climate change litigation as part of broader political mobilization strategies increases our appreciation of how climate change litigation is unfolding in practice across the Global South.

This chapter provides an analytical overview of the constellation of key actors and their roles in enabling the emergence of climate change litigation in the Global South. The actors include environmental NGOs, regulatory agencies, local and Indigenous communities, climate change activists, environmental cause lawyers located inside and outside NGOs, and funders. This broad range of actors reflects the growing complexity, polycentricity, and multi-level nature of the climate change litigation space, and climate change governance more broadly. The work of these actors often cuts across geographical boundaries and governance scales. Local communities contesting an infrastructure project, for example, will directly or indirectly have an impact on national climate change policies. Through civil society networks that crisscross the globe, grassroots communities connect with other communities as well as NGOs, volunteers, and non-profit foundations to learn and share information resources and legal strategies. These transnational networks facilitate norm diffusion and learning, as well as the flow of financial, legal, and material support towards Global South climate litigation. In short, it can be argued that a significant amount of climate change litigation in the Global South is enabled by globalization. A broader observation is that, while climate change litigation continues to take place primarily in domestic courts, it is a transnational legal phenomenon.

Following this introduction, Section 2 comprises of a comprehensive examination of some of the actors we have identified as playing key roles in the development of climate change litigation in the Global South. As mentioned above, they include global and domestic environmental NGOs, grassroots communities, prosecutors, regulatory agencies, lawyers, and non-profit foundations. The roles they

*Litigating Climate Change in the Global South*. Jolene Lin and Jacqueline Peel, Oxford University Press.
 DOI: 10.1093/oso/9780191926525.003.0005

play include the provision of legal advice, filing cases, capacity-building to develop and enhance the conditions that are conducive to climate change litigation, and funding cases and related activities. It is often the case that various actors form loose coalitions to implement a litigation strategy, with different actors playing different and sometimes multiple roles. In Section 3, we step back from the more granular analysis of 'Who is doing what?' to consider more broadly the different, prototypical modes of legal action in the Global South and how they are shaped by particular actors. Drawing on earlier work, this section advances a theoretical framework to explain these modes and their implications for the emergence of climate litigation in the Global South. Section 4 concludes with a summary of the ideas advanced in this chapter.

## 2. PLAINTIFFS, CAMPAIGNERS, AND FUNDERS

This section seeks to provide a descriptive overview of the actors involved in climate change litigation in the Global South. While environmental NGOs play a leading role in climate litigation in both the Global North and the Global South, the role that prosecutors play in Brazil, for example, is *sui generis* and forms a unique pathway for climate litigation that is rarely seen in the Global North. The role of funders is under-explored in the literature because there is a paucity of publicly available information about their activities. The discussion of their role in this section breaks new ground and provides novel insights into this relatively invisible actor and the influence funders exercise over the developing Global South case law.

### 2.1 Environmental NGOs

Since the proliferation of environmental NGOs, networks, and coalitions in the 1980s, NGOs have gained prominent recognition as political actors with important roles to play in the governance of environmental issues.[1] To take a few instances from the voluminous literature, environmental NGOs have been described as '*agitators* for environmental action, *architects* of governance solutions,

[1] Thomas Princen and Matthias Finger point out that the rapid growth of NGOs is phenomenal given that almost all environmental NGOs and networks were started in the 1980s; see Matthias Finger and Thomas Princen, *Environmental NGOs in World Politics: Linking the Local and the Global* (1st edn, Routledge 1994) 1. Also see Emilie Dupuits, 'Civil Society and NGOs as Drivers of Change in Environmental Governance' in Gustavo Sosa-Nunaz and Ed Atkins (eds), *Environment, Climate Change and International Relations* (E-International Relations Publishing 2016); Bas Arts, *The Political Influence of Global NGOs: Case Studies on the Climate and Biodiversity Conventions* (1st edn, International Books 1998); Peter Newell, *Climate for Change: Non-State Actors and the Global Politics of the Greenhouse* (1st edn, CUP 2009); Katharina Rietig, 'Public Pressure versus Lobbying—How Do Environmental NGOs Matter Most in Climate Negotiations?' (2011) Centre for Climate Change Economics and Policy Working Paper No 79, Grantham Research Institute on Climate Change and the

and *entrepreneurs* for new sorts of initiatives'.[2] Naghmeh Nasiritousi argues that they are '*shapers* of information and ideas, *brokers* of knowledge, norms and initiatives, and *doers* of implementing policies and influencing behaviours'.[3] Michele Betsill describes NGOs involved in global climate governance as '*activists* raising awareness and calling for action; as *diplomats* working with governments to craft climate policies; and as *governors* developing new mechanisms for steering society towards a low-carbon future'.[4] Sheila Jasanoff has pointed out that '[t]he place of non-governmental organizations (NGOs) in international governance seems nowhere more securely established than in the field of environmental governance' and that the intense preparatory activity in the NGO sector leading up to the 1992 United Nations Conference on Environment and Development (UNCED) showed that environmental NGOs had developed extensive skills in policymaking and implementation which added to their more traditional roles in campaigning, advocacy, and ideological consciousness raising.[5]

When it comes to lobbying or advocacy, which can take place at many levels (ie local, national, and international), NGOs adopt strategies ranging from *information* politics (the production and dissemination of relevant information) and *leverage* politics (ie putting pressure on or creating alliances with stronger actors) to *accountability* politics (monitoring and holding actors accountable). In both the Global North and the Global South, environmental NGOs use litigation (and other legal processes) as part of their information, leverage, and accountability strategies. As mentioned in Chapter 3, NGOs filed (either on their own or as part of a coalition) 70% of the cases in the Global South docket.

## Global environmental NGOs

Greenpeace, the world's largest environmental NGO network with several national and regional offices in the Global South, has filed cases in Argentina, the Philippines, Mexico, Indonesia, and Malaysia.[6] As it is a global NGO, campaigners in Greenpeace's offices in the Global South have access to resources from Greenpeace offices in the Global North to facilitate and support their climate

Environment Working Paper No 70, <www.lse.ac.uk/granthaminstitute/wp-content/uploads/2014/02/WP70_environmental-NGOs-climate.pdf> accessed 13 February 2023.

[2] Kate O'Neill, 'Architects, Agitators, and Entrepreneurs: International and Nongovernmental Organizations in Global Environmental Politics' in Regina S Axelrod and Stacy D VanDeveer (eds), *The Global Environment: Institutions, Law, and Policy* (5th edn, Sage Publishing 2019) 35.

[3] Naghmeh Nasiritousi, 'Shapers, Brokers and Doers: The Dynamic Roles of Non-State Actors in Global Climate Change Governance' (PhD thesis, Linköping University 2016) 2.

[4] Michele Betsill, 'Chapter 22: NGOs' in Karin Bäckstrand and Eva Lövbrand (eds), *Research Handbook on Climate Governance* (1st edn, Edward Elgar 2015) 251.

[5] Sheila Jasanoff, 'NGOs and the Environment: From Knowledge to Action' (1997) 18(3) Third World Quarterly 579, 579.

[6] Greenpeace International has a legal unit consisting of a team of specialized lawyers that provide legal support to regional and national Greenpeace organizations on advocacy efforts and strategic litigation. Jasper Teulings, former General Counsel of Greenpeace International, has expressed the view

change legal mobilization efforts, including the filing of lawsuits. For example, in order to build the necessary evidence base for a potential climate lawsuit premised on challenging poor enforcement of air pollution laws and the need to update air quality standards based on latest public health and scientific knowledge, Greenpeace campaigners in the Philippines spent nearly a year arranging for the transport of air quality measuring equipment from Greenpeace headquarters in Bonn (Germany) and setting up the data collection system on site.[7] The data collection would not have been possible without the equipment from Bonn.[8] Within Greenpeace in Southeast Asia, there is evidence of learning and imitation—important norm diffusion processes—amongst campaigners working in different countries, which suggests that the replication of climate change litigation and advocacy strategies is happening in a dynamic manner (as opposed to being centrally planned and coordinated).[9]

*Greenpeace Argentina and Others v Argentina and Others* (*Greenpeace Argentina v Argentina*) was filed in January 2022 by Greenpeace Argentina and

that '[w]e use strategic litigation to enforce environmental laws and hold both governments and corporations to account. We remind lawmakers of the legitimacy of environmental interests and the right to seek access to environmental justice'; see 'About Us: Legal Unit' (*Greenpeace International*, 2023) <www.greenpeace.org/international/explore/about/legal/> accessed 13 February 2023. About the organizational structure of Greenpeace, see 'About Us: Our Structure' (*Greenpeace International*, 2023) <www.greenpeace.org/international/explore/about/structure/> accessed 13 February 2023.

[7] See Interview No 6:

> In the Philippines, we have the Clean Air Act of 1999. It stipulates that air pollution standards should be reviewed at least every two years but this has not happened at all since the law came into force. So the pollution standards in the Philippines are at least two decades behind the science. Our theory in our case is that the outdated standards actually allow the expansion of coal power plants. And while there is compliance with the law, compliance with these outdated standards is actually harmful for the people.

[8] ibid.

[9] Interviews Nos 6, 8, and 9. Air pollution lawsuits were first filed in Indonesia and subsequently in Thailand and Malaysia. Campaigners in Greenpeace Philippines have been 'inspired' by their colleagues in Indonesia to launch similar litigation. See 'Thai CSOs and Local Groups File a Lawsuit to Demand Government Action against PM2.5 Pollution' (*Greenpeace Southeast Asia*, 22 March 2022) <www.greenpeace.org/southeastasia/press/45218/thai-csos-and-local-groups-file-lawsuit-to-demand-government-action-against-pm2-5-pollution/> accessed 13 February 2023; 'Green Groups File Landmark Complaint to Malaysian Human Rights Commission for Public Inquiry into Haze Pollution as a Human Rights Violation' (*Greenpeace Southeast Asia*, 7 December 2021) <www.greenpeace.org/southeastasia/press/45072/landmark-complaint-filed-to-malaysian-human-rights-commission-for-public-inquiry-into-haze-pollution-as-a-human-rights-violation/> accessed 13 February 2023; 'Seven Indonesia's State Officials Found Negligent for Jakarta Air Pollution' (*Greenpeace Southeast Asia*, 16 September 2021) <www.greenpeace.org/southeastasia/press/44679/seven-indonesias-state-officials-found-negligent-for-jakarta-air-pollution/> accessed 13 February 2023. While there was a strong public health angle to the advocacy campaign accompanying the lawsuit filed by Greenpeace Indonesia, the lawsuit was viewed internally by campaigners as part of their work on climate change justice. One of the biggest sources of air pollution in Jakarta is the emissions from the coal-fired power plants, so tackling air pollution was viewed as a strategic complement to the other strands of work to eventually shut the coal plants down. Campaigners noted that the public health angle 'made sense' as most citizens were frustrated with the poor air quality in Jakarta while climate change awareness is relatively low; see Interviews 8 and 9.

other civil society organizations, challenging the approval of offshore oil exploration projects off the coast of Buenos Aires in the Argentine Sea.[10] In the submissions to the court, the plaintiff NGOs raised several arguments about, inter alia, the risk of harm to marine biodiversity, the application of the precautionary principle, procedural irregularities in the environmental impact assessment (EIA) process, and that the projects will foster a significant reliance on fossil fuels which is inconsistent with Argentina's Nationally Determined Contribution (NDC). It is noteworthy that while the submissions in this case included nearly 125 pages of scientific data and climate change arguments, none of the decisions taken in the preliminary proceedings was made on the basis of the climate change arguments.[11] It has been suggested that this is because climate change arguments rarely feature in cases before the Argentinian courts.[12] The lack of familiarity with climate change jurisprudence renders it more likely that a judge will reach a ruling based on more familiar environmental law arguments.[13] Nonetheless, for Greenpeace Argentina and some of the other plaintiff NGOs,

On norm diffusion, see Martha Finnemore and Kathryn Sikkink, 'International Norm Dynamics and Political Change' (2005) 52(4) International Organization 887; Richard Rose, 'What Is Lesson-Drawing?' (1991) 11(1) Journal of Public Policy 3; Zachary Elkins and Beth Simmons, 'On Waves, Clusters and Diffusion: A Conceptual Framework' (2005) 598 The Annals of the American Academy of Political and Social Science 33.

[10] '*Greenpeace Argentina et al., v. Argentina et al.*' (*Grantham Research Institute on Climate Change and the Environment*) <https://climatecasechart.com/non-us-case/greenpeace-argentina-et-al-v-argentina-et-al/> accessed 22 January 2024. Also see Natalie Alcoba, 'Argentinian Activists Fight against Offshore Drilling Plans' *Al Jazeera* (11 March 2022). <www.aljazeera.com/news/2022/3/11/argentine-activists-fight-against-offshore-drilling-plans> accessed 11 August 2023. The class action environmental lawsuit sought to have the regulatory approvals for seismic exploration to take place declared unconstitutional, null, and void. Most environmental and climate change cases in Argentina are brought as class actions or *acciones de amparo*—a guarantee against the violation of rights recognized by the Constitution. For a helpful concise introduction to the *amparo*, see Allan R Brewer-Carías, 'The Latin American "Amparo": A General Overview' (Latin America Workshop on Human Rights & Legal Theory, Leitner Center for International Law and Justice, Fordham Law School, New York, 4 April 2008) <https://allanbrewercarias.com/wp-content/uploads/2008/04/1028.-Brewer.-LATIN-AMERICAN-AMPARO.-A-GENERAL-OVERVIEW.pdf> accessed 13 February 2023.

[11] Rafael Colombo, 'Argentina's Hydrocarbon Border Expansion: The Rise of Climate Litigation Claims against Offshore Fossil Fuel Exploitation' (*Sabin Center for Climate Change Law Peer Review Network on Global Climate Litigation monthly webinar series*, 5 April 2022) <https://climate.law.columbia.edu/events/argentinas-hydrocarbon-border-expansion-rise-climate-litigation-claims-against-offshore> accessed 1 March 2023.

[12] Lucas Micheloud, 'Argentina's Hydrocarbon Border Expansion: The Rise of Climate Litigation Claims against Offshore Fossil Fuel Exploitation' (*Sabin Center for Climate Change Law Peer Review Network on Global Climate Litigation monthly webinar series*, 5 April 2022) <https://climate.law.columbia.edu/events/argentinas-hydrocarbon-border-expansion-rise-climate-litigation-claims-against-offshore> accessed 1 March 2023.

[13] Gaston Medici-Colombo, 'Argentina's Hydrocarbon Border Expansion: The Rise of Climate Litigation Claims against Offshore Fossil Fuel Exploitation' (*Sabin Center for Climate Change Law Peer Review Network on Global Climate Litigation monthly webinar series*, 5 April 2022) <https://climate.law.columbia.edu/events/argentinas-hydrocarbon-border-expansion-rise-climate-litigation-claims-against-offshore> accessed 1 March 2023.

*Greenpeace Argentina v Argentina* is a climate change lawsuit filed with the intention to halt fossil fuel exploration.[14]

*Wijana, Sufarlan, Astawa and Greenpeace Indonesia v Governor of Bali* was brought by some citizens together with Greenpeace Indonesia before the Denpasar Administrative Court challenging the grant of an environmental permit for the construction of a coal-fired power plant in Celukan Bawang, northern Bali. The facts of this case were considered in Chapter 3. What is relevant to note at this juncture is that, while this case was technically a legal defeat, it is interesting that, for the NGOs and environmental lawyers involved, the case was a success in terms of building local environmental and climate change litigation capacity, strengthening the climate change advocacy movement, and drawing lessons for future litigation efforts.[15]

There are other global environmental NGOs whose involvement in climate change litigation does not entail filing cases in their own name but mainly providing advocacy, legal research, and strategic support for the campaign which surrounds the lawsuit. An example is EarthRights International, which has offices in Washington DC (the United States of America), Lima (Peru), and Thailand (Chiang Mai) EarthRights is part of a coalition of NGOs and environmental law firms working with a Karen minority community in Thailand to pursue their case, *Residents of Omkoi v Expert Committee on EIA Consideration and the Office of Natural Resources and Environmental Policy and Planning*.[16] As one interviewee explained:

> To make things manageable, there is a division of labour amongst the partners . . . most of what Earthrights has supported is a lot of training for the community. This includes giving them information about the project itself because they were not given any information during the EIA process. So, it's training about the project itself, about the EIA process and what the rights that the community has in this process, and about the administrative litigation process itself as we just filed the lawsuit recently.[17]

EarthRights also runs a residential training programme for young activists and NGO workers. Over a period of seven to eight months, the students of the Earth Rights School receive training and instruction on a broad range of topics including

[14] Laura Colombo, 'How People Power Helped Save the Argentine Sea from Oil Companies' (*Greenpeace International*, 16 February 2022) <www.greenpeace.org/international/story/52326/greenpeace-argentina-saved-argentine-sea-oil-companies/> accessed 1 March 2023.

[15] See Interview No 9. Also see Karen Beckwith, 'Narratives of Defeat: Explaining the Effects of Loss in Social Movements' (2015) 77(1) The Journal of Politics 2, on how social movements, after sometimes bitter and brutal defeats, make sense of those losses and give meaning to them.

[16] Chiang Mai Administrative Court, Black Case No Sor 1/2565, 14 April 2022.

[17] Interview No 7.

the impact of Chinese foreign investment, EIA and climate change, and the dispute resolution mechanisms of various international organizations (eg the World Trade Organization) and multilateral financial institutions (eg the Asian Infrastructure Investment Bank).[18] The EarthRights School has an active alumni base that keeps in close contact with EarthRights, and it is often through the alumni networks that potential climate change lawsuits are identified and supported.[19]

### Transnational advocacy networks

Global NGOs like Greenpeace and EarthRights International often portray their work as both local and global at the same time. It is local because these global NGOs generally do not engage in climate change litigation without local partners and support from local communities, as the lack of local knowledge and community support can seriously jeopardize any advocacy and litigation strategy. It is global because these NGOs are usually part of transnational advocacy networks (TANs) that foster exchanges of ideas, knowledge, and material resources across domestic and international spheres.

In their seminal work, *Activists Beyond Borders: Advocacy Networks in International Politics*, Margaret Keck and Kathryn Sikkink define TANs as networks of activists, distinguishable largely by the centrality of principled ideas, causes, and norms in motivating their formation.[20] Their research showed that TANs tend to form when: (1) channels between domestic groups and their governments are blocked or ineffective, making the international arena the only space for the domestic issue to gain attention; (2) activists and other actors actively promote their networks in the belief that this will further their campaigns; and (3) conferences and other forms of international contact create arenas for networks to be formed and strengthened.[21] When channels between the state and its domestic actors are blocked, the boomerang pattern of influence that is characteristic of TANs is likely to emerge: domestic NGOs bypass their state and seek support from (ie throw a boomerang to) international allies to bring pressure on the state from

[18] 'EarthRights School' (EarthRights International) <https://earthrights.org/trainings-and-networks/training/earthrights-school/> accessed 1 March 2023.

[19] It is stated on EarthRights' website that '[g]raduates of our training programs become part of a global network of defenders and go on to secure amazing victories in their communities. EarthRights maximizes the impact of our alumni by providing ongoing structured opportunities for them to network, collaborate, exchange information and resources, and receive advanced training and technical assistance through our various alumni networks'; 'Networks' <https://earthrights.org/trainings-and-networks/networks/> accessed 1 March 2023.

[20] Margaret E Keck and Kathryn Sikkink, *Activists Beyond Borders: Advocacy Networks in International Politics* (2nd edn, Cornell University Press 1998) 1. These principles and values specify criteria for determining whether actions are right or wrong, and whether outcomes are just or unjust.

[21] Keck and Sikkink (n 20) 12. The Climate Litigation Accelerator (CLX) is a good example of a global platform for activists, scholars and litigants to create and strengthen TANs. CLX describes itself as 'a global collaborative hub for research, advocacy, and strategic litigation on the climate emergency' and 'spearheads and supports climate lawsuits and other forms of advocacy'; 'Climate Litigation Accelerator Toolkit' (*CLX Climate Litigation Accelerator*) <https://clxtoolkit.com/> accessed 1 March 2023.

outside. The international allies then 'throw the boomerang back' by pressuring key state actors, corporations, and intergovernmental bodies.[22] It has been noted that, 'for the less powerful third world actors, networks provide access, leverage, and information (and often money) they could not expect to have on their own; for northern groups, they make credible the assertion that they are struggling with, and not only for, their southern partners'.[23]

The largest TAN in the Global South climate change litigation space by far is the #STOPEACOP alliance against the East African Crude Oil Pipeline (EACOP).[24] The shareholders of this mega-project to exploit oil reserves in landlocked Uganda are the French energy major Total Energies, China National Offshore Oil Cooperation (CNOOC), and the Ugandan National Oil Company. Ugandan officials have long hoped that exploitation of the oil reserves would accelerate economic development.[25] The project includes the exploitation of six oil fields, the construction of an oil refinery, water abstraction systems, storage facilities, and the

[22] Keck and Sikkink (n 20) 12–13; Leah Shipton and Peter Dauvergne, 'The Politics of Transnational Advocacy Against Chinese, Indian, and Brazilian Extractive Projects in the Global South' (2021) 30(3) The Journal of Environment & Development 240, 243.

[23] Keck and Sikkink (n 20) 12–13. There is a large body of literature on TANs. See, for further discussion, Paula Franco Moreira and others, 'South–South Transnational Advocacy: Mobilizing Against Brazilian Dams in the Peruvian Amazon' (2019) 19(1) Global Environmental Politics 77, 80; Alexandra-Maria Bocse, 'Hybrid Transnational Advocacy Networks in Environmental Protection: Banning the Use of Cyanide in European Gold Mining' (2021) 21 International Environmental Agreements: Politics, Law and Economics 285; Young Sokphea, 'Transnational Advocacy Networks in Global Supply Chains: A Study of Civil Society Organizations' Sugar Movements in Cambodia' (2017) 13(1) Journal of Civil Society 35; Susan Kay Park, 'How Transnational Environmental Advocacy Networks Socialize International Financial Institutions: A Case Study of the International Finance Corporation' (2005) 5(4) Global Environmental Politics 95; Simona Davidescu and Aron Buzogány, 'Cutting Deals: Transnational Advocacy Networks and the European Union Timber Regulation at the Eastern Border' (2021) 56(3) International Spectator 105; Peggy Spitzer Christoff and Jamie M Sommer, 'Women's Empowerment and Climate Change Adaptation in Gujarat, India: A Case-Study Analysis of the Local Impact of Transnational Advocacy Networks' (2018) 10(6) Sustainability 1920; Michael G Huelshoff and Christina Kiel, 'Swan Song: Transnational Advocacy Networks and Environmental Policy in Chile—The Case of the Cisnes de Cuello Negro' (2012) 1(2) Interest Groups & Advocacy 260; Jonathan Rosenberg, 'Transnational Advocacy and the Politics of Sustainable Development in a Small Island Developing State: An Uncertain Future for the Grenada Dove' (2018) 27(2) Journal of Environment & Development 236.

[24] 'Stop the East African Crude Oil Pipeline' (*#STOPEACOP*) <www.stopeacop.net/home> accessed 1 March 2023. Other TANs include the alliance that filed the petition that launched the *Carbon Majors Inquiry* by the Human Rights Commission of the Philippines, Fridays for Future, and Families for Climate which mobilizes parents around the world to fight for climate justice; see Annalisa Savaresi and Margaretha Wewerinke Singh, 'Historic Inquiry Holds the Carbon Majors Accountable for the Impacts of Climate Change in the Philippines' (*The Center for Climate Change, Energy and Environmental Law*, 10 May 2022) <https://sites.uef.fi/cceel/historic-inquiry-holds-the-carbon-majors-accountable-for-the-impacts-of-climate-change-in-the-philippines/> accessed 11 August 2023; Clemens Bohl and others, 'The Climate Crisis and the Fridays for Future Movement: Causes, Responsibilities and Solutions through the Lense of Framing Theory' (2021) Department of Development Studies, University of Vienna Working Paper No 10 <https://ie.univie.ac.at/fileadmin/user_upload/p_ie/INSTITUT/Publikationen/IE_Publications/ieWorkingPaper/WP10_Bohl_et_al_Final.pdf> accessed 1 March 2023; 'Mobilizing Parents, Kids and Families for Climate Justice and a Livable Future for All' (*Families for Climate*) <www.familiesforclimate.org/> accessed 1 March 2023.

[25] Tom Wilson, 'Uganda's Oil Ambitions Delayed as Deals Stall' *Financial Times* (15 September 2019) <www.ft.com/content/bbc65a9a-d4af-11e9-8367-807ebd53ab77> accessed 1 March 2023.

EACOP.[26] The EACOP will be longest heated oil pipeline in the world. Its proposed route will cut across a national park and will most likely displace communities and significantly endanger wildlife amongst other catastrophic environmental and climate change impacts.[27] In the face of scientific and policy research findings that show unequivocally that more fossil fuel production would be 'dangerously out of sync' with climate change targets,[28] the decision in 2022 by Total and its partners to forge ahead with the project has set in motion what has been described as 'a litmus test for large-scale oil development in the age of net zero'.[29]

The TAN brings together over 260 civil society organizations which are supporting the campaign to stop the construction of the EACOP 'with public mobilizations, legal actions, research, shareholder activism and media advocacy'.[30] The #STOPEACOP alliance has focused on halting the construction of the pipeline for two main reasons. First, the environmental consequences of constructing the EACOP are significant enough to galvanize environmental groups in Uganda and beyond into action. Secondly, while Total and CNOOC hold licences to extract the oil in Uganda, they can only begin drilling once the EACOP is built as it is the only way to export the oil out of landlocked Uganda. By focusing the global campaign's resources on halting the construction of the pipeline, the objective is to render the entire oil drilling project unviable so that the oil is simply kept in the ground.

The boomerang strategy is clearly in play because domestic channels between the Ugandan government and civil society on this issue are blocked. Local environmental activists and human rights defenders who express criticism of the project have faced harassment and threats.[31] Environmental NGOs have been shut down by the Ugandan authorities for alleged regulatory breaches.[32] The fact that the majority shareholder of the EACOP project is a French energy company lends itself to galvanizing French (and, more broadly, European) civil society organizations into action. The pressure to halt the project is not only directed at the state where the controversial activity is located. Significant pressure is being directed at Total, for example by lobbying investors, banks, and insurance firms around the world to

[26] Les Amis de la Terre France and Survie (members of #STOPEACOP), 'Total See You in Court' (*Total See You in Court*) <www.totalincourt.org/> accessed 1 March 2023.

[27] Leslie Hook, 'The Oil Giants Drilling among the Giraffes in Uganda' *Financial Times* (12 April 2022) <www.ft.com/content/e1670042-11bd-4c68-9bde-a599d94bd8c0> accessed 1 March 2023.

[28] 'Fossil Fuel Production "Dangerously out of Sync" with Climate Change Targets' (*UN News Audio Hub*, 20 October 2021) <https://news.un.org/en/story/2021/10/1103472> accessed 1 March 2023.

[29] Hook, 'The Oil Giants' (n 27).

[30] 'Meet the #STOPEACOP Alliance' (*#STOPEACOP*) <www.stopeacop.net/our-coalition> accessed 1 March 2023.

[31] 'Submission on the Challenges and Threats Facing Climate Defenders—To the UN Special Rapporteur on the Rights to Freedom of Peaceful Assembly and of Association' (*Human Rights Watch*, 1 June 2021) <www.hrw.org/news/2021/06/01/submission-challenges-and-threats-facing-climate-defenders> accessed 2 March 2023; 'Ugandan Human Rights Defender Detained' (*Friends of the Earth Europe*, 2 June 2021) <https://friendsoftheearth.eu/news/ugandan-human-rights-defender-detained/> accessed 2 March 2023.

[32] 'Environment Watchdogs Slam Arrest of Ugandan Activists Protesting Total Oil Project' (*RFI*, 23 October 2021) <www.rfi.fr/en/africa/20211023-environment-watchdogs-slam-arrest-of-ugandan-activists-protesting-total-oil-project> accessed 2 March 2023.

withhold funding for the project or withdraw from investing in Total itself. On 21 May 2022, for example, the #STOPEACOP alliance reported that Deutsche Bank, Citi, JPMorgan Chase, Wells Fargo, and Morgan Stanley have confirmed they will not join the project loan to finance the EACOP. Their actions have been followed by the insurer Beazley Group and the Italian export credit agency SACE.[33] Two networks of NGOs filed the lawsuits challenging the EACOP, *Friends of the Earth France and Others v Total* and *Centre for Food and Adequate Living Rights and Others v Uganda, Tanzania and the East African Community*.[34] Both networks are made up of civil society organizations which are members of the larger #STOPEACOP alliance, with the litigation amplifying the collective efforts by the transnational alliance to put a stop to fossil fuel development.

### Domestic environmental NGOs

Climate change litigation in the Global South is by no means driven primarily by global NGOs. Many domestic NGOs are typically smaller grassroots organizations which do not have legal expertise on their staff and rely on partnering with other NGOs and environmental lawyers to file cases. However, in jurisdictions with a public interest litigation track record, such as India, there are domestic NGOs which have established a track record in environmental litigation and have expanded the remit of their work to include climate change.

An example of litigation filed by grassroots NGOs is *Kallpavalli Vrishka Pempakamdarula Paraspara Sahayaka Sahakara Sangam Ltd and Others v Union of India*, a legal challenge against the construction of windmills in the Kallpavalli area in the state of Andhra Pradesh, India.[35] The applicants argued that the construction of the windmills (as part of a Kyoto Protocol Clean Development Mechanism project) caused adverse effects on the local ecology and sought their removal. The main applicant, Kallpavalli Vrishka Pempakamdarula Paraspara Sahayaka Sahakara Sangam Ltd (which literally means Kalpavalli Tree-Growers' Cooperative Society), is a federation of 10 informal Forest Protection Committees that oversees the Kalpavalli Community Conservation Area.[36] Another example is *Hanuman Laxman Aroskar and Others v Union of India*, which arose from an

[33] 'Seven Financiers Abandon TotalEnergies' EACOP Pipeline in a Week' (*#STOPEACOP*, 20 May 2022) <www.stopeacop.net/our-news/seven-financiers-abandon-totalenergies-eacop-pipeline-in-a-week> accessed 2 March 2023. It should also be noted that Marsh McLennan is arranging the insurance coverage for the pipeline, despite protests from more than a hundred of its staff citing the 'disastrous consequences' for the climate and the company's reputation; Leslie Hook and Ian Smith, 'Marsh Revealed in Oil Pipeline Project Shunned by Leading Banks and Insurers' *Financial Times* (19 May 2022) <www.ft.com/content/597a2b01-fb54-4fd3-b326-dadf52dc250a> accessed 2 March 2023.

[34] *Les Amis de la Terre v Total*, Tribunal Judiciaire de Paris, 28 February 2023; *Center for Food and Adequate Living Rights (Cefroht) v The Attorney General of the Republic of Uganda* (Reference No 39 of 2020, East African Court of Justice) <www.eacj.org/?cases=reference-no-39-of-2020-center-for-food-and-adequate-living-rights-cefroht-3-others-v-the-attorney-general-of-the-republic-of-uganda-2-others> accessed 2 March 2023. Also see discussion in Chapter 3, Section 4.

[35] *Kallpavalli Vrishka Pempakamdarula Paraspara Sahayaka Sahakara Sangam Ltd v Union of India*, Original Application No 92/2013, National Green Tribunal, 25 August 2015.

[36] 'Kalpavalli' (*The Timbaktu Collective*) <https://timbaktu.org/kalpavalli/> accessed 2 March 2023.

appeal challenging the grant of an Environmental Clearance for the development of a greenfield international airport at Mopa village in Goa.[37] The first appellant, Hanuman Laxman Aroskar, is a resident of Mopa village.[38] The second appellant, Federation of Rainbow Warriors, is a group of public-spirited citizens in Goa. The group, founded in 2014, has been 'fighting against projects which uproot local communities and destroy the state's natural richness'.[39] The general secretary of Rainbow Warriors is a Goa-based environmentalist, Abhijit Prabhudesai.[40] An engineer by training, he quit his job in 2002 and returned to his native Goa where he now works on conservation.[41]

In South Africa, the South Durban Community Environmental Alliance (SDCEA) is made up of 19 affiliate community organizations.[42] The alliance is engaged in lobbying, reporting, researching, and litigating environmental issues in South Durban. In 2021, SDCEA challenged the authorization of a gas-fired power plant (*South Durban Community Environmental Alliance & Groundwork v Minister of Forestry, Fisheries, and the Environment*)[43] and the government's approval of offshore oil and gas exploration (*South Durban Community Environmental Alliance v Minister of Environment, Forestry and Fisheries and Others*).[44] In the latter case, SDCEA filed a review application in the Pretoria High Court challenging the decision of the Minister of Forestry, Fisheries and the Environment to authorize Eni South Africa and Sasol to explore for oil and gas in Block 236 off the Kwa-Zulu Natal coast. The case has the potential to establish ground-breaking precedents regarding the legal obligations for the assessment of all future oil and gas exploration in light of climate change.

[37] (2020) 12 SCC 1.

[38] Krishnadas Rajagopal, 'SC Lifts EC Suspension on Goa Mopa Airport' *The Hindu* (17 January 2020) <www.thehindu.com/news/national/other-states/sc-lifts-ec-suspension-on-goa-mopa-airport/article30580464.ece> accessed 2 March 2023.

[39] Derek Almeida, 'In Goa, Rainbow Warriors Build on Community' *Civil Society* (26 November 2021) <www.civilsocietyonline.com/interviews/in-goa-rainbow-warriors-build-on-community/> accessed 2 March 2023.

[40] Jayashree Nandi, 'Benches Shut, NGT Forcing Petitioners to Come to Delhi' *The Times of India* (26 April 2018) <https://timesofindia.indiatimes.com/home/environment/developmental-issues/benches-shut-ngt-forcing-petitioners-to-come-to-delhi/articleshow/63918018.cms> accessed 2 March 2023; The Goan Network, 'Proposed Amendments to Forest Laws Assailed' *The Goan Everyday* (30 October 2021) <www.thegoan.net/goa-news/proposed-amendments-to-forest-laws-assailed/76272.html> accessed 2 March 2023.

[41] Raman Kirpal, 'A Goan Village Besieged by Mines is Fighting Back' (*Firstpost*, 20 December 2014) <https://www.firstpost.com/politics/a-goan-village-besieged-by-mines-is-fighting-back-94508.html> accessed 2 March 2023.

[42] 'About' (*South Durban Community Environmental Alliance*) <https://sdcea.co.za/about/> accessed 2 March 2023.

[43] *South Durban Community Environmental Alliance v Minister of Forestry, Fisheries and the Environment* (17554/2021) [2022] ZAGPPHC 741 (6 October 2022).

[44] *South Durban Community Environmental Alliance v Minister of Environment, Forestry and Fisheries* (Case No 29433/21). Copies of the application and expert affidavits can be found here: <https://naturaljustice.org/sdcea-goes-to-court-over-authorisation-granted-to-eni-and-sasol-for-offshore-oil-and-gas-exploration/> accessed 2 March 2023.

In some instances, notably in Brazil, domestic NGOs work alongside political parties to file climate change lawsuits. For example, *PSB and Others v Brazil (on deforestation and human rights)* was formally filed by seven political parties but was spearheaded by a coalition of NGOs including Artigo 19, Articulação dos Povos Indígenas do Brasil (Apib), Conectas Direitos Humanos, Conselho Nacional das Populações Extrativistas (CNS), Engajamundo, Greenpeace Brazil, Instituto Alana, Instituto Socioambiental (ISA), Observatório do Clima, and Associação Alternativa Terrazul.[45] These NGOs played a crucial role in organizing the political parties to bring the case, including creating the evidence base to support their argument that the federal government had failed to implement Brazil's Action Plan for Prevention and Control of the Legal Amazon Deforestation (PPCDAm).[46]

Some domestic NGOs become key actors in South-South cooperation when they partner with NGOs located in other Global South jurisdictions to pool their resources and expertise together in legal mobilization efforts. The legal challenge against the EACOP in the East African Court of Justice, for example, was filed by a coalition of four East African civil society organizations.[47] It can be argued that there is much to be done to foster South-South cooperation, especially for apex NGOs to build the capacity of smaller NGOs to use the law as a climate change governance tool.

## 2.2 Prosecutors and Regulatory Agencies

Climate change lawsuits have been brought by prosecutors and regulatory agencies in the Global South to enforce environmental and natural resources laws against recalcitrant polluters. There are also cases filed against government officials on allegations of corruption and maladministration that have enabled environmental destruction and further climate harm. In contrast, public prosecutors have not filed any climate change lawsuits in the Global North, with prosecutions against climate activists in certain jurisdictions falling outside the scope of climate litigation as the prosecutions do not address climate change concerns and are brought to enforce laws protecting public order and private property.[48] Some regulatory

[45] Copies of the application and decision can be found here: <http://climatecasechart.com/non-us-case/brazilian-socialist-party-and-others-v-brazil/>accessed 2 March 2023.

[46] ibid.

[47] The plaintiff NGOs are Centre for Food and Adequate Living Rights (CEFROHT), the Africa Institute for Energy Governance (AFIEGO), Natural Justice, and the Centre for Strategic Litigation (Tanzania).

[48] eg *DPP v Fraser* [2008] NSWSC 244, where the Supreme Court of New South Wales dismissed charges against two Greenpeace activists for causing malicious damage to property. The activists had trespassed into a coal loader and chained themselves to a conveyor belt inside, rendering it out of operation for almost two hours; *Ochs v District of Columbia* 258 A 3d 169 (2021), where the Court of Appeals of the District of Columbia affirmed a lower court decision to convict the defendants of crowding, obstructing, or incommoding a street. During a climate change protest, they had remained seated on

agencies in the Global North, such as the Australian Competition and Consumer Commission, have also made it a priority to combat 'greenwashing' by companies.[49] The powerful role played by the Brazilian public prosecution service in environmental and climate change litigation provides for thought-provoking discussion about whether it is a model that can be replicated elsewhere in the Global South to improve environmental and climate law enforcement.

### Prosecutor-led climate change litigation in Brazil

On 1 June 2021, international news organizations reported that the head of the Brazilian Federal Prosecution Office had sought authorization from the Supreme Court to launch a criminal investigation into allegations that Environment Minister Ricardo Salles had obstructed federal police investigations into the role that he and other officials allegedly played in allowing illegal exports of timber from the Amazon region to the United States and Europe.[50] The court authorized the investigation and granted access to Salles' home and financial records.[51] This joined a growing chorus of demands by environmental advocates and sustainability-focused investors to have Ricardo Salles removed as Environment Minister for his efforts to roll back environmental protection in Brazil, as well as earlier lawsuits filed by Brazilian prosecutors seeking Salles' dismissal on the basis of administrative malpractice that has led to countless initiatives which

the street after being warned multiple times by police officers to leave; *R v Brown* [2022] EWCA Crim 6, [2022] All ER 44, where the Court of Appeal dismissed the appellant's appeal against his conviction for public nuisance but allowed the appeal against his sentence. The appellant had glued himself to the top of a commercial aircraft to draw attention to climate change and the contribution that air travel makes to emissions. See also Mark Townsend, 'Don't Put Climate Activists on Trial, CPS Urged' *The Guardian* (30 October 2021) <www.theguardian.com/environment/2021/oct/30/dont-put-climate-activists-on-trial-cps-urged> accessed 2 March 2023. See, however, a Swiss court's ruling in favour of a dozen climate activists who staged a tennis match inside Credit Suisse's premises in November 2018 without causing property damage. The judge concluded that, in light of the imminent danger of the climate emergency, the protestors' actions were necessary and proportional. The decisions and related court documents can be found here: <https://climate-laws.org/geographies/switzerland/litigation_cases/credit-suisse-protesters-trial> accessed 2 March 2023.

[49] 'Media Release (22-294MR): ASIC Acts against Greenwashing by Energy Company' (*Australian Securities & Investments Commission*, 27 October 2022) <https://asic.gov.au/about-asic/news-centre/find-a-media-release/2022-releases/22-294mr-asic-acts-against-greenwashing-by-energy-company/> accessed 7 March 2023. See also Peter Hannam, 'ACCC Begins "Greenwashing" Crackdown on Companies' False Environmental Claims' *The Guardian* (11 October 2022) <www.theguardian.com/australia-news/2022/oct/11/accc-begins-greenwashing-crackdown-on-companies-false-environmental-claims> accessed 7 March 2023; 'Media Release (23-043MR): ASIC Launches First Court Proceedings Alleging Greenwashing' (*Australian Securities & Investments Commission*, 28 February 2023) <https://asic.gov.au/about-asic/news-centre/find-a-media-release/2023-releases/23-043mr-asic-launches-first-court-proceedings-alleging-greenwashing/> accessed 7 March 2023.

[50] Ricardo Brito, 'Brazil's Top Prosecutor Asks Supreme Court to Probe Environment Minister' *Reuters* (1 June 2021) <www.reuters.com/business/environment/brazils-top-prosecutor-asks-supreme-court-probe-environment-minister-2021-06-01/> accessed 2 March 2023.

[51] Bryan Harris, 'Brazil's Police Target Environment Minister in Smuggling Probe' *Financial Times* (20 May 2021) <www.ft.com/content/eee7efea-93ed-4f1d-832d-988dea12c96e> accessed 2 March 2023.

violate the constitutional duty to protect the environment.[52] A few weeks later, on 23 June 2021, Salles announced that he was resigning as Environment Minister.[53]

Numerous actors were involved in efforts to remove Salles almost as soon as he took office in 2019, undoubtedly motivated in part by his criminal conviction—less than a month after he was appointed Environment Minister—for altering maps to benefit mining companies during his tenure as Sao Paulo's environmental secretary in 2016. What is notable is the prominent role played by the Brazilian prosecution service in using the law to hold Salles accountable and ultimately force his resignation.

In Brazil, public prosecutors are members of the Ministério Público, literally translated as the Public Ministry but more usefully translated as the public prosecution service.[54] The prosecution service consists of a state prosecutor office in each of Brazil's 26 states and representatives of the federal prosecutor's office in each state.[55] Following the introduction of constitutional reforms in the 1980s, the Ministério Público established itself as 'the lawyer of society', that is, the legal representative of society's interests. Prosecutors work across many public interest areas including fighting political corruption, consumer protection, and environmental protection. Often graduates of the country's best law schools, they are selected through highly competitive national exams. McAllister's seminal research on the role of public prosecutors in environmental law enforcement found that the prosecutors have a reputation for their legal acumen and strong sense of public service, and they often see themselves as 'public advocates waging battle against greedy or careless businesses and corrupt governmental officials'.[56] In the

[52] 'Brazil Prosecutors Target Minister Over Amazon Destruction' (*DW Akademie*, 7 July 2020) <www.dw.com/en/brazil-prosecutors-target-minister-over-amazon-destruction/a-54076374> accessed 2 March 2023. For a concise overview of the steps Salles and his office took to roll back environmental protection, see Luís Roberto Barroso and Patrícia Perrone Campos Mello, 'How to Save the Amazon: Why the Forest Has More Value Standing Than Cut Down' (2020) 12(2) Revista de Direito da Cidade 1379, 1399.

[53] Bryan Harris and Michael Pooler, 'Resignation of Brazil Environment Minister Cheered by Activists' *Financial Times* (24 June 2021) <www.ft.com/content/84f64281-30c2-4e0c-a9c8-c9f166e5eff7> accessed 2 March 2023; Rodrigo Pedroso and Jessie Yeung, 'Brazilian Environment Minister Ricardo Salles Resigns Amid Amazon Illegal Logging Probe' *CNN* (24 June 2021) <https://edition.cnn.com/2021/06/24/americas/brazil-ricardo-salles-resigns-intl-hnk/index.html> accessed 2 March 2023.

[54] Lesley McAllister, *Making Law Matter: Environmental Protection and Legal Institutions in Brazil* (1st edn, Stanford University Press 2008).

[55] ibid.

[56] ibid 5. There is a significant body of literature on the Brazilian Ministério Público; see eg Maria Tereza Sadek and Rosângela Batista Cavalcanti, 'The New Brazilian Public Prosecution: An Agent of Accountability' in Scott Mainwaring and Christoper Welna (eds), *Democratic Accountability in Latin America* (OUP 2003); Flavianne Fernanda Bitencourt Nóbrega, 'The New Institutional Design of the Procuracy in Brazil: Multiplicity of Veto Players and Institutional Vulnerability' (2007) ALACDE Annual Papers <https://escholarship.org/uc/item/8fp6n37w> accessed 14 August 2023; Matthew Taylor and Vinícius Buranelli, 'Ending Up in Pizza: Accountability as a Problem of Institutional Arrangement in Brazil' (2007) 49(1) Latin American Politics and Society 59; Bernardo Mueller, 'The

area of environmental protection, the Ministério Público has emerged as the most powerful actor in Brazilian environmental politics, and most significant environmental cases are brought to the courts by the public prosecutors.[57]

In his empirical research on the Brazilian Ministério Público, Coslovsky argues that the prosecution service is far from being a unitary actor which shares the same deep-seated conviction to serve the public.[58] Instead, Coslovsky contends that the Ministério Público is made up of various groups of prosecutors who loosely comprise two factions, each driven by distinct modes of action and that have built their own separate but overlapping organizational policies, practices, and structures to support their mode of action. One faction can be described as being cautious and reactive, adopting a formalistic case-processing approach. They seek to conform, not to trailblaze, and to *baixar a pilha* (lower the pile of cases) that accumulates on their desks.[59] In contrast, the other faction engages in social problem-solving and 'relational regulation'.[60] Far from performing the main tasks of writing legal opinions and arguing in open court, the prosecutors in this faction seek to enforce the law by addressing root causes of non-compliant behaviour and finding solutions that are acceptable to all. They use their prosecutorial powers and professional status to reach out to other institutions to devise solutions collectively; the prosecutors in turn support and enhance the ability of these other institutions to prevent or solve environmental challenges. Coslovsky comments: 'Prosecutors refer to this practice as *negotiating deals, building alliances, assembling networks or stitching solutions together*.' (emphasis in original)[61] As McAllister has also argued:

> the Brazilian Ministério Público resolves most of its environmental cases through investigation and the threat of prosecution, using its leverage to negotiate remedial measures. As such, its approach is not entirely legalistic but resembles 'flexible enforcement', in which agencies use elements of both the cooperative and legalistic styles to respond in a calibrated way to the behaviour of the regulated entity.[62]

Fiscal Imperative and the Role of Public Prosecutors in Brazilian Environmental Policy' (2010) 32(1) Law and Policy 104; Azul América Aguilar Aguilar, 'Institutional Changes in the Public Prosecutor's Office: The Cases of Mexico, Chile and Brazil' (2012) 4(2) Mexican Law Review 261.

57 McAllister (n 55) 85; Interview No 2.

58 Salo Coslovsky, 'Relational Regulation in the Brazilian Ministério Público: The Organizational Basis of Regulatory Responsiveness' (2011) 5 Regulation & Governance 70.

59 ibid 75–78.

60 ibid 78–82.

61 ibid 75. For discussion about the excessive discretion and limited accountability of the Ministério Público, see Fábio Kerche and Thiago Gomide Nasser, 'Autonomy and Discretionary Power of the Public Prosecutor's Office in Brazil' (2008) 50(2) Dados 259; Ernani Carvalho and Natalia Maria Leitao de Melo, 'Brazilian Public Prosecution Office and Independence' (2012) APSA 2012 Annual Meeting Paper <https://papers.ssrn.com/sol3/papers.cfm?abstract_id=2106323> accessed 14 August 2023; Salo Coslovsky and Amit Nigam, 'Building Prosecutorial Autonomy from Within: The Transformation of the Ministério Público in Brazil' (2015) NYU Wagner Research Paper No 2709104 <https://papers.ssrn.com/sol3/papers.cfm?abstract_id=2709104> accessed 14 August 2023; Gustavo Andrey de Almeida Lopes Fernandes and others, 'The Failures of Horizontal Accountability at the Subnational Level: A Perspective from the Global South' (2020) 30 Development in Practice 687.

62 McAllister (n 55) 16.

The Ministério Público does not provide all the resources that prosecutors need to engage in relational regulation. To fill this gap, some prosecutors establish alliances with NGOs, community groups, social movements, and like-minded professionals in relevant government agencies.[63] Most NGOs in Brazil do not have legally trained staff and do not typically file lawsuits. Instead, they work with prosecutors to halt environmentally destructive behaviour by providing on-the-ground leads, data that can be used as evidence in court, and attracting media coverage which adds to the pressure to hold the polluters to account.[64] It is therefore not surprising to learn that many Brazilian climate change lawsuits have been filed by prosecutors and that civil society organizations are working with prosecutors to implement their climate change advocacy initiatives.[65] Several Brazilian cases discussed in Chapter 3 of this book were filed by the Ministério Público, including the *Airlines* case and the *Steel Company* case.[66]

### Prosecutor-led climate litigation in China

In their research on the prospects for climate change litigation in China, Yue Zhao and her colleagues concluded that the vast majority of cases that could be classified as climate change litigation are actions related to contractual disputes involving companies in the energy and biotechnology sectors. In these cases, 'concern for climate change does not feature even at the periphery of the argument. Instead of the plaintiff, it is the court that seeks to help implement public policies on the low-carbon economy.'[67] They suggest that, 'instead of the court-driven regulatory policy-making process that is to be found especially in the Netherlands, India and Pakistan, China follows a government-led pathway in response to climate change, leaving courts with a secondary supporting role.'[68] In this regard, the role of public prosecutors in environmental law enforcement in China is worth exploring in

[63] Coslovsky (n 58) 75; Interview No 2.

[64] Informal discussions with Brazilian public prosecutors.

[65] eg *Public Prosecutor's Office v H Carlos Scheider S/A Comercio e Industria* (2007) Special Appeal No 650.728 (Superior Court of Justice); *Public Prosecutor's Office v Oliveira* (2008) No 2008/0215494-3 (Superior Court of Justice); *Sao Paulo Public Prosecutor's Office v United Airlines* (2014) Civil Appeal No 000292010.2014.4.03.9999 (3rd Federal Regional Court); *Public Prosecutor's Office v KLM* (2020) No 2020/0001750-7 (Superior Court of Justice).

[66] See 'Overview of the Brazilian Docket' in Chapter 4 of this book.

[67] Yue Zhao, Shuang Lyu, and Zhu Wang, 'Prospects for Climate Change Litigation in China' (2019) 8(2) Transnational Environmental Law 349, 362.

[68] ibid 365. For more discussion on climate change litigation in China, see Jiangfeng Li, 'Climate Change Litigation: A Promising Pathway to Climate Justice in China' (2019) 37(2) Virginia Environmental Law Journal 132; Richard Zhang Qing, 'Towards a Breakthrough in China's Climate Change Litigation: Environmental Public-Interest Litigation Filed by NGOs' in Xiangbai He, Hao Zhang, and Alexander Zahar (eds), *Climate Change Law in China in Global Context* (1st edn, Routledge 2020); Yue Zhao, Wei Liu, and Shuang Lyu, 'Tort-Based Public Interest Litigation on Air Pollution in China: A Promising Pathway for Chinese Climate Change Litigation?' in Jolene Lin and Douglas A Kysar (eds), *Climate Change Litigation in the Asia Pacific* (CUP 2020); Zhu Yan, 'The Subordinate and Passive Position of Chinese Courts in Environmental Governance' in Lin and Kysar (eds) (ibid); Xiangbai He, 'Mitigation and Adaptation through Environmental Impact Assessment Litigation: Rethinking the Prospect of Climate Change Litigation in China' (2021) 10 Transnational Environmental Law 413.

some detail because, as we have seen in Brazil, this could be the basis for advancing climate change litigation in the country.

China's environmental protection agencies have long faced significant difficulties with enforcing environmental laws and holding polluters to account for the environmental damage they cause.[69] It is therefore not surprising that the Chinese state has sought to diversify the types of regulators and mechanisms available to combat environmental degradation.[70] The work of prosecutors in China, known as the people's procuratorates, is mainly focused on criminal prosecutions and supervision of criminal enforcement activities.[71] The Supreme People's Procuratorate is the national prosecution office that directs the work of local people's procuratorates at the provincial, municipal, and county levels. It is 'responsible and [reports] on its work' to the National People's Congress and its Standing Committee (the national legislative body).[72] In addition to oversight by the Supreme People's Procuratorate, appointments and other personnel matters in the local people's procuratorates are controlled by the people's congresses and their standing committees at the corresponding subnational level.[73] In 2015, the government commenced an experiment that would involve the country's prosecutors in tackling the country's worsening environmental problems by filing civil lawsuits against polluters and taking administrative action against government entities that were not enforcing environmental laws properly.

In July 2015, the Standing Committee of the National People's Congress introduced the procuratorate-led civil environmental public interest litigation (EPIL) pilot programme in 13 selected provinces.[74] After a two-year legal experiment, this

[69] There is voluminous literature on this topic. See eg Benjamin van Rooij, 'Implementation of Chinese Environmental Law: Regular Enforcement and Political Campaigns' (2006) 37(1) Development and Change 57; Benjamin van Rooij, 'The People vs Pollution: Understanding Citizen Action Against Pollution in China' (2010) 19 Journal of Contemporary China 55; Elizabeth C Economy, *The River Runs Black: The Environmental Challenge to China's Future* (2nd edn, Cornell University Press 2010); Alex Wang, 'The Search for Sustainable Legitimacy: Environmental Law and Bureaucracy in China' (2013) 37 Harvard Environmental Law Review 365; Denise van der Kamp, 'Blunt Force Regulation and Bureaucratic Control: Understanding China's War on Pollution' (2021) 34(1) Governance 191; Xiaoying Ma and Leonard Ortolano, *Environmental Regulation in China: Institutions, Enforcement, and Compliance* (Rowman & Littlefield 2000).

[70] See eg Jolene Lin, 'Assessing the Dragon's Choice: The Use of Market-Based Instruments in Chinese Environmental Policy' (2004) 16(4) Georgetown International Environmental Law Review 617 on attempts to introduce market-based mechanisms into Chinese environmental policy.

[71] Organic Law of the People's Procuratorates of the People's Republic of China, art 5 (2007) <www.npc.gov.cn/zgrdw/englishnpc/Law/2007-12/13/content_1384077.htm> accessed 2 March 2023; Jingjing Liu, 'China's Procuratorate in Environmental Civil Enforcement: Practice, Challenges & Implications for China's Environmental Governance' (2011) 13 Vermont Journal of Environmental Law 41, 66.

[72] Organic Law of the People's Procuratorates of the People's Republic of China, art 10 (2007) <www.npc.gov.cn/zgrdw/englishnpc/Law/2007-12/13/content_1384077.htm> accessed 2 March 2023.

[73] ibid.

[74] Supreme People's Procuratorate, Plan for the Pilot Reform Scheme of Initiating Public Interest Litigation by the People's Procuratorates (promulgated on 2 July 2015). On environmental public interest litigation in China, see eg Alex Wang and Jie Gao, 'Environmental Courts and the Development of Environmental Public Interest Litigation in China' (2010) 3 Journal of Court Innovation 37;

system was established nationwide in July 2017, and the people's procuratorates were officially given standing to sue polluters.[75] In addition, amendments to the Administrative Procedure Law also gave the procuratorates legal authority to pursue what is known as administrative EPIL. Article 25 of the Administrative Procedure Law states that when the procuratorate finds that an administrative agency has been derelict in performing its duties in the fields of environment and natural resource protection and thereby causes harm to social interests, it is empowered to issue a 'letter of prosecutorial suggestions' to the agency.[76] The aim is to secure compliance and close the case in the pre-litigation stage, but if this is not achieved the procuratorate has the power to file a lawsuit against the agency.[77]

Studies have shown that the procuratorates have little appetite to use their new powers to enforce environmental laws, let alone trailblaze the advent of climate change litigation in China. Chunyan Ding and Huina Xiao's empirical study shows that Chinese prosecutors have a strong preference for cases that involve less complicated facts, weak and small defendants, and minor environmental violations.[78] They further describe the development of the procuratorates-led EPIL system in three stages. In the pre-2015 period before the pilot programme commenced, there were sporadic episodes of procuratorates bringing lawsuits against polluters. During the 2015–17 pilot period, top-down political pressure and career

Jingjing Liu, 'Environmental Justice with Chinese Characteristics: Recent Developments in Using Environmental Public Interest Litigation to Strengthen Access to Environmental Justice' (2012) 7 Florida A & M University Law Review; Dimitri De Boer and Douglas Whitehead, 'Opinion: The Future of Public Interest Litigation in China' (*China Dialogue*, 8 November 2016) <https://chinadialogue.net/en/pollution/9356-opinion-the-future-of-public-interest-litigation-in-china/> accessed 2 March 2023; Gong Gu and An Ran, 'Progress and Obstacles in Environmental Public-Interest Litigation Under China's New Environmental Law: An Analysis of Cases Accepted and Heard in 2015' (2017) 7(2–3) Climate Law 185; Sun Qian and Jack Tuholske, 'An Exploration of and Reflection on China's System of Environmental Public Interest Litigation Comments' (2017) 47 Environmental Law Reporter News & Analysis 10497; Qing Zhang, 'A Database on Environmental Public Interest Litigation Filed by NGOs in China: 2015 to Present' (2017) SSRN Scholarly Paper ID 3065111 <https://papers.ssrn.com/sol3/papers.cfm?abstract_id=3065111> accessed 2 March 2023; Richard Zhang and Benoît Mayer, 'Public Interest Environmental Litigation in China' (2017) 1(2) Chinese Journal of Environmental Law 202; Qi Gao and Sean Whittaker, 'Standing to Sue Beyond Individual Rights: Who Should Be Eligible to Bring Environmental Public Interest Litigation in China?' (2019) 8(2) Transnational Environmental Law 327; Lei Xie and Lu Xu, 'Environmental Public Interest Litigation in China: A Critical Examination' (2021) 10(3) Transnational Environmental Law 441; Hao Zhuang and Steven A Wolf, 'Environmental Public Interest Litigation: New Roles for Civil Society Organizations in Environmental Governance in China' (2021) 7(4) Environmental Sociology 393.

[75] The Civil Procedure Law (2012) was amended in 2017 to clarify the standing of procuratorates in civil environmental public interest litigation. Procuratorates were granted standing to initiate such cases, subject to the requirement that no other eligible plaintiff has done so. Yuhong Zhao has pointed out that this was done to ensure the best use of public and procuratorial resources and allow NGOs to play a major role in civil environmental public interest litigation; Yuhong Zhao, *Chinese Environmental Law* (CUP 2021) 407.

[76] Administrative Litigation Law of the People's Republic of China (2017 Revision), art 25 <www.lawinfochina.com/display.aspx?lib=law&id=23600> accessed 2 March 2023.

[77] ibid.

[78] Chunyan Ding and Huina Xiao, 'A Paper Tiger? Prosecutorial Regulators in China's Civil Environmental Public Interest Litigations' (2021) 32 Fordham Environmental Law Journal 323. The information in the rest of this paragraph is drawn from this article.

incentives led to a spike in prosecutorial action. This then died down after the pilot programme was put on a formal legal footing—there was less political pressure and fewer incentives for procuratorates to launch cases.

A notable factor that explains the relative reluctance of Chinese procuratorates to use their powers to enforce environmental laws is the high cost of regulation that stems from interference from local government (the local people's congresses). In short, the lack of institutional autonomy that stands in sharp contrast with the Brazilian Ministério Público makes it difficult for the Chinese experiment with procuratorate-led enforcement to succeed. However, it should be noted that it is the local people's procuratorates who face constant local government interference when handling civil EPIL matters.[79] The Supreme People's Procuratorate does not face the same sort of pressure and, in fact, has publicly announced that it is actively exploring the possibility of bringing climate change litigation. In a speech delivered at a side event of the IUCN World Conservation Congress in Marseilles in September 2021, China's Deputy Chief Procurator Zhang Xueqiao said: 'To supervise and ensure the realization of the targets for "carbon peaking" and "carbon neutrality", we are actively exploring climate change procuratorial public interest litigation. We focus on urging, supervising and coordinating relevant authorities to strengthen the prevention and control of environmental impacts from high energy-consuming and high-emission construction projects, and we are exploring preventive public interest litigation.'[80] It is also notable that a ruling issued on 24 March 2021 in Deqing County (Zhejiang province) has been officially recognized by the Supreme People's Court as the first successful procuratorate-led public interest litigation on climate change.[81] This is a rapidly developing space which many observers are keenly watching because, with the necessary political support from the highest levels of government, procuratorate-led litigation can play an important role in holding to account companies and government bodies that fail to comply with China's national plans to reach peak carbon by 2030 and carbon neutrality by 2060.[82]

[79] ibid 365.

[80] ClientEarth Communications, 'China's Procurators Eye Public Interest Litigation on Climate Change' (*ClientEarth*, 18 September 2021) <www.clientearth.org/latest/latest-updates/news/china-s-procurators-may-soon-bring-climate-litigation/> accessed 2 March 2023.

[81] ClientEarth Communications, 'China's First Climate Change Public Interest Case Could Herald New Era' (*ClientEarth*, 10 May 2021) <www.clientearth.org/latest/latest-updates/opinions/china-s-first-climate-change-public-interest-case-could-herald-new-era/> accessed 2 March 2023.

[82] 'China's Achievements, New Goals and New Measures for Nationally Determined Contributions (Unofficial Translation)' (*United Nations Framework Convention on Climate Change*) <https://unfccc.int/sites/default/files/NDC/2022-06/China%E2%80%99s%20Achievements%2C%20New%20Goals%20and%20New%20Measures%20for%20Nationally%20Determined%20Contributions.pdf> accessed 2 March 2023; for discussion, see Leslie Hook, 'China Pledges to be Carbon-neutral by 2060' *Financial Times* (23 September 2020) <www.ft.com/content/730e4f7d-3df0-45e4-91a5-db4b3571f353> accessed 2 March 2023; Michal Meidan, 'Unpacking China's 2060 Carbon Neutrality Pledge'

### Regulatory agency-led litigation

In a handful of cases brought by the Indonesian government through the Ministry of Environment and Forestry (MoEF) against timber companies and oil palm concession holders for illegal logging and causing peatland fires, the plaintiff MoEF sought compensation for the environmental damage resulting from the defendants' illegal conduct as well as costs to restore the natural resource to its prior state.[83] The climate change dimension lies in the MoEF's claim requiring the defendant to pay the costs of restoring the damaged carbon sink such as a forest or a mangrove area (restoration costs) or compensation for the release of greenhouse gases (GHGs) due to the destruction of the carbon sink. These cases also represent a novel approach to deterring illegal deforestation and peatland fires which are a major source of GHG emissions in Indonesia.[84] Tackling deforestation and peatland destruction is critical from a climate change perspective because they are vital carbon sinks.[85] The destruction of tropical forest cover results in an average of 4.8 gigatons of carbon dioxide emissions annually, 'causing more emissions every year than 85 million cars would over their entire lifetime'.[86]

(*Oxford Institute for Energy Studies*, December 2020) <www.oxfordenergy.org/wpcms/wp-content/uploads/2020/12/Unpacking-Chinas-carbon-neutrality-pledge.pdf> accessed 2 March 2023; Nicholas Stern and Chunping Xie, 'China's New Growth Story: Linking the 14th Five-Year Plan with the 2060 Carbon Neutrality Pledge' (*Grantham Research Institute on Climate Change and the Environment*, March 2021) <www.lse.ac.uk/granthaminstitute/wp-content/uploads/2021/03/Chinas-new-growth-story_Linking-the-14th-Five-Year-Plan-with-the-2060-carbon-neutrality-pledge.pdf> accessed 2 March 2023; Edmund Downie, 'Getting to 30-60: How China's Biggest Coal Power, Cement, and Steel Corporations Are Responding to National Decarbonization Pledges' (*Columbia University Center on Global Energy Policy*, 25 August 2021) <www.energypolicy.columbia.edu/research/report/getting-30-60-how-china-s-biggest-coal-power-cement-and-steel-corporations-are-responding-national> accessed 2 March 2023.

[83] *Minister of Environment v PT Selatnasik Indokwarsa and PT Simpang Pesak Indokwarsa* (Decision No 105/PDT/G/2009/PN.JKT.UT) (District Court of North Jakarta, 2010); *Minister of Environment v PT Kalista Alam* (Decision No 12/PDT.G/2012/PN.MBO) (District Court of Meulaboh, 2013); *Minister of Environment v PT Merbau Pelalawan Lestari* (Decision No 157/PDT.G/2013/PN.PBR) (District Court of Pekanbaru, 2014); *Minister of Environment v PT Bumi Mekar Hijau* (Decision No 51/PDT/2016/PT.PLG) (High Court of Palembang, 2016); *Minister of Environment v PT Jatim Jaya Perkasa* (Decision No 108/PDT.G/2015/PN.JKT.UTR) (District Court of North Jakarta, 2016); *Minister of Environment v PT Waringin Agro Jaya* (Decision No 456/PDT.G-LH/2016/PN.JKT.SEL) (District Court of South Jakarta, 2017).

[84] Daisy Dunne, 'The Carbon Brief Profile: Indonesia' (*Carbon Brief*, 27 March 2019) <www.carbonbrief.org/the-carbon-brief-profile-indonesia> accessed 2 March 2023.

[85] Yude Pan and others, 'A Large and Persistent Carbon Sink in the World's Forests' (2011) 333(6045) Science 988; Virginia Gewin, 'How Peat Could Protect the Planet' (February 2020) 578(7794) Nature 204; Nancy L Harris and others, 'Global Maps of Twenty-first Century Forest Carbon Fluxes' (2021) 11 Nature Climate Change 234; Lorna I Harris and others, 'The Essential Carbon Service Provided by Northern Peatlands' (2022) 20(4) Frontiers in Ecology and the Environment 222. The Paris Agreement supports the use of biological sinks in climate change mitigation. Article 5 includes an explicit call to developed and developing countries to conserve and enhance forests and other biological carbon reservoirs.

[86] David Gibbs, Nancy Harris, and Frances Seymour, 'By the Numbers: The Value of Tropical Forests in the Climate Change Equation' (*World Resources Institute*, 4 October 2018) <https://www.wri.org/insights/numbers-value-tropical-forests-climate-change-equation> accessed 2 March 2023

The first case in which the government sought compensation for GHG emissions was *Minister of Environment v PT Selatnasik Indokwarsa and PT Simpang Pesak Indokwarsa*.[87] The defendants were two mining companies who carried out illegal activities within their permitted areas, including the clearing of protected forests for road access to the mining location, conducting open pit mining, building infrastructure within a natural reserve, and constructing an illegal port.[88] The lawsuit was filed on grounds of negligence and strict liability. The plaintiff Ministry argued that amongst the various environmental impacts was the release of up to 359 tonnes of carbon per hectare of land. Based on the Ministry's calculations that the restoration costs were 90,000 Indonesian rupiahs per tonne of carbon and the damaged area was 208 hectares, the total restoration costs for the carbon release amounted to 6.7 billion Indonesian rupiahs (approximately US$480,000).[89] The case proceeded in a convoluted manner through the courts. Eventually, the Supreme Court decided in favour of the plaintiff Ministry and awarded costs in its favour. The defendant companies were ordered to pay restoration costs for the environmental damage amounting to 32.3 billion Indonesian rupiahs (approximately US$2.3 million).[90]

Another interesting case is *Minister of Environment v PT Kalista Alam*.[91] In the aftermath of serious peatland fires in 2013, this was one of several lawsuits filed with the aim of deterring companies from burning as a way of clearing land for oil palm plantations as well as to seek compensation from these companies for the environmental damage caused by the fires. The plaintiff Ministry argued that the defendant company had not only failed to prevent and control fires but it had also intentionally drained and burned the peatland in its area. It claimed that the fires had caused serious environmental damage including the loss of hydrological function of the peatland, biodiversity loss, soil damage, and damage due to the release of carbon into the atmosphere (contributing to climate change).[92] In relation to carbon, the plaintiff Ministry specified two heads of damages: one being compensation for the GHGs released into the atmosphere because of the fires and the other being costs to restore the peatland's loss of GHG absorption capacity.[93] The Supreme Court ruled in favour of the plaintiff Ministry and awarded costs in its favour.[94]

Andri Wibisana and Conrado Cornelius have argued that these lawsuits offer 'a new and unique approach to climate change litigation' because the regulator

[87] *Minister of Environment v PT Selatnasik Indokwarsa and PT Simpang Pesak Indokwarsa* (Decision No 105/PDT/G/2009/PN.JKT.UT) (District Court of North Jakarta, 2010).

[88] *Minister of Environment v PT Selat Nasik Indokwarsa and PT Simpang Pesak Indokwarsa* (Decision No 499 K/PDT/2012) (Supreme Court, 2012) 5–6.

[89] ibid 19–20.

[90] *Minister of Environment v PT Selat Nasik Indokwarsa and PT Simpang Pesak Indokwarsa* (Decision No 109 PK/PDT/2014) (Supreme Court, 2014) 36–39.

[91] *Minister of Environment v PT Kalista Alam* (Decision No 12/PDT.G/2012/PN.MBO) (District Court of Meulaboh, 2013).

[92] ibid 38–48.

[93] *Minister of Environment v PT Kalista Alam* (n 91) 43–44.

[94] *Minister of Environment v PT Kalista Alam* (Decision No 651 K/PDT/2015) (Supreme Court, 2015).

is putting a price on carbon and seeking to internalize the externality costs of emitting GHGs via the enforcement of natural resources laws concerning illegal logging and peatland fires.[95] In other words, these are peripheral cases wherein climate change is one of a number of environmental issues being addressed through litigation. Wibisana and Cornelius also suggest that '[t]he Indonesian approach is unique because it offers a solution for tort-based climate lawsuits, by focusing on the cost of emission reduction rather than on climate change damage. The focus allows the plaintiff to avoid a very difficult, if not impossible, task of proving causation between the defendant's conduct and a particular climate-related damage suffered by the plaintiff.'[96]

It should be noted that while the defendant companies have been found negligent and ordered to pay compensation in most of these lawsuits, enforcement of the verdicts has been poor. More than a year after the Supreme Court rulings, none of the companies paid compensation because of delays in enforcing the verdicts (the Ministry has to wait months for the official record of the verdict to be available) and the absence of official procedures for the enforcement of forestry-related judicial decisions.[97] This points to the enduring difficulties of environmental law enforcement in most developing countries, which reduces the effectiveness of climate change litigation as a governance tool and in fact might lead to the conclusion that the costs outweigh the benefits of pursuing legal action against polluters. This is a recurrent issue that has been canvassed at length in the earlier chapters. It suffices to recall one of our over-arching arguments that climate change litigation is indeed a costly process and by no means a silver bullet. Nonetheless, litigation, alongside other policy interventions and social action, can lead to better climate regulation and hold governments and corporations accountable for their actions (and inaction).[98]

[95] Andri G Wibisana and Conrado M Cornelius, 'Climate Change Litigation in Indonesia' in Lin and Kysar (eds) (n 68) 248.

[96] ibid 249. See also Andri G Wibisana, 'The Many Faces of Strict Liability in Indonesia's Wildfire Litigation' (2019) 28(2) Review of European, Comparative & International Environmental Law 185; Asian Development Bank, *Climate Change, Coming Soon to a Court Near You: Climate Litigation in Asia and the Pacific and Beyond* (Asian Development Bank 2020) 144–46. For discussion about climate change litigation in Indonesia more broadly, see Agung Wardana, 'Governing through Courts? A Gloomy Picture of Climate Litigation in Indonesia' (*Völkerrechtsblog*, 21 March 2022) <https://voelkerrechtsblog.org/governing-through-courts/> accessed 2 March 2023.

[97] Hans Nicholas Jong, 'Government Wins in Forest Fire Case' *The Jakarta Post* (9 February 2017) <www.thejakartapost.com/news/2017/02/09/government-wins-in-forest-fire-case.html> accessed 2 March 2023; Ahmad Syamsudin, 'Greenpeace: Indonesian Firms Failed to Pay Billions in Fines Linked to Forest Fires' *Benar News* (15 February 2019) <www.benarnews.org/english/news/indonesian/greenpeace-report-02152019165335.html> accessed 2 March 2023; Stephen Wright and Niniek Karmini, 'Indonesia's Pulp and Palm Oil Polluters Owe Millions in Fines' *The Sydney Morning Herald* (17 February 2019). <www.smh.com.au/world/asia/indonesia-s-pulp-and-palm-oil-polluters-owe-millions-in-fines-20190215-p50y20.html> accessed 2 March 2023.

[98] The Indonesian government has been sued in a citizen lawsuit concerning its failures to prevent the devastating fires in 2015: Greenpeace Southeast Asia, 'President Jokowi Must Accept Fires

To the best of our knowledge, environmental regulatory agencies elsewhere in the Global South have not sought to use tort law to compel polluters to pay compensation and restoration costs for causing climate change through illegal activities such as deforestation. While these Indonesian cases are a mere handful, they highlight an interesting avenue for the emergence of climate change litigation through actors responsible for enforcing environmental and natural resources laws.

## 2.3 Youths and Children

That the burden of climate change falls unequally on different groups of people is most clearly exemplified in the case of youths and children.[99] Not only are today's young people going to have to deal with long-term climate change impacts longer than adults (including threats to their health, and access to food, water, and shelter), they are worse affected by climate change impacts in the present.[100] Yet, at the same time as being most affected by climate change and least responsible for the current state of global GHG emissions, young people are also least able to influence climate policy responses and international climate change negotiations. Most political systems have a voting age of 18 years and the international climate change negotiations are not especially accessible to young people.[101]

Verdict and Show He is Serious About Ending Forest Fires and People's Suffering' (*Greenpeace*, 23 July 2019) <www.greenpeace.org/southeastasia/press/2830/president-jokowi-must-accept-fires-verdict-and-show-he-is-serious-about-ending-forest-fires-and-peoples-suffering/> accessed 2 March 2023; Hans Nicholas Jong, 'Top Court Holds Indonesian Government Liable Over 2015 Forest Fires' (*Mongabay*, 23 July 2019) <https://news.mongabay.com/2019/07/top-court-holds-indonesian-government-liable-over-2015-forest-fires/> accessed 2 March 2023; Michael Taylor, 'Indonesians Hope Milestone Ruling Will Dampen Haze-fire Risks' *Reuters* (23 July 2019) <www.reuters.com/article/us-indonesia-forest-haze-idUSKCN1UI1DB> accessed 2 March 2023.

[99] 'Children' usually refers to human beings under the age of 18 years. For example, art 1 of the Convention on the Rights of the Child defines a child to be 'every human being below the age of eighteen years unless under the law applicable to the child, majority is attained earlier'; United Nations General Assembly (UNGA) Res 44/25 (20 November 1989): Convention on the Rights of the Child. The term 'youth' is often taken to include those up to their mid-20s but there is no clear definition; see eg the United Nation's *World Youth Report*, which refers to young people as those between the ages of 15 and 24 years. For our purposes, we use the term 'young people' to refer collectively to children and youths up to the age of 25. Larissa Parker and her colleagues define young people to loosely encompass children (those below the age of 18 years) as well as youths (between the ages of 15 and 35), noting that the cases they survey often involves plaintiffs in a wide range of age groups. The *Canada ENJEU* case set an upper age limit of 35 years, for example. See Larissa Parker and others, 'When the Kids Put Climate Change on Trial: Youth-Focused Rights-Based Climate Litigation Around the World' (2022) 13(1) JHRE 64, 65 (fn 4).

[100] eg higher temperatures result in increased mortality amongst children, especially those who live in areas at risk of drought. The climate crisis also increases the vulnerability of children to trafficking and forced labour. For details, see UN Office of the High Commissioner for Human Rights, 'Analytical Study on the Relationship between Climate Change and the Full and Effective Enjoyment of the Rights of the Child: Report of the Office of the United Nations High Commissioner for Human Rights' A/HRC/35/13 (4 May 2017) <https://digitallibrary.un.org/record/1298393?ln=en> accessed 2 March 2023.

[101] National delegations to the annual UNFCCC Conference of the Parties (COP) meetings generally do not include youths and children. The YOUNGO is the youth constituency of the UNFCCC.

More broadly, social and legal norms tend to define children and non-adult youths as lacking agency and maturity.[102] Children and youths are commonly portrayed as victims in need of protection rather than individuals capable of exercising rights and holding political opinions. Charities seeking donations often use images of children to elicit sympathy.[103] United Nations (UN) monitoring mechanisms and official discourse tend to emphasize the protection of children (and their vulnerability).[104]

Against this background of marginalization which has created frustration and agitation, many young people nonetheless have worked to combat climate change. The prominence of young climate activists rose significantly in 2018 when Greta Thunberg made global headlines as a lone protestor outside the Swedish parliament and swiftly became the figurehead of a global youth climate justice movement.[105] In September 2019, students around the world organized a coordinated global strike under the Fridays for Future (FFF) campaign, which mobilized

It consists of many youth-led organizations and individuals working in climate change-related fields. The YOUNGO constituency gained full recognition at COP-17 in Durban, South Africa and has since served as the official conduit for youth participation in the UN climate change talks as well as a global network of youth and youth-focused groups working on climate change; see <https://youngoclimate.org> accessed 22 January 2024. Some youth activists, however, lament that these efforts are simply to 'fill the youth quota' and do not genuinely promote intergenerational cooperation involving youths. The COPs are also perceived to be public relations events where little real action is achieved; Xiye Bastida, Disha Ravi, and Ayisha Siddiqa, 'Youth in Focus: Youth Activism's Role in Pushing Climate Action Forward' (*NYU Climate Litigation Accelerator*, 11 April 2022) <www.youtube.com/watch?v=OoSa6_jY4P4> accessed 2 March 2023. On voting, an increasing number of countries have allowed 16- and 17-year-olds to vote. For an interesting exploration on lowering the voting age, see Tommy Peto, 'Too Immature to Vote? A Philosophical and Psychological Argument to Lower the Voting Age' (2020) 13(1) Erasmus Law Review 60.

[102] AL Liou and Ioana Literat, '"We Need You to Listen to Us": Youth Activist Perspectives on Intergenerational Dynamics and Adult Solidarity in Youth Movements' (2020) 14 International Journal of Communication 4462. In Dutch civil law, for example, a child does not have legal standing and cannot be an independent party to civil court proceedings. A child has to be represented by a parent or guardian. When there is a conflict of interests between the child and the legal representative(s), the court can appoint a guardian *ad litem* to represent the child. For discussion, see Mariëlle Bruning and Jiska Peper, 'Giving Children a Voice in Court? Age Boundaries for Involvement of Children in Civil Proceedings and the Relevance of Neuropsychological Insights' (2020) 13(1) Erasmus Law Review 31; and more broadly, the 2020 Erasmus Law Review Issue 1, which is a special issue on age limits in law.

[103] Primary school children have asked organizations, including Save the Children, to end the use of emotive images of children in their advertisements; Nicola Slawson, 'Schoolkids Challenge Charities' "undignified" images of children' *The Guardian* (22 March 2016) <www.theguardian.com/voluntary-sector-network/2016/mar/22/schoolchildren-campaign-stop-charities-emotive-images-advertising> accessed 2 March 2023. Social media companies have also exploited images of children to elicit sympathy and garner profits from donor generosity; Hannah Gelbart, Mamdouh Akbiek, and Ziad Al-Qattan, 'Tiktok Profits from Livestreams of Families Begging' *BBC* (12 October 2022) <https://www.bbc.com/news/world-63213567> accessed 14 August 2023.

[104] See eg the United Nations Children's Fund (UNICEF)'s flagship publication 'The State of the World's Children' <www.unicef.org/reports/state-of-worlds-children> accessed 2 March 2023.

[105] Charlotte Alter, Suyin Haynes, and Justin Worland, 'TIME 2019 Person of the Year: Greta Thunberg' *Time* (23 December 2019) <https://time.com/person-of-the-year-2019-greta-thunberg/> accessed 2 March 2023; 'The Greta Effect' *The Economist* (19 August 2019) <www.economist.com/graphic-detail/2019/08/19/the-greta-effect> accessed 2 March 2023

some seven million participants across 6,000 protest events in 185 countries.[106] Today, youths are some of the most prominent and outspoken voices in climate activism.[107]

As part of their climate change mobilization efforts, young people have filed or participated in lawsuits to stop projects and activities that will worsen climate change and to compel governments to undertake more ambitious climate action. Youth climate activists have been described as being 'set on a path to interrogate and attempt to bring to account, by any means possible, every lagging State, diplomatic negotiation, and fossil fuel major, using all appropriate judicial and quasi-judicial avenues and other public fora'.[108] Larissa Parker and her colleagues have argued that youth-focused cases can be seen as examples of disruptive dissent which seek to transform the norms and structures that are responsible for the climate crisis.[109] By April 2023, 11 youth-led cases had been filed across the Global South including in India, Pakistan, Uganda, Peru, and Mexico. Three cases have been filed in a supranational forum by a youth plaintiff from a Global South jurisdiction. For the sake of clarity, the term 'youth-led' refers to cases filed by or involving individual children and youths, an NGO led by youths, or a class of children and young people. In the following section, we elaborate on some of the youth-led climate lawsuits that have been filed in the Global South.

[106] Joost de Moor and others (eds), *Protest for a Future II: Composition, Mobilization and Motives of the Participants in Fridays for Future Climate Protests on 20–27 September 2019 in 19 Cities Around the World* (Swedish Research Council for Sustainable Development 2020) <www.diva-portal.org/smash/get/diva2:1397070/FULLTEXT01.pdf> accessed 2 March 2023.

[107] See eg Blaise Hope, 'Get to Know the World's Top Five Youth Climate Activists' *Sustainability* (2 March 2022) <https://sustainabilitymag.com/sustainability/faces-of-change-the-top-five-youth-climate-activists-named-greta-emissions-change-pledge> accessed 2 March 2023; 'Vanessa Nakate'(*United Nations Office of the Secretary-General's Envoy on Youth*) <www.un.org/youthenvoy/vanessa-nakate/> accessed 2 March 2023; 'The Youth Advisory Group on Climate Change' (*United Nations Climate Action*) <www.un.org/en/climatechange/youth-in-action/youth-advisory-group> accessed 2 March 2023; Akane Okutsu, 'Youth Climate Activists Spread Greta Thunberg's Message in Asia' *Nikkei Asia* (2 October 2021) <https://asia.nikkei.com/Spotlight/Environment/Climate-Change/Youth-climate-activists-spread-Greta-Thunberg-s-message-in-Asia> accessed 2 March 2023; Alex Jackson, 'The Young Female Climate Activists Driving Change in Asia' *The Japan Times* (27 June 2021) <www.japantimes.co.jp/news/2021/06/27/asia-pacific/climate-activists-asia/> accessed 2 March 2023. Nicole Rogers argues that the child has made a transition from victim to empowered actor in climate change discourse; see Nicole Rogers, 'Victim, Litigant, Activist, Messiah: The Child in a Time of Climate Change' in Benjamin J Richardson (ed), *From Student Strikes to the Extinction Rebellion* (Edward Elgar 2020); and Nicole Rogers, *Law, Fiction and Activism in a Time of Climate Change* (Routledge 2020). Aoife Daly argues that there are three main reasons behind the prominence of children and young people in climate activism. The first is the grasp of climate science that the youth activists display. The second is their competence in using digital platforms to promote online activism which transcends borders. The third is their success in using social media to garner attention and support. See Aoife Daly, 'Climate Competence: Youth Climate Activism and Its Impact on International Human Rights Law' (2022) 22(2) Human Rights Law Review 1, 9–12.

[108] Ingrid Gubbay and Claus Wenzler, 'Intergenerational Climate Change Litigation: The First Climate Communication to the UN Committee on the Rights of the Child' in Ivano Alogna, Christine Bakker, and Jean-Pierre Gauci (eds), *Climate Change Litigation: Global Perspectives* (Brill 2021) 343, 347

[109] Parker and others (n 99) 78.

### A brief overview of some of the youth-led climate lawsuits

One of the earliest youth-led lawsuits in the Global South is *Rabab Ali v Federation of Pakistan*.[110] Rabab Ali, who was seven years old at the time when the case was filed, sued Pakistan's government for violating citizens' constitutional rights to life, liberty, dignity, information, and equal protection before the law, and for violating the public trust doctrine in relation to Pakistan's atmosphere and climate.[111] Represented by her father who is an environmental lawyer, Rabab Ali's petition focused on the government's approval of a plan to develop coalfields in the Thar desert region in the southeast corner of Pakistan, one of the world's largest coal reserves. For years, Pakistan used the Thar coal reserves as a bargaining chip in global climate change negotiations. It argued that since it was not mining its large coal deposits, it should receive easier access to international climate finance.[112] However, to end the energy crisis that has caused frequent power cuts for years, the government decided to develop the coalfields which contain an estimated 175 billion tonnes of watery, low-energy coal that for a long time was deemed too costly to mine.[113] The China-Pakistan Economic Corridor is the main source of investment for developing coalfields and multiple coal-fired power plants in Pakistan.[114]

The plaintiff argued that exploitation of the Thar coalfields would release approximately 327 billion tons of $CO_2$, more than one thousand times Pakistan's previous estimate for annual GHG emissions.[115] In addition, coal mining would worsen air pollution and water quality in the region.[116] It is also expected to displace thousands of Thari people from their land in violation of their right to property, dignity, and equal protection before the law.[117] The petition argued that Pakistan could potentially use renewable energy to power all of its energy needs, including in the transport, industrial, and agriculture sectors.[118] Ali also sought

[110] *Ali v Federation of Pakistan*, Constitutional Petition No _/I of 2016.

[111] ibid. '7-Year-Old Files Climate Change Lawsuit with the Supreme Court of Pakistan' (*EcoWatch*, 6 April 2016) <www.ecowatch.com/7-year-old-files-climate-change-lawsuit-with-the-supreme-court-of-paki-1891078139.html> accessed 2 March 2023.

[112] Rina Saeed Khan, 'Pakistan Races to Tap Virgin Coal Fields to Meet Energy Crunch' *Reuters* (27 February 2017) <www.reuters.com/article/us-pakistan-coal-climatechange-feature-idUSKBN16612G> accessed 2 March 2023.

[113] Khan (n 112).

[114] News Desk, '"Good News": Huge Coal Reserve Discovered in Thar' *The Express Tribune* (31 January 2022) <https://tribune.com.pk/story/2341297/good-news-huge-coal-reserve-discovered-in-thar> accessed 2 March 2023; Ministry of Planning, Development, & Special Initiatives: CPEC Authority, 'Energy Projects under CPEC' (*CPEC Authority Office*) <http://cpec.gov.pk/energy> accessed 2 March 2023.

[115] *Ali v Federation of Pakistan*, Petition, [5].

[116] Azhar Lashari, 'Development and Catastrophe' *The News on Sunday* (10 November 2019) <www.thenews.com.pk/tns/detail/568851-development-catastrophe> accessed 2 March 2023.

[117] Zofeen Ebrahim, 'Locals Urge Pakistani Government to Drop CPEC Coal Mining Plans' (*The Third Pole*, 15 July 2020) <www.thethirdpole.net/en/regional-cooperation/thar-coal-project/> accessed 2 March 2023; Shujauddin Qureshi and Zeenia Shaukat, 'Thar Coal Project and Local Community Documenting Views and Experiences of Stakeholders' (*National Commission for Human Rights*, December 2019) <www.boell.de/sites/default/files/2019-12/Thar_Coal_Project.pdf> accessed 2 March 2023.

[118] *Ali v Federation of Pakistan* (n 110), Petition, [34].

protection for mangroves for sequestering carbon and protecting against sea-level rise.[119] The case remains pending more than six years after being filed, which is arguably not surprising given the strong governmental support for coal development in Pakistan and the geopolitical significance of the China-Pakistan Economic Corridor project.[120]

*Future Generations v Ministry of the Environment*[121] is one of the best-known youth-led climate cases because it was a legal victory for the plaintiffs and a watershed decision on the rights of future generations. Inspired by the plaintiffs in *Juliana v United States*, 25 youths and a supporting NGO called Dejusticia sued the Colombian government for the deforestation of the Amazon, which, they argued, violates their constitutional rights to a healthy environment, life, health, food, and water.[122] The lawsuit was filed as a tutela, which is a legal mechanism under the Colombian Constitution that allows for an expedited hearing of cases alleging violations of fundamental rights. The claimants' plea to the court was to order the state to act in various ways, including the development and implementation of a national action plan to reduce deforestation and an intergenerational agreement for the reduction of deforestation in the Colombian Amazonian rainforest.[123]

On 5 April 2018, the Supreme Court of Colombia ruled in favour of the youth plaintiffs. The court also declared that the Colombian Amazon Forest has legal personhood, that the Colombian government has a duty to protect it and has not done enough to prevent deforestation and increasing GHGs.[124] The court ordered the government to create and implement plans to halt deforestation, to address climate mitigation and adaptation, and to develop an 'intergenerational pact for the life of the Colombian Amazon' with the plaintiffs, affected communities, and research and scientific organizations. This Supreme Court decision has been hailed as a watershed for a few reasons. First, the Supreme Court discussed issues such

119 ibid, The Questions of Law xxxvii, [42] and [43].

120 See eg David Sacks, 'The China-Pakistan Economic Corridor—Hard Reality Greets BRI's Signature Initiative' (*Council on Foreign Relations*, 30 March 2021) <www.cfr.org/blog/china-pakistan-economic-corridor-hard-reality-greets-bris-signature-initiative> accessed 2 March 2023; Zenel Garcia and Umer Rahman, 'What the China-Pakistan Economic Corridor Tells Us About the Belt and Road Initiative' *The Diplomat* (28 April 2022) <https://thediplomat.com/2022/04/what-the-china-pakistan-economic-corridor-tells-us-about-the-belt-and-road-initiative/> accessed 2 March 2023. It is interesting to note that while the constitutional challenge against the Thar coalfields project has stalled, money-laundering allegations have invited judicial scrutiny of the project and could potentially be a pathway for delaying the project: Our Correspondent, 'SC Orders NAB to Probe Thar Coal Project' *The Express Tribune* (1 December 2021) <https://tribune.com.pk/story/2331888/sc-orders-nab-to-probe-thar-coal-project> accessed 2 March 2023.

121 *Demanda Generaciones Futuras v Minambiente* (2018) STC4360-2018, Number 11001-22-03-000-2018-00319-01.

122 'Colombia' (*Our Children's Trust*) <www.ourchildrenstrust.org/colombia-global-summary> accessed 2 March 2023.

123 César Augusto Rodríguez Garavito, 'Acción de Tutela' (*Dejusticia*, 29 January 2018) <https://www.dejusticia.org/wp-content/uploads/2018/01/TutelaCambioClim%C3%A1tico.pdf?x54537> accessed 2 March 2023.

124 *Demanda Generaciones Futuras v Minambiente* (n 121).

as intergenerational equality and solidarity, accountability for climate change, and human dependence on the environment—topics which are rarely addressed by the court.[125] Secondly, the court's decision embraced the idea of how current generations should respect and protect the rights of future generations, which includes responsible use of natural resources.[126] Finally, for a court that rarely relies on international law in its jurisprudence, the Supreme Court in this case held that the state had legal obligations pursuant to the Paris Agreement.[127] The obligations included the reduction of GHG emissions and adopting effective policies to fight deforestation. It should be noted that, at the time of writing, the success of this case has been 'somewhat tainted by the Colombian government's failure to execute the Court's orders promptly and effectively'.[128] This of course raises questions about the impact of litigation and its 'value proposition', which are important issues we will return to in the concluding chapter of this book.

On 5 December 2019, 15 youth plaintiffs from the state of Baja California in Mexico, between 17 and 23 years of age, filed a lawsuit against the Mexican federal government seeking compliance with its constitutional obligations and proper implementation of the national General Law on Climate Change.[129] In the petition, the plaintiffs point out that the national climate change law was passed in 2012 but there are no regulations or public policies to put the law into effect. More broadly, it was argued that the governmental inaction amounted to a failure to respect and protect the plaintiffs' right to a healthy environment pursuant to the Mexican Constitution. While an earlier court decision in 2022 dismissed the case on the basis that the plaintiffs lacked standing, this decision was overturned on appeal to the Collegiate Court in September 2023. The case was remitted by the Collegiate Court to the Mexican Supreme Court for a decision on the merits of the youths' claims. At the time of writing, the Supreme Court's decision is pending.[130]

125 Paola Andrea Acosta Alvarado and Daniel Rivas-Ramírez, 'A Milestone in Environmental and Future Generations' Rights Protection: Recent Legal Developments before the Colombian Supreme Court' (2018) 30(3) Journal of Environmental Law 519.

126 ibid.

127 ibid.

128 Parker and others (n 99) 81.

129 'Fifteen Youths File Federal Climate Lawsuit in Mexico' (*Defensa Ambiental del Noroeste (DAN) and Our Children's Trust*, 2 September 2020) <https://static1.squarespace.com/static/571d109b04426270152febe0/t/5f4fcaa80194e649f7f1014a/1599064747861/2020.09.02+Jo%CC%81venes+v.+Gobierno+de+Me%CC%81xico+-+Press+Release.pdf> accessed 2 March 2023. On the General Law on Climate Change, see Michael B Gerrard and Anne Siders, 'Mexico's General Climate Change Law' (*Mexico-US Climate Law Network, Environmental Law Institute*, 2012) <https://scholarship.law.columbia.edu/cgi/viewcontent.cgi?article=4132&context=faculty_scholarship> accessed 2 March 2023; Israel Solorio, 'Leader on Paper, Laggard in Practice: Policy Fragmentation and the Multi-Level Paralysis in Implementation of the Mexican Climate Act' 21(9) Climate Policy 1.

130 For timeline of events in this lawsuit, see 'Jovenes *v* Gobierno de México' (*Climate Change Case Chart*) < https://climatecasechart.com/non-us-case/youth-v-government-of-mexico/> accessed 2 March 2023.

Recent international and regional cases have begun to grow from increasing domestic litigation, given the requirement of international organizations that domestic avenues must usually be exhausted.[131] It is notable that youths and children are utilizing international human rights accountability mechanisms and regional human rights courts to address climate change. These climate change-related applications are typically highly visible because of the prominence of the international/regional bodies and the compelling nature of the underlying core argument that these applications raise—that international law is the last beacon of hope for these youth plaintiffs seeking climate justice and that states have international legal obligations to protect young people from the harmful consequences of climate change.

In *Sacchi and Others v Argentina and Others*, Greta Thunberg and 15 youths (then aged between 8 and 17 years) filed a complaint to the Committee on the Rights of the Child (CRC), which is an adjudicatory body for the Convention on the Rights of the Child.[132] The *Sacchi* petitioners brought suit against five countries—Argentina, Brazil, France, Germany, and Turkey.[133] They alleged that the states parties' acts and omissions 'recklessly [caused] and [perpetuated] life-threatening climate change'[134] and violated their Convention rights. The youth plaintiffs suffered through heatwaves and droughts, had medical conditions worsened by air pollution, and had their Indigenous practices threatened.[135] The petitioners asked that the countries acknowledge that climate change is a human rights crisis and amend their laws to ensure that GHG emissions are reduced to the lowest possible level.[136]

Diverging from the standard test for determining extraterritorial jurisdiction (which is whether there was state control over the petitioners), the CRC

[131] See eg Alexia Solomou, 'European Convention on Human Rights—Exhaustion of Domestic Remedies—Effectiveness of Domestic Remedies' (2010) 104(4) American Journal of International Law 628; Lilian Chenwi, 'Exhaustion of Local Remedies Rule in the Jurisprudence of the African Court on Human and Peoples' Rights' (2019) 41(2) Human Rights Quarterly 374; Ikenna Ugboaja, 'Exhaustion of Local Remedies and the FSIA Takings Exception: The Case for Deferring to the Executive's Recommendation' (2020) 87(7) University of Chicago Law Review 1937.

[132] United Nations Convention on the Rights of the Child (unedited version), Decision adopted by the Committee on the Rights of the Child under the Optional Protocol to the Convention on the Rights of the Child on a communications procedure in respect of Communication No 104/2019, CRC/C/88/D/104/2019. An Optional Protocol to the Convention allows individual children to submit communications to the CRC alleging violations of their Convention rights; GA Res 66/183, UN Doc A/RES/66/138 (27 January 2012).

[133] *Sacchi v Argentina*, Communication of Petitioners (23 September 2019) [30]. A copy of the petition can be found here: <https://climatecasechart.com/wp-content/uploads/non-us-case-documents/2019/20190923_Communication-No.-1042019-Argentina-Communication-No.-1052019-Brazil-Communication-No.-1062019-France-Communication-No.-1072019-Germany-Communication-No.-1082019-Turkey_petition.pdf> accessed 7 March 2023. Another notable youth-led case, *Duarte Agostinho v Portugal*, No 39371/20, §§ 1, 2, 3, 8, and 14, December 2020, is not discussed here as it does not involve Global South actors and institutions.

[134] *Sacchi v Argentina* (n 133) [24].

[135] ibid [97]–[111], [112]–[114], and [134]–[158].

[136] ibid [33].

relied on the causality-based test proposed by the *Sacchi* petitioners and set out in the Inter-American Court of Human Rights' (IACHR) *Advisory Opinion on the Environment and Human Rights*.[137] Applying that test, the CRC held that victims of transboundary environmental damage, including damage caused by climate change, were within the human rights jurisdiction of a GHG-emitting state if the petitioners' harms were caused by the act or omission of that state and were 'reasonably foreseeable' consequences of the emissions allowed by that state's policies.[138] However, the CRC found the petition to be inadmissible because of the petitioners' failure to exhaust domestic remedies.[139] Nonetheless, this petition is groundbreaking because the CRC held that a state party can be held responsible for the negative impacts of its failure to mitigate climate change on the rights of children both within and outside its territory. The adoption of the new causality-based test also represents a significant expansion of extraterritorial jurisdiction which opens up opportunities for similar legal interventions in the future.[140]

## 2.4 Philanthropic funders

The conventional wisdom in social movements studies suggests that organizations must be able to mobilize material (as well as symbolic) resources to pursue collective action.[141] It follows that groups that are better resourced will be better equipped to pursue climate change litigation and are likely to be more successful when they do so.[142] In other words, the pursuit of social change or a 'rights revolution' requires a 'support structure' made up of organizations that are committed to achieving social change as well as organizations providing legal and financial

[137] *The Environment and Human Rights*, Advisory Opinion OC-23/17, Inter-American Court of Human Rights Series A No 23 (15 November 2017).

[138] *Sacchi v Argentina* (n 133), Communication of Petitioners, [261].

[139] Cali describes the Committee's decision as being a 'handy illusion'. It is an illusion because there is no domestic law precedent in Turkey that would give the youth plaintiffs a substantive hearing in the courts, including the Constitutional Court, so there is no chance to exhaust domestic remedies. It is handy because it allows the Committee to avoid backlash from states for entertaining such communications, 'whilst at the same time telling the children that the Committee is willing to take on the climate crisis, but only when children litigate and lose domestically': Başak Çali, 'A Handy Illusion? Interpretation of the "Unlikely to Bring Effective Relief" Limb of Article 7(e) OPIC by the CRC in Saachi et al.' (*EJIL:Talk!*, 1 November 2021) <www.ejiltalk.org/a-handy-illusion-interpretation-of-the-unlikely-to-bring-effective-relief-limb-of-article-7e-opic-by-the-crc-in-saachi-et-al/> accessed 2 March 2023.

[140] For a critical view on such a 'radical expansion of extraterritorial jurisdiction', see 'Sacchi *v* Argentina: Committee on the Rights of the Child Extends Jurisdiction over Transboundary Harms; Enshrines New Test' (2022) 135 Harvard Law Review 1981.

[141] See eg John D McCarthy and Mayer N Zald, 'Resource Mobilization and Social Movements: A Partial Theory' (1977) 82(6) American Journal of Sociology 1212; Jeff A Larson, 'Social Movements and Tactical Choice' (2013) 7(10) Sociology Compass 866; Paul Almeida, *Social Movements: The Structure of Collective Mobilization* (University of California Press 2019) 63–100.

[142] McCarthy and Zald (n 141).

resources.[143] Acknowledging that the mobilization of resources is rarely straightforward in developing countries,[144] it is beyond the scope of this chapter to provide detailed analysis of the mobilization of financial resources for climate change litigation. This section instead seeks to provide some insights into the rapidly growing climate change philanthropy space which is scaling up climate change litigation in the Global South based on the broad belief that litigation is an important pathway for unlocking meaningful cuts in GHG emissions as well as enhancing adaptation to climate change impacts. It is noteworthy that the growing interest in cases seeking compensation for climate damage has also garnered the attention of professional litigation funders who seek to make money from these climate cases.[145] Our discussion in this chapter does not address the role of these professional litigation funders as, to the best of our knowledge, they have not entered the Global South climate litigation space in a meaningful way at this nascent stage. Instead, the costs of climate litigation in the Global South have often been met by local and global charities as well as philanthropic organizations.

There are two main actors with distinct roles in the climate change philanthropy space. Philanthropic organizations are, broadly speaking, non-profit and non-governmental entities that use donated assets and income to give back to society through supporting charitable causes such as provision of social services in low-income communities.[146] Philanthropic organizations are also frequently referred to as private foundations or non-profit foundations. While many major philanthropic organizations typically support charitable causes at home, their philanthropy is becoming more global and, in some cases, they play an increasingly prominent role both in the scale of their financial giving and in their ability to influence the global development agenda. The world's largest private philanthropic organization is the Bill & Melinda Gates Foundation which disbursed about US$5

[143] Charles R Epp, *The Rights Revolution: Lawyers, Activists, and Supreme Courts in Comparative Perspective* (University of Chicago Press 1998) 17. Galanter argues that well-resourced actors tend to come out ahead in litigation efforts because of their ability to be 'repeat players' in the courtroom: Marc Galanter, 'Why the "Haves" Come Out Ahead: Speculations on the Limits of Legal Change' in Art Hinshaw, Andrea Kupfer Schneider, and Sarah Rudolph Cole (eds), *Discussions in Dispute Resolution: The Foundational Articles* (online edn, Oxford Academic 2021). In her 2006 study on environmental groups in Spain and Germany, Börzel concluded that '[t]he EU's legal institutions only increase opportunities for participation for those individuals and groups who possess court access and sufficient resources to use it. In other words, it is mostly the "haves" who benefit—those actors that already command considerable resources that enable them to broadly participate in political and legal processes.' Tanja A Börzel, 'Participation through Law Enforcement: The Case of the European Union' (2016) 39(1) Comparative Political Studies 128, 147.

[144] See eg Mark Aspinwall, 'Legal Mobilization Without Resources? How Civil Society Organizations Generate and Share Alternative Resources in Vulnerable Communities' (2021) 48(2) Journal of Law and Society 202.

[145] For an excellent discussion, see Camilla Hodgson, 'The Money Behind the Coming Wave of Climate Litigation' *The Financial Times* (5 June 2023) <www.ft.com/content/055ef9f4-5fb7-4746-bebd-7bfa00b20c82> accessed 25 July 2023.

[146] Joseph C Kiger, 'Philanthrophic Foundation' (*Encyclopedia Britannica*, updated 4 January 2016) <www.britannica.com/topic/philanthropic-foundation> accessed 11 August 2023.

billion a year before the onset of the global Covid-19 pandemic, mostly to health and development projects.[147] In 2021, the foundation boosted its spending to US$6.7 billion as it spent heavily on Covid vaccines, tests, and treatments.[148] A list of the top 10 philanthropic organizations, according to their global development contributions, shows that 9 out of 10 are in the United States (US) and one is in the United Kingdom (UK).[149] Apart from the Bill & Melinda Gates Foundation, other foundations on the list include Open Society Foundation, Ford Foundation, the William and Flora Hewlett Foundation, and the Children's Investment Fund Foundation.[150] These philanthropic organizations give grants to projects and individuals across Latin America, Africa, and Asia and their focus areas include democratic governance, reproductive health, environment and nature conservation, improving children's access to education, and climate change.

While there is growing interest amongst philanthropic organizations in climate change, including strategic legal interventions and litigation, the overall percentage of global philanthropic giving dedicated to climate change remains small. It is notable that climate change philanthropic funding has traditionally been directed to supporting scientific and technical solutions, with a focus on giving to major environmental NGOs in the Global North and on the reduction of GHG emissions.[151] ClimateWorks Foundation, in a 2021 report on funding trends for climate philanthropy, found that, in 2020, philanthropic giving for climate change mitigation was between US$6 and $10 billion.[152] This figure represented a growth from 2019

[147] 'Imagine What's Possible by Prioritizing Health Innovation' (*Bill & Melinda Gates Foundation*) <www.gatesfoundation.org/> accessed 2 March 2023. Joshua Chaffin and Andrew Edgecliffe-Johnson, 'The Political Awakening of Melinda French Gates' *Financial Times* (11 July 2021) <www.ft.com/content/6c6cf979-3bd3-4efd-ad57-2e4779e468c8> accessed 2 March 2021; the foundation is described as 'not just bigger than others in the sector but virtually a category unto itself'. The foundation's annual giving of US$5 billion is nearly 10 times the amount of the next largest American foundation.

[148] Andrew Edgecliffe-Johnson and Joshua Chaffin, 'Bill & Melinda Gates Foundation Plans Massive Giving Expansion' *Financial Times* (26 January 2022) <www.ft.com/content/45bb12a7-24ee-4f6e-b839-aafb9e46500d> accessed 2 March 2023.

[149] Noel Salazar, 'Top 10 Philanthropic Foundations: A Primer' (*Devex*, 1 August 2011) <www.devex.com/news/top-10-philanthropic-foundations-a-primer-75508> accessed 2 March 2023. Devex is a key media platform for the global development community and its subscribers include the United Kingdom's House of Commons, the World Economic Forum, the United Nations, and the Bill & Melinda Gates Foundation; 'Devex Delivers Insider Reporting that Drives the Global Development Agenda' (*Devex*) <https://pages.devex.com/about-devex> accessed 2 March 2023.

[150] Salazar (n 149).

[151] Matthew C Nisbet, 'Strategic Philanthropy in the Post-Cap-and-Trade Years: Reviewing U.S. Climate and Energy Foundation Funding' (2018) 9(4) Wiley Interdisciplinary Reviews: Climate Change 524. His study finds that funding favoured strategies pursued by 'insider' groups rather than 'outsider' groups that specialize in grassroots organizing. There was also a strong focus on promoting renewable energy technologies and energy efficiency although, following the failure of the 2010 federal cap and trade bill, philanthropists shifted to focus on state- or municipal-level mitigation and adaptation actions and to the needs of low-income/minority communities. Significant funding was also devoted to mobilizing public opinion and to opposing the fossil fuel industry but, overall, the bias towards particular technological solutions remained.

[152] Helene Desanlins and others, 'Funding Trends 2021: Climate Change Mitigation Philanthropy' (*Climateworks Foundation*, 7 October 2021) <www.climateworks.org/report/funding-trends-2021-climate-change-mitigation-philanthropy/> accessed 7 March 2023.

figures but was still less than 2% of overall global philanthropic giving for 2020. Of this US$6–10 billion, the report estimated that approximately US$100 million was directed to governance, diplomacy, and legal activities, with the latter including strategic litigation efforts.[153] Within this category of funding, most of the money still flows to the Global North (US$43 million across Europe and the United States), with only a fraction going to countries in the Global South (eg US$2.6 million to India, US$2.1 million to Brazil, and US$0.2 million for Africa as a region).[154] However, there is evidence of increasing interest among philanthropic funders in a more holistic approach that takes climate justice into account in decisions to fund projects and initiatives.[155] Philanthropic organizations involved in funding climate litigation are part of this broader trend, even though they remain small in number.[156] It can be argued that with greater knowledge of the role that litigation is playing to put pressure on corporations and governments to take concrete measures to address climate change and that '[w]hat donors invest in strategic litigation can yield a big result and offer a very good return',[157] philanthropic organizations will increase their funding of climate change litigation. This prospect is particularly relevant in the Global South where the shift away from carbon-intensive activities and fossil fuels is beset with delays, ambiguous policy signals, and doubt about the viability of achieving economic development without reliance on fossil fuels.[158]

The other key actor in the climate change philanthropy space is the 'regrantor'. A 'regrantor' refers to an organization that receives grants from philanthropic organizations and makes onward grants to entities engaged in climate change litigation and broader legal mobilization efforts. A 'regrantor' therefore does not bring litigation in its own name and usually plays a relatively invisible role in

[153] ibid.

[154] ibid.

[155] Seema Shah, 'Centering Equity and Justice in Climate Philanthropy' (*Candid Learning*, 25 May 2022) <https://learningforfunders.candid.org/wp-content/uploads/sites/2/2022/06/Climate-Justice-Funders-Guide.pdf> accessed 7 March 2023.

[156] Examples include Grata Fund (Australia), Urgenda, Sunrise Project, Children's Investment Fund Foundation (CIFF), and Arcadia. CIFF is an early mover amongst philanthropic organizations in providing funding for strategic climate litigation. It has made grants amounting to US$83.6 million in the area of climate litigation. This includes funding ClientEarth in Europe, the Center for Climate Integrity, the Institute for Governance and Sustainable Development, and the Foundation for International Law for the Environment (FILE); 'Grant Portfolio' (*Children's Investment Fund Foundation*) <https://ciff.org/grant-portfolio/> accessed 2 March 2023.

[157] Martin O'Brien, 'Making the Case for Strategic Litigation' in Susan Hansen, *Atlantic Insights: Strategic Litigation* (The Atlantic Philanthropies 2018) <www.atlanticphilanthropies.org/insights/insights-books/strategic-litigation> accessed 2 March 2023, 11.

[158] David Pilling, 'Can Africa Grow Without Fossil Fuels?' *Financial Times* (1 June 2022) <www.ft.com/content/1e8c12fe-4823-41a1-8069-b6150876427d> accessed 2 March 2023; Chun Sheng Goh, 'Building Indonesia's "Green" New Capital Could See Coal Use Surge (Analysis)' (*Mongabay*, 2 August 2022) <https://news.mongabay.com/2022/08/building-indonesias-green-new-capital-could-see-coal-use-surge-analysis/> accessed 2 March 2023; Alan Smith, 'Climate Graphic of the Week: Asia-Pacific Remains Stuck on Coal for Half Its Power Needs' *Financial Times* (2 July 2022) <www.ft.com/content/f58cf05a-dbaa-4359-9ae7-a1da37449f48> accessed 2 March 2023.

supporting and enabling climate change litigation. An example of one of the largest regranting entities globally with a focus on climate litigation is the Foundation for International Law for the Environment (FILE). FILE describes itself as a 'not-for-profit philanthropic organisation working to accelerate legal action on climate change'.[159] Its portfolio of climate litigation includes more than 250 cases across 49 countries that span both the Global North and South.[160] FILE has a team of professional lawyers, grants managers, and strategists who work with over 50 organizations or grantees to support their efforts to use legal intervention to tackle climate change and biodiversity loss. Apart from (re)making grants to organizations to provide vital financial support for litigation, FILE also provides communications, field-building, and research support to amplify the impact of their grantees' work.[161]

A 'regrantor' like FILE sees itself as playing an intermediary role in connecting the world of philanthropy with that of climate change litigation. Many philanthropic organizations may not be willing to donate to climate change litigation efforts because they do not have the in-house expertise to evaluate funding proposals and to decide on questions of the type of cases to fund and where, for example. Regranting organizations, in contrast, usually possess the legal expertise and strategic litigation experience to answer such questions and therefore ensure that proposals which seek funding for climate change litigation are properly evaluated. Philanthropic organizations are therefore more willing to donate to a 'regrantor' based on the understanding that the 'regrantor' is a charitable or non-profit entity that has the technical and legal expertise to make decisions to support organizations in bringing climate change lawsuits.[162]

There are also regranting organizations that do not focus on climate litigation but support litigation alongside other programmes and initiatives. An example is the European Climate Foundation (ECF), which mentions on its website that, in responding to gaps in the French, German, British, and Spanish responses to climate change, it takes 'a broad approach to the types of interventions [it] funds—from studies, litigation, and political caucusing to activism and media engagement—because all of these activities build towards an ambitious and thoughtful movement' to address the climate emergency.[163]

[159] 'We bring together the power of people and the law to compel governments and corporations to take action on climate change.' (*Foundation for International Law for the Environment*) <www.filefoundation.org/> accessed 2 March 2023.

[160] ibid.

[161] ibid.

[162] Interview 10.

[163] 'Spain' (*European Climate Foundation*) <https://europeanclimate.org/expertises/spain/> accessed 2 March 2023; 'Germany' (*European Climate Foundation*) <https://europeanclimate.org/expertises/germany/> accessed 2 March 2023; 'France' (*European Climate Foundation*) <https://europeanclimate.org/expertises/france/> accessed 2 March 2023; 'United Kingdom' (*European Climate Foundation*) <https://europeanclimate.org/expertises/united-kingdom/> accessed 2 March 2023. Over 80% of the European Climate Foundation's funding is 'regranted' to support the work of their partners. In 2021,

It can be argued that 'regrantors' are far from being passive actors in the climate change litigation arena. 'Regrantors' can play an important role in shaping the types of litigation undertaken through their decisions to fund certain cases and not others. The nature of certain types of cases also renders them more dependent on funding made available by 'regrantors'. An example would be cases that challenge governments' macro-level climate policy responses, also known as framework litigation. It has been pointed out that '[t]he future of framework litigation is likely to be contingent on the continued availability of significant philanthropic funding, given that most of these cases do not seek monetary damages and so are less likely to attract the attention of commercial litigation funders, in contrast with other fields of climate litigation'.[164]

'Regrantors' can also influence the geography of climate change litigation by directing their financial resources towards certain jurisdictions or focusing their attention on a particular region. For example, Tara Climate Foundation is a 'regrantor' with a mission to support organizations working to accelerate Asia's energy transformation.[165] Its regional focus is East, Southeast, and South Asia excluding India and China.[166] Tara was initially established as an initiative of the ECF and, in 2022, became an independent foundation based in Singapore.[167] The African Climate Foundation is 'the first African led strategic climate change grantmaking foundation on the continent' which 'was established to provide a mechanism through which philanthropies can contribute to Africa's efforts to address climate change'.[168] As most regranting organizations draw their philanthropic support from the Global North, there is the risk that climate change litigation seeded in the Global South with such philanthropic backing may reflect Northern priorities. Some common strategies adopted by 'regrantors' to reduce such risks include

it made 1,177 grants to 713 grantee organizations. It received funding from, inter alia, Bloomberg Philanthropies, Ikea Foundation, and Climateworks Foundation; 'Funders | Grantees' (*European Climate Foundation*) <https://europeanclimate.org/funding-grantmaking/> accessed 2 March 2023.

[164] Catherine Higham, Joana Setzer, and Emily Bradeen, 'Challenging Government Responses to Climate Change through Framework Litigation' (*Grantham Research Institute on Climate Change and the Environment* and *Centre for Climate Change Economics and Policy*, 7 September 2022) 11 <www.lse.ac.uk/granthaminstitute/publication/challenging-government-responses-to-climate-change-through-framework-litigation/> accessed 2 March 2023.

[165] 'Accelerating Asia's Clean Energy Transformation' (*Tara*) <www.taraclimate.org/> accessed 2 March 2023.

[166] ibid.

[167] ibid. Tara's website does not provide any information on whether it has made grants to support climate change litigation. However, Tara advertised on a public platform in August 2022 that it was seeking to hire a senior lawyer in Asia who would, amongst other things, 'support FILE's Head of Asia Pacific Strategy and team to develop FILE's legal strategy across the region and build and strengthen the network of climate-focused lawyers in the region through funding, developing compelling legal strategies, and building capacity and knowledge across the region'; on file with authors.

[168] 'News & Analysis' (*The African Climate Foundation*) <https://africanclimatefoundation.org/> accessed 2 March 2023.

engagement and collaboration with a wide range of local stakeholders, grassroots organizations, regional experts, and local law firms.[169]

For those outside the system, it is difficult to get a sense of how much Global South climate litigation is being externally funded by philanthropic organizations, for what purposes, and how this might be shaping the goals and outcomes of the cases that are brought. There is a paucity of publicly available data on philanthropic funding support for climate litigation—beyond headline data like the number of cases supported—and strategy documents and evaluations of programmes remain confidential. From the little that is publicly available, for example on FILE's website, we can surmise some key themes, including a focus on funding cases that bring about systemic change. FILE's multiple areas of support—from grant-making to provision of in-house legal expertise, communications and research support, and 'field-building' activities [170]—point to its broader influence over the landscape of climate litigation than purely acting as a 'regrantor'. On its website, FILE describes itself as having these additional roles, using terms like 'connector', 'incubator', 'field-builder', 'co-strategist', and 'protector'.[171] This has implications, not just for the cases that may emerge in the Global South, but also for the kinds of impacts they are directed to producing, a subject to which we return in Chapter 6.

## 2.5 Lawyers

The law is often described as a profession, referring to 'any type of work which needs special training or a particular skill, often one which is respected because it involves a high level of education'.[172] The eminent American jurist, Roscoe Pound, highlighted that the term *profession* 'refers to a group pursuing a learned art as a common calling in the spirit of public service—no less a public service because it may incidentally be a means of livelihood'.[173] This idea of rendering service for the benefit of others and to improve society is intended to lie at the crux of the legal profession although the rise of big multinational law firms and the nature of commercial legal practice today has led many to question whether law is, in reality, a

[169] This is mentioned on the websites of FILE, Tara, and the European Climate Foundation.
[170] Foundation for International Law for the Environment (n 159).
[171] ibid.
[172] *Cambridge International Dictionary of English*, 1995. The word *profession* is derived from the Latin *professio* or *professionem* which means to make a public declaration. The term evolved to describe an occupation that requires new entrants to take an oath professing their dedication to the ideals associated with that calling; Edward D Re, 'The Profession of the Law' (2000) 15(2) Journal of Civil Rights and Economic Development 109, 110.
[173] Roscoe Pound, *The Lawyer from Antiquity to Modern Times* (1st edn, West Publishing Company 1953) 5.

business.[174] A rich body of literature on lawyers arguing strategic human rights cases has highlighted how these lawyers identify themselves with the cause.[175] Similarly, many lawyers in the climate litigation space are motivated by the desire to make a difference by using their specialist knowledge to mobilize the law to address climate change.

The existence of public interest litigation frameworks in South Asia (Sri Lanka, Bangladesh, India, Nepal, and Pakistan) reduces barriers such as high costs and complex court procedures (although, of course, there are still tremendous practical hurdles limiting the access of the poor to the courts). This also makes it easier for motivated lawyers to file lawsuits in their own name in pursuit of what they believe is the public interest.[176]

Nepal's first climate lawsuit was filed by an environmental lawyer, Padam Bahadur Shrestha, who petitioned the Supreme Court to direct the government to enact climate change legislation as well as implement existing climate mitigation and adaptation policies.[177] The Supreme Court ruled in Padam Bahadur Shrestha's favour and held that Nepal, being a signatory to the UN Framework Convention on Climate Change and the Paris Agreement, was legally obliged to enact domestic laws 'which are parallel to international standards'.[178] This landmark decision has been hailed as one of a handful of judgments which clarify that climate mitigation and adaptation is the responsibility of the state according to the principle of *parens patriae*.[179] Although a climate change Act has yet to be enacted in Nepal, the government responded to the Supreme Court's decision by passing the Environment Protection Act of 2019 and the Forests Act of 2019, both aimed at addressing the need for climate mitigation and adaptation measures.[180] Padam Bahadur Shrestha

174 Champ S Answers, 'The Law. A Business or a Profession?' (1908) 17(8) Yale Law Journal 602; Mark A Cohen, 'Law Is a Profession and an Industry—It Should Be Regulated That Way' *Forbes* (29 March 2018) <www.forbes.com/sites/markcohen1/2018/03/29/law-is-a-profession-and-an-industry-it-should-be-regulated-that-way/?sh=4738347f6598> accessed 3 March 2023; Norman Bowie, 'The Law: From a Profession to a Business' (1988) 41(4) Vanderbilt Law Review 741.

175 Austin Sarat and Stuart A Scheingold (eds), *Cause Lawyers and Social Movements* (Stanford University Press 2006).

176 See eg Michael G Faure and AV Raja, 'Effectiveness of Environmental Public Interest Litigation in India: Determining The Key Variables' (2010) 21(2) Fordham Environmental Law Journal 239, 269: 'Citizens are willing to invest a lot of time and money in bringing cases in the public interest even though the results will potentially and subsequently allow a large number of third parties to free ride and benefit at the litigants' expense. Given the high number of cases that have effectively been brought, usually by a small group of individuals, one can argue that altruistic motives have provided sufficient incentives to bring these cases in the public interest.'

177 *Advocate Padam Bahadur Shrestha v The Office of the Prime Minister and Council of Ministers, Singhadurbar, Kathmandu*, Decision No 10210, NKP, Part 61, Vol 3 (2018).

178 ibid [12].

179 Selvaraj Puwanitha, 'Case Review of *Advocate Padam Bahadur Shrestha v The Office of the Prime Minister and Council of Ministers, Singhadurbar, Kathmandu and Others*' (2021) 1 South Asian Environmental Law Reports 123 <https://ejustice.lk/south-asia-environmental-law-report-volume-1/> accessed 3 March 2023.

180 The Environment Protection Act 2019, Act No 9 of 2019; The Forests Act 2019, Act No 12 of 2019; Dawn, 'Nepal to Draft Laws to Address Climate Change' *Asia News Network* (29 January 2019) <https://asianews.network/nepal-to-draft-laws-to-address-climate-change/> accessed 3 March 2023.

is vocal about his commitment to the pursuit of climate justice. On his personal website, he writes: 'With global warming and climate change plaguing Nepal and the world at large, I have been working to advocate for judicial, legislative and policy interventions to tackle the looming crisis with a deep sense of responsibility and conviction.'[181]

In India, Gaurav Kumar Bansal, who has been described as a 'Supreme Court advocate, wildlife activist and conservationist', has filed several environmental public interest lawsuits and one which has been included in our analysis as a climate lawsuit.[182] In *Gaurav Kumar Bansal v Union of India*,[183] the petitioner asked the National Green Tribunal to direct the central and state governments to show the steps taken by them to implement the National Action Plan on Climate Change.[184] It has also been due to Gaurav Kumar Bansal's filing of public interest lawsuits that the Delhi High Court directed the Delhi Police to provide gender awareness training for its officers[185] and that the Supreme Court has asked Uttarakhand and six other states to explain their failure to implement the Disaster Management Act.[186] Motivated by the desire to address injustice and environmental challenges, Gaurav Kumar Bansal often waives his fees as his clients are too poor to afford them. In an interview, he recalled a judge asking how he managed financially given the type of cases he generally took up and shared his reply to the judge: 'I told him that I earned enough to get my four *rotis* and *dal*. What else does a man really need?'[187]

There are also lawyers who do not bring cases in their own names but provide legal representation in climate change lawsuits. What is distinctive about these

[181] Advocate Padam Bahadur Shrestha (*Advocate Padam Bahadur Shrestha Environment Lawyer*) <https://padamshrestha.com/> accessed 3 March 2023

[182] Vineet Upadhyay, 'IFS Officer and Magsaysay Award Recipient Chaturvedi Withdraws from Corbett Probe' *The New Indian Express* (9 November 2021) <www.newindianexpress.com/nation/2021/nov/09/ifs-officer-and-magsaysay-award-recipient-chaturvedi-withdraws-from-corbett-probe-2381368.html> accessed 3 March 2023; *Gaurav Kumar Bansal v Union of India*, National Green Tribunal, 25 March 2019; *Gaurav Kumar Bansal v Union of India* MANU/GT/0326/2020, National Green Tribunal, 23 December 2020.

[183] *Guarav Kumar Bansal v Union of India*, National Green Tribunal, Original Application No 498 of 2014, Orders of the Tribunal, 23 July 2015 <www.indiaenvironmentportal.org.in/files/state%20action%20plan%20on%20climate%20change.pdf> accessed 3 March 2023.

[184] In its final order, the Tribunal directed states that had yet to finalize their action plans in accordance with the National Plan to prepare and submit these plans to the Ministry of Environment, Forest and Climate Change for approval expeditiously; ibid.

[185] ibid.

[186] ibid.

[187] Soni Mishra, 'Sparkle, Without an Uncle' *The Week* (14 June 2015) <www.theweek.in/theweek/specials/young-legal-professionals.html> accessed 3 March 2023. India has a long history of environmental public interest litigation (EPIL) stalwarts, the most prominent being MC Mehta. Lavanya Rajamani describes the existence of a 'clique of public interest litigators and petitioners' who have performed a valuable role as public watchdogs but raised concerns about whether 'their leverage with the Court may result in converting "one strain of opinion into policy, while annihilating all others"'; Lavanya Rajamani, 'Public Interest Environmental Litigation in India: Exploring Issues of Access, Participation, Equity, Effectiveness and Sustainability' (2007) 19(3) Journal of Environmental Law 293, 305.

lawyers is that they are activist lawyers working with civil society organizations, environmentalists, and other stakeholders to pursue strategic litigation as part of a wider social mobilization effort to address climate change. These lawyers do not simply act on their clients' instructions and stay behind the scenes. To the contrary, these lawyers are publicly vocal about their beliefs about the climate change emergency and their commitment to use their legal expertise to pursue the societal changes they believe are necessary. They speak to international media about the cases they work on, participate in global networks, and use social media to raise awareness.

One such lawyer is Melinda Janki, 'the environmental lawyer taking on Big Oil in Guyana'.[188] Melinda Janki is an international human rights and environmental lawyer. She started her legal career in London where she advised international financial institutions, governments, and corporations.[189] Her past work includes drafting Guyana's environmental impact assessment laws, the drafting of the Escazú Agreement, and acting for the WaiWai in the creation of a protected area over their territory of 2,300 square miles.[190] As stated on her personal website, Melinda Janki's work is currently focused on challenging a massive ExxonMobil-led oil and gas project off Guyana's coast, where the company is planning to extract over nine billion barrels of oil and trillions of cubic feet of gas.[191]

Melinda Janki is representing two Guyanese citizens who have filed the first constitutional climate lawsuit in the Caribbean to challenge fossil fuel production on the grounds that it worsens climate change and violates human rights, including the right to a healthy environment and the rights of future generations.[192] Quadad

[188] Melinda Janki, 'Introducing Melinda Janki, the Environmental Lawyer Taking on Big Oil in Guyana' (*YouTube*, 2 July 2021) <www.youtube.com/watch?v=7Avp__oCVqM> accessed 3 March 2023.

[189] 'Melinda Janki' (*Melinda Janki*) <https://melindajanki.wordpress.com/> accessed 3 March 2023.

[190] ibid. The Escazú Agreement refers to the United Nations, Regional Agreement on Access to Information, Public Participation and Justice in Environmental Matters in Latin America and the Caribbean. The creation and management of Kanashen, home to the Wai Wai and the only Indigenous-owned territory in Guyana's protected area system, has attracted global attention. See eg Gaulbert Sutherland, 'Wai-Wais Ink Historic Forest Conservation Deal (Guyana)' (*Institute for Agriculture & Trade Policy*, 3 October 2007) <www.iatp.org/news/wai-wais-ink-historic-forest-conservation-deal-guyana> accessed 3 March 2023; WWF Factsheet, 'Significant Stories: Empowering Communities Through Participatory MRV in Guyana' (*WWF Forest and Climate Programme*, 2016) <https://wwfint.awsassets.panda.org/downloads/guyana_significantstories__online.pdf> accessed 3 March 2023; Gemma Handy, 'Guyana tribe goes hi-tech to protect its land' (*BBC*, 19 March 2017) <www.bbc.com/news/world-latin-america-39282922> accessed 3 March 2023; Jessica George-Joseph, 'When Indigenous Lands Become Part of a National Protected Area System: The Unique Case of the Kanashen Protected Area of Southern Guyana' *IUCN World Conservation Congress Marseille News* (27 August 2021) <https://www.iucncongress2020.org/newsroom/all-news/when-indigenous-lands-become-part-national-protected-area-system-unique-case> accessed 3 March 2023.

[191] Melinda Janki (n 188).

[192] An excellent case summary of *Thomas & De Freitas v Guyana* is available here: <www.justiceinstituteguyana.org/wp-content/uploads/2021/08/Thomas-De-Freitas-v-AG-case-summary.pdf> accessed 3 March 2023; Cate Bonacini, 'Guyanese Citizens File Climate Case Claiming Massive Offshore Oil Project is Unconstitutional' (*Centre for International Environmental Law*, 21 May 2021) <www.ciel.org/news/guyana-consitutional-court-case-oil-and-gas/> accessed 3 March 2023; 'Guyana: Citizens

de Freitas, an Indigenous youth, and Troy Thomas, a scientist and university lecturer, filed their constitutional challenge on 21 May 2021.[193] Melinda Janki is also acting for the plaintiffs in two other lawsuits that challenge the legality of oil exploration permits issued to Esso Exploration (a subsidiary of ExxonMobil)[194] as well as the Guyanese Environmental Protection Agency's decision to grant Esso Exploration a modified permit that allows it to flare gas in exchange for payment of a fee.[195] She has also written legal letters to ExxonMobil's shareholders to scrutinize the company's investments in Guyana, alerting them to the risks of stranded assets and loss of shareholder value due to ongoing litigation and the inevitable transition to renewable energy.[196] Melinda Janki regularly gives interviews in global newspapers and publishes updates on her cases, thereby raising the profile of the three cases challenging oil and gas production plans in Guyana and which currently make up the country's climate litigation docket.[197]

Finally, there are the NGOs that are staffed by lawyers. They usually do not bring cases in their own name but provide legal advice to the individuals and communities that wish to file climate change lawsuits. For example, Dejusticia, also known as the Center for Law, Justice and Society, is a non-profit legal organization based in Colombia. It is 'dedicated to the strengthening of the rule of law and the promotion of social justice and human rights in Colombia and the Global South'. Its work includes 'litigating in the most impactful forums'[198] and that includes its legal representation of the youth plaintiffs in *Future Generations v Ministry of the Environment and Others*.[199] Like many other actors involved in climate litigation, Dejusticia does not enter this space without significant experience and an established track record

File Climate Case about Exxonmobil Massive Project Concerning Its Social and Environmental Impacts' (*Business & Human Rights Resource Centre*, 24 May 2021) <www.business-humanrights.org/en/latest-news/guyana-citizens-file-climate-case-about-exxonmobil-massive-project-concerning-its-social-and-environmental-impacts/> accessed 3 March 2023.

193 The Supreme Court adjourned the matter until June 2022. At the time of writing, there are no further updates.

194 *Thomas v Environmental Protection Agency* (2020-HC-DEM-CIV-FDA-460). The Supreme Court ultimately accepted a settlement agreement between Thomas, the Environmental Protection Agency, and Esso.

195 *Henry v Environmental Protection Agency* (2021-HC-DEM-CIV-FDA).

196 Melinda Janki, 'Open Letter to Members of the Coalition United for a Responsible Exxon (CURE) and to Other Concerned Exxon Shareholders' (*Oil and Gas Governance Network*, 24 May 2021) <www.oggn.org/wp-content/uploads/2021/06/Final_Guyana-letter-to-Exxon-shareholders-with-intl-endorsements_sent-26May2021-to-CURE-Exxon.pdf> accessed 6 March 2023.

197 See eg Melinda Janki, 'Guyana: A First Step Towards Climate Litigation' (*Climate Law: A Sabin Centre Blog*, 29 December 2020) <https://blogs.law.columbia.edu/climatechange/2020/12/29/guyana-a-first-step-towards-climate-litigation/> accessed 6 March 2023; Anastasia Moloney, 'Climate Lawsuits Snowball as South Americans Seek a Healthy Environment' *Reuters* (1 June 2021) <www.reuters.com/article/us-climate-change-litigation-latam-idUSKCN2DD25E> accessed 6 March 2023; Antonia Juhasz, 'The Quest to Defuse Guyana's Carbon Bomb' (*Wired*, 20 December 2022) <www.wired.com/story/the-quest-to-defuse-carbon-bomb-guyana/> accessed 6 March 2023.

198 'About Us' (*Dejusticia*) <www.dejusticia.org/en/about/> accessed 6 March 2023.

199 *Demanda Generaciones Futuras v Minambiente* STC4360-2018 (Supreme Court, Republic of Colombia).

in environmental litigation or strategic litigation in other areas such as human rights and constitutional law.[200] As one of Dejusticia's founders, César Rodríguez Garavito, who is also the Founding Director of the Climate Litigation Accelerator at New York University School of Law, has argued, the Global South's

> route to climate litigation is not serendipitous, or the result of the absence of specialized climate change legislation that litigants would otherwise have used in framing their cases. Instead, it is a route whose tracks were firmly laid over the last three decades through public interest law practice, research, and judicial activism regarding constitutional rights in general and socioeconomic rights (SERs) in particular.[201]

Another example is the NGO known as LIFE, founded by lawyers Ritwick Dutta and Rahul Choudhary in 2005. A veteran environmental lawyer who has litigated over 350 cases, one of the earliest cases Dutta is known for is one against the mining company Vedanta, in which he represented the Dongria Kondh tribals seeking a ban on bauxite mining in the Niyamgiri hills in southwest Odisha.[202] LIFE 'aims to protect the environment through the creative use of law and legal processes in India'.[203] The organization's focus is on working alongside affected communities; 'For this reason, LIFE is never the litigant. Instead, the organisation sees itself as providing legal and scientific support for grassroots groups fighting on the frontlines of environmental protection.'[204]

LIFE describes itself as 'a national level public interest environmental law group which applies a unique combination of law, science and advocacy to protect biodiversity hotspots, flora and fauna, clean up air and water and protect vulnerable communities'. It also seeks to ensure that India 'adopts and implements a low

[200] For more information about Dejusticia's other cases and focus areas, see 'Litigation' (*Dejusticia*) <www.dejusticia.org/en/how-we-work/litigation/> accessed 6 March 2023.

[201] César Rodríguez-Garavito, 'Human Rights: The Global South's Route to Climate Litigation' (2020) 114 American Journal of International Law Unbound 40. For details about the Climate Litigation Accelerator, see 'Climate Litigation Accelerator' (*Centre for Human Rights & Global Justice*) <https://chrgj.org/focus-areas/climate-and-environment/climate-litigation-accelerator/> accessed 6 March 2023.

[202] Megha Mandavia, 'Environment Lawyer Ritwick Dutta's Moves Have India Inc See Red' *The Economic Times* (23 December 2013) <https://economictimes.indiatimes.com/industry/indl-goods/svs/metals-mining/environment-lawyer-ritwick-duttas-moves-have-india-inc-see-red/articleshow/27763912.cms> accessed 6 March 2023; Prerna Singh Bindra, 'Ritwick Dutta | The case for our forests' (*mint*, 9 August 2014) <www.livemint.com/Leisure/5hzY1gykB3DD9VgvZTOSgJ/Ritwick-Dutta--The-case-for-our-forests.html> accessed 6 March 2023.

[203] Right Livelihood Laureates, 'Awarded 2021: Legal Initiative for Forest and Environment (LIFE)' (*Right Livelihood*) <https://rightlivelihood.org/the-change-makers/find-a-laureate/legal-initiative-for-forest-and-environment-life/> accessed 19 August 2023. See also Mayank Aggarwal, 'Interview: Renewable Energy Goals Can't Be Met by Excluding People or Making Them Hostile, Say Founders of LIFE' *Mongabay* (25 October 2021) <https://india.mongabay.com/2021/10/interview-renewable-energy-goals-cant-be-met-by-excluding-people-or-making-them-hostile-say-founders-of-life/> accessed 19 August 2023; Bindra, 'Ritwick Dutta | The case for our forests' (n 202).

[204] Right Livelihood Laureates (n 203).

carbon path of development and strengthening climate resilience'.[205] It has been focusing on India's environmental impact assessment process, specifically public consultation[206] and social and environmental impacts. LIFE provides scientific and legal support to affected communities and also provides capacity-building training for forest department officers. LIFE's legal research has also been used to support petitions filed before the National Green Tribunal, eg *Chandra Bhal Singh v Union of India*,[207] which was about effective implementation of the Biological Diversity Act. Ritwick Dutta has represented several plaintiffs and appellants in the cases considered part of India' climate litigation docket, including *Vimal Bhai and Others v Ministry of Environment and Forests*,[208] *Pandey v Union of India*,[209] and *Society for Protection of Environment and Biodiversity and Others v Union of India*.[210] He served as *amicus curiae* in *Court on its Own Motion v State of Himachal Pradesh*.[211]

A final example is the environmental law NGO, ClientEarth, that is well known in the Global North for bringing some of the most ground-breaking climate lawsuits against governments and companies. For example, *ClientEarth v Enea* is the first climate risk case in the world, brought by ClientEarth to challenge the construction of a major coal power plant in Poland.[212] As a shareholder in Enea, ClientEarth filed a legal challenge against the company's decision to pass a resolution authorizing the Ostroleka C coal power plant. A second example is ClientEarth's landmark climate case challenging the UK government's net zero strategy and the government's breach of its legal duties under the 2008 Climate Change Act.[213] The case was filed in January 2022. On 18 July 2022, the High Court ruled in ClientEarth's favour. It concluded that the government's net zero strategy was inadequate, that the government had breached the Climate Change Act, and that the minister who was responsible for authorizing the strategy did not have the

205 'About Us' (*Legal Initiative for Forest and Environment (LIFE)*) <https://thelifeindia.org.in/about-us/> accessed 6 March 2023. Also see Ritwick Dutta, 'Climate Change in the Courts: An Environmental Lawyer's Viewpoint' (2022) 55(3) Contributions to Indian Sociology 157.

206 On public consultation generally, see MP Ram Mohan and Himanshu Pabreja, 'Public Hearings in Environmental Clearance Process: Review of Judicial Intervention' (2016) 51(50) Economic and Political Weekly 68.

207 Original Application 347/2016 (IA No 185/2020, IA No 206/2020, and IA No 344/2020), IA No 355/2020, IA No 356/2020, IA No 357/2020, and IA No 410/2020, National Green Tribunal, 16 December 2020.

208 Appeal No 5 of 2011 (challenging the grant of a Forest Clearance permit for a hydroelectricity dam project), National Green Tribunal, 14 December 2011.

209 Original Application No 187/2017, National Green Tribunal, 15 January 2019.

210 Original Application 677/2016 (MA No 148/2017) (consolidation of five cases challenging the exemption of certain construction projects from environmental impact assessment laws), National Green Tribunal, 8 December 2017.

211 Original Application 488/2014 (MA No 1085/2015), National Green Tribunal, 1 August 2017.

212 The judgment in Polish and an English translation are available online at Sabin Centre for Climate Change Law: 'ClientEarth v. Enea' (*Sabin Center for Climate Change Law*) <https://climatecasechart.com/non-us-case/clientearth-v-enea/> accessed 19 August 2023.

213 *Friends of the Earth v Secretary of State for Business, Energy and Industrial Strategy* [2022] EWHC 1841 (Admin).

legally required information on how carbon budgets would be met if the strategy were to be implemented.[214] In October 2022, the UK government announced that it would not seek to appeal the High Court's ruling.[215]

ClientEarth describes itself as 'a charity that uses the power of the law to protect people and the planet . . . we work on laws throughout their lifetime, from the earliest stages to implementation. And when those laws are broken, we go to court to enforce them.'[216] ClientEarth's work in the Global North has a strong focus on litigation alongside other legal interventions. In contrast, ClientEarth's work in the Global South focuses on the earlier phases of the lifecycle of the law. This includes stakeholder engagement to support effective implementation of climate change laws, and training lawyers, public prosecutors, and judges.[217] In some instances, ClientEarth lawyers provide legal support to local communities and lawyers in their climate litigation efforts. An example would be the legal challenge that landslide victims in the Bududa district of Uganda filed against the government alleging that, by failing to act on the known landslide risks, the Ugandan government had violated their rights to life, to property, and to a clean and healthy environment.[218]

## 2.6 Indigenous Peoples

'Indigenous peoples', also sometimes referred to as 'first nations', 'tribal peoples', 'aboriginal', and 'native' peoples, have an estimated population of 476.6 million worldwide, comprising 6.2% of the global population.[219] There is no fixed, universal definition of 'Indigenous peoples' as it varies both across different countries[220] and

[214] ibid [28]–[59], [69]–[99], [147]–[154], [212]–[222], and [231]–[260].

[215] Adam Vaughan, 'Government Drops Appeal Over Net-Zero High Court Ruling' *The Times* (13 October 2022) <www.thetimes.co.uk/article/government-drops-appeal-over-net-zero-high-court-ruling-szrqlpfn2> accessed 6 March 2023.

[216] Alice Garton, Marcin Stoczkiewicz, and Peter Barnett, 'Briefing—Ostrołęka C: Energa's and Enea's Board Members' Fiduciary Duties to the Companies and Shareholders' (*ClientEarth*, 20 September 2018) <http://blogs2.law.columbia.edu/climate-change-litigation/wp-content/uploads/sites/16/non-us-case-documents/2018/20180920_Not-Available_na-1.pdf> accessed 6 March 2023, 10.

[217] 'China' (*ClientEarth*) <www.clientearth.org/our-global-reach/china/> accessed 6 March 2023; 'Asia & the Pacific' (*ClientEarth*) <www.clientearth.org/our-global-reach/asia-the-pacific/> accessed 6 March 2023.

[218] *Tsama William v Attorney General* (in the High Court of Uganda at Mbale, Misc Cause No 024 of 2020); ClientEarth Communications, 'Landslide Victims Take Ugandan Government to Court' (*ClientEarth*, 22 October 2020). <www.clientearth.org/latest/latest-updates/news/landslide-victims-take-ugandan-government-to-court/> accessed 6 March 2023.

[219] International Labour Organization, *Implementing the ILO Indigenous and Tribal Peoples Convention No 169: Towards an Inclusive, Sustainable and Just Future* (*International Labour Organization*, 2019) 13; Rodolfo Stavenhagen, 'Human Development Report Office Occasional Paper: Background Paper for HDR 2004 (2004/14): Indigenous Peoples in Comparative Perspective: Problems and Policies', United Nations Development Programme (2014) 1.

[220] Stavenhagen (n 219); Richard Madden and others, 'Indigenous Identification: Past, Present and a Possible Future' (2019) 35(1) Statistical Journal of the IAOS 23; Linzhu Wang, 'The Definition of

also across different international organizations.[221] Adopted by the UN General Assembly in September 2007, the UN Declaration on the Rights of Indigenous Peoples (UNDRIP) is the most comprehensive international instrument on the rights of Indigenous peoples.[222] The UNDRIP does not have a formal definition of Indigenous peoples. This absence of a formal definition was intentional given the diversity of Indigenous peoples in various countries.[223] The UN ultimately decided on self-identification as the basis for recognizing Indigenous peoples.[224] According to the UN Permanent Forum on Indigenous Issues (UNPFII), the identification of Indigenous peoples should be based on the following seven factors: (1) self-identification as Indigenous peoples at the individual level and accepted by the community as their member; (2) historical continuity with pre-colonial and/or pre-settler societies; (3) strong link to territories and surrounding natural resources; (4) distinct social, economic, or political systems; (5) distinct language, culture, and beliefs; (6) form non-dominant groups of society; and (7) resolve to maintain and reproduce their ancestral environments and systems as distinctive peoples and communities.[225]

Indigenous lands cover a quarter of the world's land surface area and overlap with some 40% of terrestrial protected areas and ecologically intact landscapes.[226] These lands are typically vulnerable to economic exploitation and environmental degradation, which threaten Indigenous rights and the integrity of Indigenous

Indigenous Peoples and its Applicability in China' (2015) 22(2) International Journal on Minority and Group Rights 232.

[221] See Douglas Sanders, 'Indigenous Peoples: Issues of Definition' (1999) 8 Irish Journal of Counselling and Psychotherapy 4.

[222] GA Res 61/295, UN Doc A/RES/61/295 (13 September 2007).

[223] Report of the Working Group on Indigenous Population on its fifteenth session, UN Doc E/CN.4/Sub.2/1997/14, [129] (13 August 1997).

[224] GA Res 61/295 (n 222) art 8.

[225] United Nations Permanent Forum on Indigenous Issues, 'Indigenous Peoples, Indigenous Voices: Factsheet' (*United Nations*) <www.un.org/esa/socdev/unpfii/documents/5session_factsheet1.pdf> accessed 6 March 2023. Kingsbury has pointed out that the requirement of 'historical continuity with pre-invasion and pre-colonial societies' may render the definition underinclusive for states which do not have a European colonial history; Benedict Kingsbury, '"Indigenous Peoples" in International Law: A Constructivist Approach to the Asian Controversy' (1998) 92(3) American Journal of International Law 414, 420. However, proponents of this requirement argue that the requirement of historical continuity is essential to secure goals such as decolonization and self-determination for indigenous peoples; see Irene Watson and Sharon Venne, 'Talking up Indigenous Peoples' Original Intent in a Space Dominated by State Interventions' in Elvira Pulitano (ed), *Indigenous Rights in the Age of the UN Declaration* (CUP 2012) 105.

[226] Stephen T Garnett and others, 'A Spatial Overview of the Global Importance of Indigenous Lands for Conservation' (2018) 1 Nature Sustainability 369. It should be noted that the overlap between Indigenous lands and protected areas has also caused controversy where Indigenous peoples have been displaced. See Stan Stevens (ed), *Indigenous Peoples, National Parks, and Protected Areas: A New Paradigm Linking Conservation, Culture, and Rights* (University of Arizona Press 2014) 1–6; Britta Sjöstedt, 'Rights of Indigenous Peoples and Environmental Protection in *Jus Post Bellum*' in Carsten Stahn and Jens Iverson (eds), *Just Peace After Conflict: Jus Post Bellum and the Justice of Peace* (OUP 2020).

lands.[227] In the face of encroachment on their rights and intrusion on their ways of life, Indigenous peoples have sought to defend their rights through a variety of methods including advocacy campaigns, protests, and litigation.[228] Coupled with the fact that many Indigenous peoples have eco-centric worldviews and cultural beliefs that are deeply intertwined with protecting nature, Indigenous peoples have been characterized as environmental stewards of their lands.[229] Tragically, a 2021 report by campaign group Global Witness estimated that over 40% of the 200 fatal attacks against environmental activists annually are inflicted on Indigenous peoples.[230]

There are several examples of Indigenous peoples who have used litigation to complement other efforts (including protest and advocacy) to protect their lands from environmentally destructive economic activities such as oil drilling and illegal logging.[231] One of the most prominent examples of Indigenous peoples-led litigation to protect the environment is the *Case of the Mayagna (Sumo) Awas Tingni Community v Nicaragua* filed by the Awas Tinggi Community in Nicaragua to the IACHR.[232] The community sought to challenge a logging grant that the

[227] Jérémie Gilbert, 'Indigenous Peoples and Litigation: Strategies for Legal Empowerment' (2020) 12(2) Journal of Human Rights Practice 301.

[228] Open Society Justice Initiative, 'Strategic Litigation Impacts: Indigenous Peoples' Land Rights' (*Open Society Foundations*, 2017) <www.justiceinitiative.org/publications/strategic-litigation-impacts-indigenous-peoples-land-rights> accessed 7 March 2023; see, in particular, discussions on protests and other forms of engagement in Section 5 'Impacts on Attitudes and Behaviours'.

[229] Christopher J O'Bryan and others, 'The Importance of Indigenous Peoples' Lands for the Conservation of Terrestrial Mammals' (2021) 35 Conservation Biology 1002; Monica Gratani and others, 'Indigenous Environmental Values as Human Values' (2016) 2(1) Cogent Social Sciences 1 <www.tandfonline.com/doi/full/10.1080/23311886.2016.1185811> accessed 6 March 2023. For example, the *Maori* people of New Zealand believe that they share a kinship with nature. This is guided by the belief of *kaitiakitanga* or a duty to protect their present environment as a display of reverence to their ancestors and to safeguard their future; 'Indigenous People and Nature: A Tradition of Conservation' (*United Nations Environment Programme*, 26 April 2017) <www.unep.org/news-and-stories/story/indigenous-people-and-nature-tradition-conservation> accessed 6 March 2023. The Woodlands Cree (also known as Nēhîthâwâk) in Canada believed that they 'personify' the land 'as our Mother Earth', walking on the dust created from their ancestors' bones; Herman Michell, 'Nēhîthâwâk of Reindeer Lake, Canada: Worldview, Epistemology and Relationships with the Natural World' (2005) 34(1) The Australian Journal of Indigenous Education 33, 38. However, one ought not to oversimplify and presume that all Indigenous peoples hold the belief of the stewardship of nature at the risk of doubting the 'authenticity' of Indigenous peoples who do not hold such beliefs; Noah Theriault, 'Environmental Politics and the Burden of Authenticity' in James F Eder and Oscar L Evangelista (eds), *Palawan and its Global Connections* (Ateneo de Manila University Press 2014) 363.

[230] Ali Hines, 'Decade of Defiance' (*Global Witness*, 29 September 2022) <www.globalwitness.org/en/campaigns/environmental-activists/decade-defiance/> accessed 6 March 2023. This group was recognized in 'Indigenous People and Nature: A Tradition of Conservation' (*United Nations Environment Programme*, 26 April 2017) <www.unep.org/news-and-stories/story/indigenous-people-and-nature-tradition-conservation> accessed 6 March 2023.

[231] The UN Environment Programme's Law and Environment Assistance Platform has documented more than a hundred environmental cases filed by or on behalf of Indigenous peoples as of January 2023. See 'Legislation and Case-Law' (*United Nations Environment Programme and Montevideo Environmental Law Programme (LEAP)*) <https://leap.unep.org/knowledge/legislation-and-case-law?search_api_fulltext=indigenous%20peoples&sort_by=aggregated_date&sort_order=DESC&f%5B0%5D=legislative_cases_content_type%3Acase_law> accessed 6 March 2023.

[232] *Case of the Mayagna (Sumo) Awas Tingni Community v Nicaragua* (Merits, Reparations and Costs), Inter-American Court of Human Rights Series C No 79 (Judgment of 31 August 2001).

Nicaraguan government intended to issue to commercially develop 62,000 hectares of tropical forest.[233] The plaintiffs argued that the government failed to protect the community's property rights over its ancestral lands and natural resources.[234] The IACHR held in favour of the Indigenous plaintiffs that Nicaragua had violated their rights to judicial protection and property under the American Convention on Human Rights.[235] Legal challenges against logging have also been brought in Papua New Guinea, where Indigenous landowners challenged land acquisition for palm oil development on the basis that there was no proper consent and consultation,[236] and in Belize where Indigenous peoples claimed customary land tenure over land that was authorized for oil drilling and logging.[237] A case in India brought on behalf of the Monpa community also successfully challenged the construction of a hydroelectric dam on the basis that the environmental impact assessment conducted for the proposed project had been flawed.[238]

It is noteworthy that the recognition of legal personhood of natural entities such as rivers is attributable in part to the efforts of Indigenous peoples.[239] A high-profile example to date is the granting of legal personality to the Whanganui River in New Zealand as part of a settlement claim.[240] In Colombia, a historic lawsuit brought by Indigenous and Afro-Colombian communities led to Ruling T-622 in 2017—a landmark decision by the Colombian Constitutional Court granting legal rights to the Atrato river in Colombia's Chocó region.[241] One of the court's orders was for the formation of a commission of guardians of the river. Ever since the issuance of the ruling, the communities have come up with creative ways to engage state institutions, the church, youths, and Indigenous organizations in their uphill battle to implement the ruling in the face of climate change, illegal mining, and lax enforcement of environmental laws (that allows the degradation of the Chocó bio-region to continue).[242]

[233] ibid [6].
[234] ibid [2].
[235] ibid [173].
[236] *Maniwa v Malijiwi* [2014] PGNC 25 (Papua New Guinea).
[237] *The Maya Leaders Alliance v The Attorney General of Belize* [2015] CCJ 15 (AJ), CCJ Appeal No BZCV2014/002, BZ Civil Appeal No 27 of 2010, 30 October 2015.
[238] *Save Mon Region Federation v Union of India*, MA No 104 of 2012 (Arising out of Appeal No 39 of 2012), National Green Tribunal, 14 March 2013.
[239] Elizabeth Macpherson, Julia Torres Ventura, and Felipe Clavijo Ospina, 'Constitutional Law, Ecosystems, and Indigenous Peoples in Colombia: Biocultural Rights and Legal Subjects' (2020) 9(3) Transnational Environmental Law 521. See also Mihnea Tănăsescu, 'Rights of Nature, Legal Personality, and Indigenous Philosophies' (2020) 9(3) Transnational Environmental Law 429 (on how the recognition of legal personality is strategic rather than genealogical for indigenous peoples).
[240] Te Awa Tupua (Whanganui River Claims Settlement) Act 2016 (New Zealand).
[241] *Atrato River Decision* T-622/16 of November 10, 2016 (Sentencia T-622/16 de Noviembre 10, 2016) (Colombia).
[242] For an excellent discussion, see Brian Roewe, 'In Colombia's Chocó Region, "River Guardians" Protect the Rights of the Rio Atrato' (*Earthbeat*, 28 January 2022) <www.ncronline.org/news/earthbeat/colombias-choc-region-river-guardians-protect-rights-rio-atrato> accessed 6 March 2023; Philipp Wesche, 'Rights of Nature in Practice: A Case Study on the Impacts of the Colombian Atrato River

Climate change disproportionately impacts Indigenous peoples.[243] For instance, climate change worsens the socio-economic conditions of Indigenous peoples and is likely to threaten their access to land and natural resources.[244] It is thus unsurprising that Indigenous peoples have been active in litigating climate change issues in both the Global North and the Global South.[245] In fact, it has been argued that the initiative which kickstarted the global movement of climate change litigation stemmed from an Indigenous peoples-led case. The Inuit people, led by Shelia Watt-Cloutier and on behalf of all Inuit people in the Arctic regions, sought to compel the US to join the Kyoto Protocol in a petition to the IACHR.[246] They argued that the US's emissions amounted to a breach of human rights accorded to the Inuit people as it led to the loss of permafrost, and consequently, the loss of the ways of life of the Inuit people, given its adverse effects on housing, communication, safety, and food gathering.[247] While the IACHR's response was disappointing, this petition can be viewed as the birth of the movement of seeking legal protection of rights accorded by international treaties in domestic and international courts.[248]

In the Global South docket, there are a number of notable cases that have been brought by Indigenous peoples on their own or in partnership with other plaintiffs. In *Baihua Caiga and Others v PetroOriental SA*, Indigenous plaintiffs and environmental and human rights NGOs in Ecuador filed for a constitutional injunction against a Chinese oil company. They argued that the company's gas flaring

Decision' (2021) 33(3) Journal of Environmental Law 531; Diego Cagüeñas and others, 'The Atrato River and Its Guardians: Ecopolitical Imagination for Weaving New Rights' (2020) 56(2) Revista Colombiana de Antropología 169; Paola Villavicencio Calzadilla, 'A Paradigm Shift In Courts' View On Nature: The Atrato River And Amazon Basin Cases In Colombia' (2019) 15(1) Law, Environment and Development Journal 49.

[243] Laura Westra, *Environmental Justice and the Rights of Indigenous Peoples: International and Domestic Legal Perspectives* (Earthscan 2008).

[244] Eugenia Recio and Dina Hestad, 'Indigenous Peoples: Defending an Environment for All' (*International Institute for Sustainable Development*, 22 April 2022) <www.iisd.org/articles/deep-dive/indigenous-peoples-defending-environment-all> accessed 6 March 2023.

[245] For a discussion on selected Global North cases, see Sam Bookman, 'Indigenous Climate Litigation in Anglophone Settler-Colonial States: Context, Cases and Caution' (*Verfassungsblog*, 25 March 2022) <https://verfassungsblog.de/indigenous-climate-litigation-in-anglophone-settler-colonial-states/> accessed 6 March 2023. Global North cases include: *Nature Conservation Council v New South Wales (NSW) Nature Conservation Council of NSW v Minister for Water, Property and Housing*, 2021/00282599 (Australia); *Mataatua District Maori Council v New Zealand*, 2016 WAI 2607 (New Zealand); *Pabai Pabai and Guy Paul Kabai v Commonwealth of Australia*, VID622/2021 (Australia); *Lho'imggin v Her Majesty the Queen*, 2020 FC 1059 (Canada); *Kang v KSURE and KEXIM* (2022) (South Korea); *SkeenaWild Conservation Trust v Government of Canada* (2016), Court File No T-1836-16 (Canada).

[246] See discussion in Randall S Abate, *Climate Change and the Voiceless: Protecting Future Generations, Wildlife, and Natural Resources* (CUP 2019) 17–42.

[247] *Petition to the Inter American Commission on Human Rights Seeking Relief from Violations Resulting From Global Warming Caused by Acts and Omissions of the United States* (7 December 2005) 2–3.

[248] Elizabeth A K Warner and Randall S Abate, 'International and Domestic Law Dimensions of Climate Justice for Arctic Indigenous Peoples' (2013) 43(113) Ottawa Law Review 113, 136.

activities contribute to climate change and violate their constitutional rights, such as the rights to nature, a healthy environment, food, water, health, land, and life. They sought an injunction from the court to prohibit gas flaring and for the defendants to pay reparations by financing adaptive climate policies. The case was held to be inadmissible as the plaintiffs failed to demonstrate how the defendant had violated their constitutional rights.[249]

The plaintiffs in *Thomas & De Freitas v Guyana* were an Indigenous youth from the Rupununi region of Guyana and a scientist who brought the case on their behalf and in the public interest.[250] They argued that the state of Guyana had violated their constitutional rights to a healthy environment, sustainable development, and the rights of future generations by approving oil exploration licences to a local subsidiary of ExxonMobil. The case is pending at the time of writing.[251]

In *Hearing on Climate Change Before the Inter-American Commission on Human Rights*, 15 NGOs from South America called for the IACHR to recognize the impacts of climate change on the rights of Indigenous peoples, women, children, and rural communities.[252] One of the NGOs, Pachamama Alliance (*Fundación Pachamama*) in Ecuador, supports Indigenous peoples in fighting for their rights and aids them in capacity-building.[253] In March 2022, the IACHR and the Office of the Special Rapporteur on Economic, Social, Cultural and Environmental Rights (REDESCA) jointly published Resolution No 3/21 'Climate Emergency: Scope of Inter-American Human Rights Obligations'.[254] The resolution recognizes that climate change is a human rights emergency, constituting one of the greatest threats to the full enjoyment of human rights by present and future generations including the right to life, food, housing, health, water, and the right to a healthy environment.[255] The resolution also reinforces the Escazú Agreement's procedural rights

[249] *Baihua Caiga v PetroOriental SA* (Judgement No 22201202000469, First Instance, Entry Number 1).

[250] *Thomas & De Freitas v Attorney-General of Guyana* (2021) 2021-HC-DEM-CIV-FDA. A concise summary of this case is available at Justice Institute Guyana, 'Thomas & De Freitas v Attorney-General of Guyana Summary' (*Justice Institute Guyana*) <www.justiceinstituteguyana.org/wp-content/uploads/2021/08/Thomas-De-Freitas-v-AG-case-summary.pdf> accessed 6 March 2023. Also see 'Legal Challenge to Halt Oil Production: High Court Grants Esso's Request to Join Case Despite Opposition by Applicants' *Guyana Times* (28 September 2021) <https://guyanatimesgy.com/legal-challenge-to-halt-oil-production-high-court-grants-essos-request-to-join-case-despite-opposition-by-applicants/> accessed 6 March 2023.

[251] *Thomas & De Freitas v Attorney-General of* Guyana (n 250)

[252] 'Historic Hearing! Organizations Warn the IACHR of the Impacts of the #ClimateCrisis on Human Rights' (*Dejusticia*, 22 May 2020) <www.dejusticia.org/en/historic-hearing-organizations-warn-the-iachr-of-the-impacts-of-the-climatecrisis-on-human-rights/> accessed 6 March 2023.

[253] 'An Alliance for the Amazon and the World' (*Pachamama Alliance*) <https://pachamama.org/amazon> accessed 6 March 2023.

[254] Inter-American Commission on Human Rights (IACHR) and Special Rapporteur for Economic, Social, Cultural and Environmental Rights (REDESCA), 'Resolution No. 3/2021 Climate Emergency: Scope of Inter-American Human Rights Obligations' (Adopted by the *IACHR* on December 31st, 2021)' (*IACHR and REDESCA*) <https://media.business-humanrights.org/media/documents/Res-3-21_EN.pdf> accessed 7 March 2023.

[255] ibid 8–18.

guarantees, calling on states to 'ensure that climate norms, policies and actions are constructed, updated and/or reexamined in a transparent and participatory manner'.[256] The resolution establishes the obligation to 'delimit and demarcate the collective ancestral territory', where states should avoid 'granting concessions for projects' without consultation and consent.[257] Indigenous communities seeking to challenge governmental decisions to grant concessions and licences can rely on this provision to support their claims.[258]

Finally, *Envol Vert and Others v Casino* is not strictly within the Global South docket but we have included discussion of this case in our analysis of 'transnational cases' in earlier chapters of this book, which describes cases filed in Global North jurisdictions with the involvement of Global South plaintiffs.[259] In this case, an international coalition of 11 NGOs sued the French supermarket chain Casino for causing environmental and human rights damage through its involvement in the cattle industry in Brazil and Colombia.[260] Some of the NGOs are Indigenous peoples' associations, namely Coordinator of the Indigenous Organizations of the Brazilian Amazon (COIAB), the National Organization of Indigenous Peoples of the Colombian Amazon (OPIAC), the Federation of Indigenous Peoples of Pará (FEPIPA), and the Federation of Indigenous Peoples and Organizations of Mato Grosso (FEPOIMT).[261] The case is pending at the time of writing.

As climate litigation in the Global South continues to grow apace, it can be expected that Indigenous peoples will become an increasingly significant actor in using litigation to advance climate protection in tandem with upholding their cultural and territorial rights.

## 3. MODES OF CLIMATE LITIGATION IN THE GLOBAL SOUTH

Section 2 has outlined the variety of key actors that have enabled the emergence of climate change litigation in the Global South, including global and local environmental NGOs, prosecutors, regulatory agencies, youths and children, philanthropic funders and regrantors, activist lawyers, public interest environmental law

256 ibid 8.

257 ibid 47.

258 Maria Antonia Tigre, 'Inter-American Commission on Human Rights' First Resolution on the Climate Emergency: Implications for Climate Litigation' (*Climate Law: A Sabin Centre Blog*, 11 April 2022) <https://blogs.law.columbia.edu/climatechange/2022/04/11/guest-commentary-inter-american-commission-on-human-rights-first-resolution-on-the-climate-emergency-implications-for-climate-litigation/> accessed 6 March 2023.

259 *Envol Vert v Casino* (2021) (Saint-Étienne Judicial Court, France).

260 ibid.

261 'Casino Sue in Court' (*Envol Vert*, 3 March 2021) <https://envol-vert.org/en/news/2021/03/casino-in-court/> accessed 6 March 2023.

firms, and Indigenous peoples. Some of these actors are familiar from the landscape of Global North climate litigation, particularly NGOs, philanthropic donors, lawyers, and Indigenous, youth, and child plaintiffs. Others, such as prosecutors, regulatory agencies, and the transnational advocacy networks of NGOs and regranting organizations, do not feature as prominently in climate litigation in the Global North, which suggests that their involvement is more characteristic of the Global South experience of climate litigation.

A key insight that emerges from our analysis of actors involved in climate litigation and legal mobilization efforts in the Global South is the complexity of the emerging landscape. Climate litigation in the Global South is clearly now a multi-actor space, with cases brought, funded, and supported by a diversity of groups beyond environmental NGOs. The case law examples included in this chapter and earlier chapters reveal that different actors involved in Global South climate litigation typically do not work in silos. Loose coalitions involve several actors in a single case, with each actor playing different and sometimes overlapping roles. The next level of cooperation in some of these coalitions is to activate legal mobilization efforts in a number of jurisdictions in a coordinated fashion to achieve a strategic end. For example, in seeking to halt the construction of a coal-fired power plant in Indonesia that is financed by a consortium of South Korean and Japanese banks, a litigation strategy could entail an EIA challenge on grounds that the approving authority in Indonesia did not adequately take climate change considerations into account alongside shareholder activism in South Korea and Japan. If the banks concerned had made public commitments to phase out coal financing, legal letters and further proceedings can be brought to hold the banks accountable for greenwashing.[262] This example simply highlights the potential for transnational coalitions to accelerate decarbonization and adaptation in the Global South by bringing multiple actors to work together across jurisdictional boundaries.

In this section of the chapter, we take a step back from the complexity of these actor interactions on the ground to consider if there are general observations that can be made about the prototypical roles different actors are playing in Global South climate litigation. Building on earlier work, we suggest a typology of five dominant modes of climate litigation, which we have labelled 'grassroots activists', 'hero litigators', 'farmers', 'engineers', and 'enforcers' (see Table 5.1).[263]

[262] For definitions and discussions on greenwashing, see Cathie Armour, 'What Is "Greenwashing" and What Are its Potential Threats?' (*Australian Securities & Investments Commission*, 7 July 2021) <https://asic.gov.au/about-asic/news-centre/articles/what-is-greenwashing-and-what-are-its-potential-threats/> accessed 7 March 2023; United Nations' High-Level Expert Group on the Net Zero Emissions Commitments of Non-State Entities, 'Integrity Matters: Net Zero Commitments by Businesses, Financial Institutions, Cities and Regions' (*United Nations*, 8 November 2022). <www.un.org/sites/un2.un.org/files/high-level_expert_group_n7b.pdf?_gl=1*xvdbd9*_ga*Njc3ODcyMDQxLjE2NjIzNTk2Mjc.*_ga_TK9BQL5X7Z*MTY3ODE2MDgwMi4zLjAuMTY3ODE2MDgwMi4wLjAuMA> accessed 7 March 2023.

[263] This table is adapted from Jolene Lin and Jacqueline Peel, 'The Farmer or the Hero Litigator? Modes of Climate Litigation in the Global South' in César Rodríguez-Garavito (ed), *Litigating the*

**Table 5.1** Prototypical roles of actors in Global South climate litigation

| Role | Actors |
|---|---|
| Grassroots Activist | Domestic NGOs representing local community interests.<br>Public interest environmental law firms or youth and child plaintiffs may also play this role.<br>May build from the foundation of strong public interest legal tradition in some countries. |
| Hero Litigator | A dominant figure—the activist lawyer or law firm—drives the litigation strategy and process.<br>Inspired to fight for climate justice on behalf of the community.<br>Actors playing a 'hero litigator' role will often be the 'public face' of the litigation. |
| Farmer | Philanthropic foundations and other NGOs who provide funding or other support to local lawyers and NGOs to 'seed' new climate litigation.<br>Work as part of multi-actor networks. |
| Engineer | Generally global NGOs or transnational advocacy networks who seek to replicate the success of a particular legal strategy in other jurisdictions.<br>Philanthropic organizations may also seek to play this role to achieve systemic impacts from funding interventions. |
| Enforcer | Prosecutors or regulatory agencies bringing lawsuits to enforce local laws.<br>Local NGOs may engage with enforcement agencies to support these actions. |

## 3.1 Actor Roles in Global South Climate Litigation

The role of grassroots activists in climate litigation is generally to represent local community interests, including those of particular communities like Indigenous peoples, or local environmental concerns such as pollution of air and water. By connecting local issues with global concerns, grassroots activists are advancing a mode of climate litigation that is quite distinctive of the Global South experience, ie one that is focused on the enforcement of natural resources and environmental laws to address fundamental drivers of climate change such as deforestation and expansion of coal mines.

The Adivasi, or Indigenous, communities and other forest-dwelling communities across the coal-rich state of Odisha in India, for example, have long challenged the opening and expansion of coal and mineral mines as violations of their rights over forest land under the Forest Conservation Act and the Forest Rights Act.[264] These laws recognize the role of forest-dwelling communities as legitimate

*Climate Emergency: How Human Rights, Courts, and Legal Mobilization Can Bolster Climate Action* (CUP 2022) 197.

[264] One of the best-known examples would be the Dongria Kondh's struggle against the mining company, Vedanta, which involved local, national, and international campaigning and litigation; Ankur

stewards of the land.[265] As Arpitha Kodiveri has argued, these communities have reframed the concept of 'stewardship' over time from protecting their land rights towards a broader understanding in which they steward the forests on behalf of the global community in the fight against climate change.[266] By rearticulating their identities as 'climate warriors', the Adivasi and other forest-dwelling communities seek to internationalize what would otherwise be a very local issue and, in doing so, are able to challenge the expansion of the coal-mining sector.[267]

The principal actors fulfilling a grassroots activist role are often domestic NGOs, although public interest environmental law firms or Indigenous, youth, and child plaintiffs may also work in this way. While early Global South climate cases following the grassroots activist pathway showed little evidence of collaboration between the local NGOs and plaintiffs involved and other organizations (either in the Global South or the Global North), this picture is rapidly changing. Coalitions of domestic NGOs are now more likely to engage with global NGOs in TANs on major cases, such as the challenge to the EACOP. There are also emerging efforts to build South-South cooperation to share knowledge, resources, and expertise to strengthen cases and their outcomes across Global South jurisdictions. Philanthropic organizations, particularly those with regional regranting functions, are also taking more interest in 'grassroots' climate litigation, helping to play a role in building the capacity of local NGOs and the public interest environmental law 'ecosystem' in Global South countries to support the emergence of a robust climate jurisprudence.

The role of 'hero litigators' is that of the

> lawyer-activist who is passionate about the use of litigation and other legal tools to champion climate justice. She is a dominant figure who has a high-profile role in relation to the litigation, often raising publicity for the case (and climate litigation more broadly) through press conferences and appearances on television programs. The Hero Litigator drives the litigation strategy and process.[268]

The role of the hero litigator in Global South climate litigation can be seen in several of the examples of activist lawyers and law firms described earlier. Some plaintiffs, such as young climate activists, may also be described as 'hero litigators'.

Paliwal, 'Dongria Kondhs in Odisha Win against Vedanta as Supreme Court Bans Mining in Niyamgiri' (*Land Conflict Watch*, 22 October 2016) <www.landconflictwatch.org/conflicts/tribals-in-niyamgiri-protest-against-bauxite-mining-by-vedanta-limited> accessed 7 March 2023.

265 The Scheduled Tribes and Other Traditional Forest Dwellers (Recognition of Forest Rights) Act, 2006, ss 3–5; The Forest (Conservation) Act, 1980, s 2 and related Compliance of the Schedule Tribes and Other Traditional Forest Dwellers (Recognition of Forest Rights) Act, 2006.

266 Arpitha Kodiveri, 'From Being Adivasi to Becoming Climate Warriors: Transformation in the Politics of Recognition and Legal Mobilization in India's Coal Mining Areas' (on file with authors).

267 ibid 13.

268 Lin and Peel in César Rodríguez-Garavito (ed), *Litigating the Climate Emergency* (n 263).

What distinguishes the 'hero litigator' role from that of 'grassroots activist' is the central part that the actor plays 'front of house' in the litigation rather than just contributing 'behind the scenes'. Actors playing a 'hero litigator' role often are the 'public face' of the litigation and can help to champion certain climate justice arguments or novel causes of action. Increasingly, as climate justice organizations and funders grow their focus on the Global South, hero litigators are collaborating with broader-based coalitions or campaigns in which legal mobilization and litigation is just one part.

The 'farmer' is a role that is often fulfilled by philanthropic funders and global NGOs who are seeking to 'seed' climate change litigation beyond the Global North. These actors may focus on providing funding support to Global South climate litigation, directly or through regranting arrangements, or take 'softer' approaches such as providing assistance with advocacy, legal research, and strategic support for the campaign which surrounds a lawsuit.

By definition, 'farmers' do not operate as the sole actor in climate litigation but rather as part of multi-actor networks that include domestic NGOs, community plaintiffs, activist lawyers, local public interest law firms, and sometimes prosecutors. Funding networks supporting Global South climate litigation are in existence and expanding, but their operations, objectives, and strategies, as discussed earlier, are often not publicly disclosed. More evidence is available about the role played by global NGOs in seeding Global South climate cases, either in their own name or as part of NGO coalitions.

It can be argued that, in addition to influencing the type of climate litigation cases that are pursued in the Global South, the role of 'farmers' in the climate litigation landscape is fulfilling a range of other functions, such as promoting the diffusion of norms and knowledge (including from the local branches of global NGOs back to the headquarters and amongst various branches); fostering the capacity of courts, lawyers, and NGOs in Global South countries to undertake cases; and more generally building the 'field' of climate litigation globally.

The role of 'engineers' in climate litigation is related to that of 'farmers' but it is more specifically focused on replicating a particular legal strategy or successful case in other jurisdictions. A good example is that of *Urgenda v the Netherlands*.[269] Following the success of the litigation, which compelled the Dutch government to increase the ambition of its climate change policies and created the political momentum for the passage of climate change legislation, the Urgenda Foundation proactively supported similar litigation in Belgium, Germany, Ireland, and the UK.[270]

[269] ECLI:NL:HR:2019:2007, Number 19/00135.

[270] Urgenda's case theory is heavily influenced by Roger Cox, whose book explicitly endorses a transplant model to climate litigation. See Roger Cox, *Revolution Justified: Why Only the Law Can Save Us Now* (Planet Prosperity Foundation 2012); Roger Cox, 'A Climate Change Litigation Precedent: *Urgenda Foundation v The State of the Netherlands*' (2016) 34(2) Journal of Energy and Natural Resources Law 143, 161.

An engineering function in Global South climate litigation could be fulfilled by global NGOs or TANs. There is also the potential for philanthropic organizations to embrace this model, given the desire of these organizations to achieve 'systemic' change with their funding interventions.

There is, however, limited evidence in the Global South docket of an 'engineering' approach, at least in the sense of Global North organizations directing the roll-out of a particular litigation template in the Global South. What we are observing instead is the adoption of a more nuanced approach, with significant efforts by external NGOs or funders to engage with local lawyers and NGOs. Legal strategies are being adapted and carefully planned over years to take into account political context, and the institutional and legal structures in the Global South.[271] This is encouraging because local wisdom and knowledge play a critical role in ensuring that litigation and broader legal mobilization support the quest for climate justice, prevent the creation of bad precedents which can cause significant setbacks and generate backlash,[272] and avoid the pitfalls of inciting reprisal and retaliation against complainants (and their families and supporters).[273] There is also potential for 'engineers' to foster South-South cooperation which may provide an avenue for successful Southern cases to serve as models for litigation in other similarly placed jurisdictions in the Global South.

Finally, Global South climate litigation has prominent examples of actors playing an 'enforcer' role. In this role, actors such as prosecutors and regulatory agencies are engaged in the enforcement of climate, environmental, planning, or natural resource laws which have climate-relevant outcomes. It can be argued that because the dominant focus of climate litigation in the Global South has been on ensuring existing laws are implemented by governments and properly enforced, the enforcer role is one that is distinctive to the Global South climate litigation experience.

[271] On the importance of political context, see eg Scott Cummins and Deborah Rhode, 'Public Interest Litigation' (2009) 36(4) Fordham Urban Law Journal 603, 609 ('in order to evaluate the social change potential of litigation in a given circumstance, it is necessary to examine the conditions—political, economic, cultural and organizational—within which a lawsuit operates').

[272] Michael J Klarman has noted that political backlash is 'especially likely when a court decision not only contravenes public opinion but has supporters who are less intensely committed than are its opponents'; Michael J Klarman, *From the Closet to the Altar: Courts, Backlash and the Struggle for Same-Sex Marriage* (OUP 2013) 172.

[273] Report of the Special Rapporteur on the situation of human rights defenders, UN Doc A/71/281 (3 August 2016), chs IV(D) and V <https://documents-dds-ny.un.org/doc/UNDOC/GEN/N16/247/09/PDF/N1624709.pdf?OpenElement>; Matthew Hale, 'The Critical Role of Environmental Rights Defenders—and the Risks They Face' (*Freedom House*, 10 November 2022) <https://freedomhouse.org/article/critical-role-environmental-rights-defenders-and-risks-they-face>; Julieta Biegner, 'Environmental Defenders Under Threat Around the World' (*Global Witness*, 5 November 2019) <www.globalwitness.org/en/blog/environmental-defenders-under-threat-around-world/>; 'Submission on the Challenges and Threats Facing Climate Defenders: To the UN Special Rapporteur on the Rights to Freedom of Peaceful Assembly and of Association' (*Human Rights Watch*, 1 June 2021) <www.hrw.org/news/2021/06/01/submission-challenges-and-threats-facing-climate-defenders>. All accessed 7 March 2023.

The examples of this role in action, particularly in Brazil, suggest that even if government prosecutors or agencies are the primary actors involved, their work is supported by collaboration with NGOs and public interest law firms. This is true even in China, where prosecutor-led climate litigation is at a nascent stage and tied to state priorities. Non-state actors including experts and NGOs assist in the enforcement role through the provision of technical information and assisting with capacity-building and judicial training efforts.

## 3.2 Fluid, Multi-Actor Coalitions

The above typology of typical roles played by actors in Global South climate litigation provides a shorthand description of the ways in which different individuals, groups, and organizations are mobilizing for legal action. In this regard, it provides some insights into the questions raised at the outset of this chapter about how various actors are shaping climate change litigation in the Global South and how this leads to Global South climate litigation taking on different characteristics and approaches compared to the Global North. The typology also has some predictive value in the sense that it provides a framework for evaluating where we might see further climate litigation emerge, based on the characteristics and conditions in different jurisdictions which might cohere with one or more of these prototypical models.

That the reality of actor involvement in Global South climate litigation is more complex is not surprising and is testament to the rapid development of this field over recent years. While more localized litigation efforts remain important—drawing on the roles of grassroot activists, youth plaintiffs, and hero litigators—these efforts are increasingly embedded within broader networks that may include local and global NGOs, specialist law firms, prosecutors and regulatory agencies, regional regrantors, and large philanthropic funders. This landscape of fluid, multi-action coalitions as well as densely networked TANs provides significant potential for coordination of actions towards broader, systemic, and transformative goals, including those linked to international climate change laws such as the Paris Agreement. The coalitions and networks also create important pathways for sharing of resources, knowledge, strategy, tactics, and approaches that can strengthen the capacity of local actors while also helping to build the wider field of climate litigation. It can be argued that these transnational networks and activities are likely to continue to develop as a distinctive feature of Global South climate litigation.

## 4. CONCLUSION

This chapter has sought to provide an analytical overview of the various actors that are involved in change litigation in the Global South including youths, NGOs, Indigenous peoples, activist lawyers, and funders. A key observation that has emerged from our discussion is the extent to which these actors are enmeshed in rapidly growing transnational networks of knowledge-sharing, strategy-building, and financial support. It remains to be seen whether these networks facilitate convergence in some of the trends we observe in climate change litigation across the Global North and the Global South. This would not be entirely surprising given the active commitment of many of the actors discussed in this chapter towards building transnational partnerships and serving as connectors to support the use of litigation to tackle the climate emergency. In the next chapter, we turn our attention to assessing the value of all this work. In other words, 'What can we say about the impact of litigation in galvanizing action on climate change?'

# 6
# Assessing the Impact of Global South Climate Litigation

## 1. INTRODUCTION

The previous chapters have illustrated how climate litigation is emerging and accelerating across the Global South in ways that have distinct differences from the experience of climate litigation in the Global North. This case law is still very young, with the majority of cases filed since 2019.[1] Notwithstanding the 'youth' of this jurisprudence, there is growing scholarly, practitioner, and funder interest in Global South climate litigation and the part it might play in strengthening local and global climate governance.[2]

In this concluding chapter, we turn to consider the early impacts emerging from Global South climate litigation to date and potential areas of future development for this body of jurisprudence. We talk about 'impact' in terms of the outcomes achieved by the litigation that extend beyond courtroom findings, as well as likely trajectories of future outcomes. Impact in this sense is a measure of 'whether the outcomes of these cases actually help to address the problem of climate change in a meaningful way'.[3] The pathways by which impact is achieved may involve 'direct' effects, where litigation outcomes produce legal, behavioural, or policy change, or 'indirect' effects which are more diffuse in nature and implicate changes in the attitudes of key actors, such as governments, corporations, civil society, and the public.[4] The kinds of impacts generated may be 'material'—entailing 'tangible changes in the conduct of groups or individuals'—or 'symbolic'—consisting

[1] eg in Jacqueline Peel and Jolene Lin, 'Transnational Climate Litigation: The Contribution of the Global South' (2019) 113(4) American Journal of International Law 679, we identified 34 Global South climate cases. As earlier chapters note, this body of jurisprudence encompasses 128 cases from Global South countries as of April 2023.

[2] As an example, see the Global Network for Human Rights and the Environment project compiling a Global South climate litigation database of scholarship at <https://gnhre.org/?page_id=14314> accessed 7 May 2023.

[3] Jacqueline Peel and Hari M Osofsky, 'Climate Change Litigation' (2020) 16(1) Annual Review of Law and Social Science 21, 32.

[4] See Jacqueline Peel and Hari M Osofsky, *Climate Change Litigation: Regulatory Pathways to Cleaner Energy* (CUP 2015).

*Litigating Climate Change in the Global South*. Jolene Lin and Jacqueline Peel, Oxford University Press.
 DOI: 10.1093/oso/9780191926525.003.0006

of 'changes in ideas, perceptions, and collective social constructs relating to the litigation's subject matter'.[5]

Given both the novelty of Global South climate litigation and the nascent state of knowledge on the assessment of impact from climate cases,[6] any evaluation of outcomes from the litigation is necessarily provisional. The assessment provided in this chapter draws on our knowledge of the case law, information derived from interviews with actors participating in the litigation, and our understanding (often obtained through our practice and scholarly networks) of the diversity of actors involved and broader ecosystem of global climate litigation. This involves a qualitative analysis, rather than being a data-driven, quantitative exercise. Further, as Joana Setzer, Nicola Silbert, and Lisa Vanhala discuss in the context of questions about the 'effectiveness' of climate litigation, these inquiries do not yield a single answer:

> what is seen as effective will depend on the goals or objectives that are established by a group of actors at the outset of the legal process (and potentially re-established along the way).[7]

Indeed, as is clear from our interviews conducted with litigation participants and in the reaction to particular cases, there is often no consensus view about what litigation is designed to achieve, as different stakeholders may have different answers to this question and there may also be contrasting views on whether a case was a 'success' or beneficial depending on the 'eye of the beholder'. In the same manner, what impacts are valued by and valuable for litigants and other stakeholders will be heavily influenced by context, with the result that we would expect this overall landscape—as for other dimensions of climate litigation—to look different in the Global South compared to the Global North.

This chapter begins in Section 2 with a discussion of different understandings of climate litigation's impact and the frameworks that have been developed for assessing such impact. Available frameworks have been adapted from other areas of strategic or impact litigation, such as human rights litigation, which offer many relevant lessons for examining the impact achieved by climate litigation.[8] We offer some suggestions on how these frameworks (and concepts of impact) might need to be adjusted for the task of assessing the impacts of climate litigation in

[5] César Rodríguez-Garavito and Diana Rodríguez-Franco, *Radical Deprivation on Trial: The Impact of Judicial Activism on Socioeconomic Rights in the Global South* (CUP 2015) 19–20.

[6] Joana Setzer and Lisa Vanhala, 'Climate Change Litigation: A Review of Research on Courts and Litigants in Climate Governance' (2019) 10(3) WIREs Climate Change 1.

[7] Joana Setzer, Nicola Silbert, and Lisa Vanhala, 'The Effectiveness of Climate Change Litigation' in F Sindico and others (eds), *Edward Elgar Research Handbook on Climate Change Litigation* (Edward Elgar forthcoming).

[8] Nicola Silbert, 'In Search of Impact: Climate Litigation Impact through a Human Rights Litigation Framework' (2022) 13(1) Journal of Human Rights and the Environment 272.

the Global South, stressing the need for consideration of the context in which a particular case or series of cases is brought when conducting this evaluation. For instance, outcomes commonly sought in Global North climate litigation might be less relevant in some Global South contexts. To take one example, media coverage of a case is often viewed as important in many Global North climate litigation campaigns for raising public awareness and keeping the issue of climate change on the public or political agenda in order to induce policy change.[9] However, there are many Global South jurisdictions where such a publicity-seeking strategy is likely to backfire and promote adversarial confrontation between the government and civil society actors. This could then lead to negative impacts, such as the use of national security laws to reduce societal space for environmental activism and advocacy.[10]

Section 3 turns to an analysis of some of the (early) impacts we are seeing from Global South climate cases. It has not been possible, given the now significant volume of Global South climate litigation and the recently decided or pending nature of many cases, to provide a comprehensive, overarching evaluation of the litigation's impact. Instead, for this analysis, we have sought to provide a 'snapshot' of such impact as it exists at the time of writing by selecting a sample of some of the longest-running and/or high-profile examples of climate litigation in Global South countries, which tend to be the ones where we have the most evidence to hand.[11] This evidence is drawn from a documentary review of post-case developments, information provided by interviewees and that drawn from our own networks, and insights offered by relevant literature considering these cases.[12] We acknowledge that a bias towards analysis of impact from high-profile cases pervades the climate litigation literature more generally and can lead to lower-profile cases with more incremental, cumulative effects being overlooked.[13] However, at present, capture of the impacts from this latter category of cases is constrained by limitations of the available data.

Section 4 of the chapter builds from the analysis of impacts in the previous section, as well as the existing trends in the case law summarized in Chapters 3 and 4, to consider potential areas for future development of Global South climate litigation. Our hypothesis guiding this discussion is that climate litigation in the Global

[9] Peel and Osofsky, 'Climate Change Litigation' (n 3).

[10] See eg Guilherme Pratti, 'Brazil Fights Itself: Climate Litigation Against Incoming Backfire' (*Opinio Juris*, 3 June 2022) <http://opiniojuris.org/2022/06/03/brazil-fights-itself-climate-litigation-against-incoming-backfire/> accessed 10 March 2023, describing the Bolsonaro government's efforts to undermine environmental and climate frameworks in Brazil 'from the inside' at the same time as civil society and advocates fought to keep these (in)actions under scrutiny.

[11] We acknowledge that the picture of impact that this 'snapshot' yields may well change at different time points, such that the view of impact five years out from a case decision may look very different from that after a 10- or 20-year period, and so on.

[12] Hence, we are not attempting to apply the impact assessment frameworks described in Section 2, an exercise which would be a substantial study in and of itself, most likely requiring a multi-year time frame and dedicated funding to assemble a more comprehensive data set.

[13] Setzer and Vanhala (n 6).

South has different characteristics from that in the Global North and is likely to develop in different ways in coming years. We therefore consider the extent to which future trajectories that have been identified in the Global North context might be replicated or departed from in the Global South.

Section 5 concludes with a summary of ideas advanced in the chapter and the book as a whole.

## 2. FRAMEWORKS FOR ASSESSING IMPACT

Climate litigation, as a species of public interest litigation, is rarely concerned only with securing outcomes for the parties involved. This raises salient questions about the broader impact of this litigation, including its contribution to climate governance and the implementation of international treaties, such as the Paris Agreement. As we noted in the book's introduction, in its Sixth Assessment Report, the Intergovernmental Panel on Climate Change (IPCC) addressed, for the first time, the question of climate litigation's implications for climate governance, concluding that '[c]limate litigation is growing and can affect the outcome and ambition of climate governance'.[14]

As the ecosystem of climate litigation actors widens to include more complex coalitions, including non-governmental organizations (NGOs) and funders supporting strategic legal interventions to advance climate action,[15] the impact achieved by cases is also attracting more attention. Despite this, definitions of impact applicable to climate litigation, and agreed frameworks for assessing the impact of climate cases, remain remarkably thin on the ground.[16] Those that do exist have often drawn on the experience of assessing impact in other areas, such as development projects or human rights litigation. In this section of the chapter, we consider how the impact of climate litigation can be understood, available frameworks for assessing the impact of climate litigation, and how such frameworks might need to be adjusted for the Global South context.

### 2.1 Defining Climate Litigation's Impact

In studies considering the impact of litigation—particularly strategic rights litigation—the notion of impact employed is generally a broad one. Summarizing a

[14] Intergovernmental Panel on Climate Change, *Climate Change 2022: Mitigation of Climate Change* (Contribution of Working Group III to the Sixth Assessment Report of the Intergovernmental Panel on Climate Change, April 2022) (hereafter WGIII Report) 1358.

[15] For a discussion of the role of these actors, see Chapter 5.

[16] The authors are aware through their consultancy work of some such frameworks that have been developed but which are not available in the public domain.

range of such literature, Malcolm Langford finds that litigation's impact 'constitutes the total influence or effect of a decision', whether that is net positive in terms of the goals sought or net negative as a result of the litigation's unintended, adverse consequences.[17] Impact in this sense stretches beyond the (final) judgment in a case and its direct implementation or enforcement to encompass what might be termed 'beyond courtroom' effects.[18] In this chapter, we understand the impact of Global South climate litigation in a similarly broad way. We focus on the outcomes of the cases beyond the courtroom and whether they make a meaningful contribution (or not) to addressing the problem of climate change.

### Types of impact

While the impact achieved by any particular case or course of litigation is usually highly context-dependent,[19] literature considering the nature of litigation's impact often distinguishes between different types or categories of impact.

For instance, as highlighted above, impact can be net positive for addressing climate change or net negative, although the latter category of outcomes is often not extensively assessed.[20] Negative impacts might include political or other backlash following the decision, or stem from litigation by corporate interests to resist the effects of the decision, or simply the failure of government defendants to carry out any enforcement action. While there is a smattering of examples, 'anti-regulatory' litigation seen in the United States has not been a prominent feature of the Global South climate case law,[21] possibly because powerful business groups have other avenues through lobbying and the like to achieve their commercial ends. More common is the situation (evidenced by the *Earthlife Africa* case discussed further in Section 3) of a project authorization overturned in litigation being subsequently reissued by governmental authorities, as well as judgments languishing indefinitely with no enforcement action being taken.[22] Situations can also arise, however, where litigation does not lead to enforcement action but there are other beneficial

[17] Malcolm Langford, 'The Impact of Public Interest Litigation: The Case of Socio-Economic Rights' (2021) 27(3) Australian Journal of Human Rights 505, 514.

[18] We note that this terminology was developed by Sophoi consultants in a 'Sensemaking and Learning Review of FILE's Climate Legal Portfolio', see online at <www.sophoi.co.uk/projects>.

[19] Oran R Young, 'Effectiveness of International Environmental Regimes: Existing Knowledge, Cutting-edge Themes, and Research Strategies' (2018) 108(50) Proceedings of the National Academy of Sciences of the United States of America (PNAS) 19853.

[20] For an example of negative consequences in a Global North context see Peel and Osofsky 'Climate Change Litigation' (n 3) for discussion of US anti-regulatory cases. A similar phenomenon is noted by Joana Setzer and Catherine Higham in their *Global Trends in Climate Change Litigation: 2023 Snapshot* (June 2023).

[21] There are some examples in Brazil, discussed in Chapter 4, as well as the *DG Khan* case in Pakistan, which involved a company lawsuit against the government arguing a constitutional rights violation.

[22] Forestry-related climate litigation in Indonesia, discussed in Chapter 3, provides a good example here with no effort to collect the fines levied in judgments even several years after the decision.

outcomes from the case such as that its overall impact is viewed as one of 'winning by losing'.[23]

Another distinction commonly drawn is between litigation's direct effects and its indirect effects. César Rodríguez-Garavito and Diana Rodríguez-Franco's analysis of socio-economic rights litigation in Latin America judges an intervention to have had direct effects if it has 'produced an observable change in the conduct of those it directly targets'.[24] Indirect effects, on the other hand, are more diffuse and represent the larger flow-on consequences of litigation. They may include the effects of the judgment in catalysing wider law and policy reform, in providing a legal precedent that is utilized in future litigation, or in creating a 'shadow of liability' that presages future claims.[25] Indirect effects of litigation also often comprise the political and 'symbolic' dimensions of impact as, for instance, where the case helps shape a social movement or elevates the public standing or political salience of an issue. César Rodríguez-Garavito and Diana Rodríguez-Franco distinguish symbolic effects of this kind from 'material' ones that are more tangible in nature.[26]

Jacqueline Peel and Hari Osofsky, focusing on the regulatory impact of climate litigation in Australia and the United States,[27] also apply a framework of direct and indirect effects. They provide a conceptual mapping of different pathways by which litigation might achieve regulatory impact, validated by their interviews with litigation participants and targets.[28] Peel and Osofsky treat climate litigation impacts as direct 'when litigation results in a formal change in climate change law and policy', whether through judicial interpretation of constitutional doctrine or statutes, or through development of the common law (Australia and the United States both being common law jurisdictions).[29] Indirect pathways, by contrast, involve 'behavioural and norm change that influences choices of key actors' such as corporations, governments, environmental organizations, and individuals.[30] The authors give examples where cases provide 'a stimulus for regulatory change through increasing costs or reputational risks associated with particular projects

[23] Brian Preston, 'The Influence of the Paris Agreement on Climate Litigation: Legal Obligations and Norms (Part I)' (2021) 33(1) Journal of Environmental Law 1.

[24] Rodríguez-Garavito and Rodríguez-Franco (n 5) 18.

[25] ibid 18–20.

[26] See Rodríguez-Garavito and Rodríguez-Franco (n 5) 20. The authors present a two-by-two table with a categorization of impacts of litigation as: (1) material, direct, eg designing public policy as ordered by a ruling; (2) material, indirect, eg forming coalitions of activists to influence the issue under consideration; (3) symbolic, direct, eg defining or reframing the problem in a new way such as seeing climate change as a human rights issue; and (4) symbolic, indirect, eg transforming public opinion about the problem's urgency and gravity.

[27] See Peel and Osofsky, 'Climate Change Litigation' (n 3) 32–34. The authors construe the 'regulatory' impact of litigation in broad terms to extend beyond rulemaking to include the wider influence of a case on the behaviours of different actors: see ibid 32–34.

[28] ibid 36 and Figure 2.1.

[29] ibid 37.

[30] ibid 47.

or business practices, raising public awareness of the climate change problem or generating shifts in public opinion or social norms'.[31]

More recently, Peel, together with co-authors Alice Palmer and Rebekkah Markey-Towler, completed a review of the legal literature published between 2000 and 2021 concerning the impact of climate litigation.[32] While noting that much of the literature's discussion of impact was anecdotal and unsystematic, the report's authors identified different types of impact from litigation discussed in the scholarship on climate case law.[33] In their analysis, direct effects discussed in the literature included:

- Changes or proposed amendments to laws, regulations, or other policies following the litigation;
- Changes to the interpretation of a law or its scope of application following the decision, eg recognizing that 'pollutants' covers greenhouse gases (GHG) or broadening standing rules to allow civil society to challenge government climate policy; and
- Behavioural changes of the litigation target in the wake of a decision, eg where a company issues a new corporate emissions policy to comply with a judgment.

On the other hand, indirect effects discussed in the literature (which the authors noted was less common than for direct effects), included:

- Changes in behaviour by litigation targets or similar entities not directly required by the ruling but which could plausibly be traced back to the litigation, eg a pension fund regulator proposing to adopt new guidelines for reporting climate risk following complaints from members about inadequate disclosures by some funds;
- The role of litigation in setting a benchmark (or precedent) for future litigation or inspiring other litigants to bring similar cases;
- Increasing costs and risks, including reputational risks, due to the litigation or credible litigation threats; and
- Broader systemic changes potentially stemming from litigation, including changing social norms or raising awareness of climate issues.

Another way to classify the types of impact from climate litigation is based on spatial factors (ie where impacts are felt) and temporal factors (ie over what time

[31] ibid.

[32] Jacqueline Peel, Alice Palmer, and Rebekkah Markey-Towler, 'Review of Literature on Impacts of Climate Litigation' (Report, Children's Investment Fund Foundation, 27 May 2022).

[33] ibid 22–23.

frame). Spatial elements of impact usually refer to impacts occurring in different systems—social, legal, and political—or in different geographies (acknowledging that the legal cultures and socio-political contexts of countries variably considerably and may limit the impact that can be achieved through litigation). Given the long lead-time of litigation, with cases often taking many years to reach final resolution and for impacts to manifest fully, temporal elements of impact are also highly relevant. An assessment of climate litigation's impact might therefore distinguish between immediate effects and intermediate and longer-term effects.[34] Moreover, intermediate effects a few years out from a judgment might initially seem positive but should not be over-emphasized if, in the longer term, they produce limited tangible gain for climate action.[35]

### Role of context

One of the key findings on the impact of litigation that emerges from other literatures is that the types of impacts that can be expected from cases will be 'highly conditional', ie dependent on the prevailing conditions and context.[36] In particular, the broader socio-political context matters, with studies consistently finding that stronger governance and adherence to the rule of law facilitates litigation seeking the implementation of constitutional rights.[37] This is borne out by the analysis in Chapter 4 of the conditions that have facilitated the development of climate litigation in 'frontrunner' Global South jurisdictions like India, Brazil, and South Africa. Such conditions include good access to justice, existing laws and institutions to support climate claims, protection of environmental rights, and a well-developed civil society—all characteristics generally associated with the environmental rule of law.[38]

The important role played by context may also explain why litigation in some circumstances and some jurisdictions does not or will not have the impacts that might be expected by advocates. For instance, institutional factors, such as the power of local courts to enforce judgments and craft effective remedies, as well as the capacity of the state or other targeted actors to implement rulings or be transformed by them may often be determinative of the capacity of climate litigation to have meaningful impact.[39] Beyond institutional structures, resourcing constraints (human, economic, budgetary) and the extent of cohesiveness of civil society coalitions may also influence litigation outcomes. In addition, Malcolm Langford

[34] Sandra Botero and Daniel Brinks, 'A Matter of Politics: The Impact of Courts in Social and Economic Rights Cases' in Malcolm Langford and Katherine Young (eds), *Oxford Handbook on Economic and Social Rights* (OUP 2022).

[35] Langford, 'The Impact of Public Interest Litigation' (n 17).

[36] ibid.

[37] ibid.

[38] United Nations Environment Programme, *Environmental Rule of Law* (First Global Report, January 2019).

[39] Langford, 'The Impact of Public Interest Litigation' (n 17); Young (n 19).

identifies the potential for litigation to displace other effective strategies for producing social change if actors are lulled into a false sense of security by a court 'win' that leads to a lack of follow-up implementation.[40]

In studies of climate litigation's impact focusing on the Global North, a common finding has been to recognize that litigation is *a* tool but not *the* tool for producing social and policy change on climate.[41] This points to the importance of the other strategies that surround and support litigation, such as policy advocacy, media, and linked campaigns or social movements. Some of the most 'successful' Global North cases have built a considerable campaign and communications apparatus alongside litigation efforts to amplify their impact. Whether this approach offers the same benefits in a Global South context is considered further below.

In sum, using the courts to drive social and policy transformation towards a given end is a highly conditional strategy. As yet, even in the Global North where the climate litigation jurisprudence is of longer standing, there is little evidence that climate cases generate significant long-term impacts, and their contribution may often be more on the indirect and symbolic side rather than in giving rise to direct, material effects. In this sense, courts and climate litigation may function most effectively as a site for agenda setting, issue framing, social mobilization, and generating political pressure, which in turn builds the momentum towards material outcomes on mitigation and adaptation.

## 2.2 Frameworks for Assessing the Impact of Climate Litigation

Given the complexities of the types of impact from climate litigation and their context-dependence, it is unsurprising to find that no clear frameworks have emerged for assessing the impact of litigation. In the broader social science literature, there is no consensus on approaches for setting a baseline from which impact can then be measured, given the lack of feasibility of standard scientific approaches such as conducting a randomized controlled experiment. Nonetheless, as Joana Setzer and Lisa Vanhala outline, scholarship in socio-legal studies, political science, and sociology offers a useful foundation upon which to begin to assess the impact of cases and court decisions.[42]

Some methodologies use a counterfactual approach to see what changes have occurred since a legal intervention, or closely related situations may be used as a proxy to test whether a plausible causal link can be posited between litigation and particular outcomes.[43] However, the danger with this backwards-looking

[40] Langford, 'The Impact of Public Interest Litigation' (n 17).
[41] Peel and Osofsky, 'Climate Change Litigation' (n 3).
[42] Setzer and Vanhala (n 6).
[43] See Ottar Maestad, Lise Rakner, and Octavio L Motta Ferraz, 'Assessing the Impact of Health Rights Litigation: A Comparative Analysis of Argentina, Brazil, Colombia, Costa Rica, India and South

approach is that it cannot clearly establish causation between the intervention and an outcome following its conclusion. This is because the method generally does not account for 'greater advances that might have occurred if an alternative strategy or tactic was adopted', or whether the changes 'might have occurred simply in any case due to other exogenous factors'.[44]

A middle way forward used in analysis is to consider the 'before and after' of a particular legal intervention and whether there are reasonable explanations that the litigation, as opposed to other available strategies, delivered or contributed to the observed result.[45] Teasing out these different factors may be done by introducing and then eliminating independent variables that either constrain or facilitate different outcomes.[46] Quantitative approaches for undertaking this task, such as multivariate regression analysis, have been applied in some studies of the impact of court decisions. For instance, Christopher Berry used quantitative and multivariate analysis to measure whether funding for schools in the United States increased following court orders issued in certain states between 1970 and 2003, finding that the presence of a positive judgment led to 'roughly $450 per pupil in additional revenue'.[47] However, these techniques are far less effective at measuring non-quantifiable impacts, such as field-building, mobilization of civil society coalitions, or changes in public attitudes—the kinds of impacts often sought in climate litigation.[48]

An alternative qualitative method, which has been applied in some studies of climate litigation impact,[49] is 'process-tracing' that offers a framework for analysing causal relationships between antecedents and outcomes.[50] This technique

Africa' in Alicia Ely Yamin and Siri Gloppen (eds), *Litigating Health Rights: Can Courts Bring More Justice to Health* (Harvard University Press 2011) 279.

[44] Langford, 'The Impact of Public Interest Litigation' (n 17).

[45] ibid 514–15. This could include examining strategies such as active mobilization and/or protest, law, and policy reform and media advocacy.

[46] ibid.

[47] Christopher Berry, 'The Impact of School Finance Judgments on State Fiscal Policy' in Martin R West and Paul E Peterson (eds), *School Money Trials: The Legal Pursuit of Educational Adequacy* (Washington DC, Brookings Institution 2007) 213.

[48] Setzer and Vanhala (n 6) 13. It is possible to do media analysis studies as a way of testing how an issue is perceived by the public and whether that changes over time in conjunction with particular litigation. However, the extent of media commentary may often be a function of the case or campaign strategy and available communication resources, something that is more often a feature of Global North (or transnational) cases than for those brought in the Global South.

[49] The handful of published examples include Lisa Vanhala and Cecilie Hestbaek, 'Framing Climate Change Loss and Damage in UNFCCC Negotiations' (2016) 16(4) Global Environmental Politics 111 and Sébastien Jodoin, Annalisa Savaresi, and Margaretha Wewerinke-Singh, 'Rights-Based Approaches to Climate Decision-Making' (2021) 52 Current Opinion in Environmental Sustainability 45. See also Charles Beauregard and others, 'Climate Justice and Rights-based Litigation in a Post-Paris World' (2021) 21(5) Climate Policy 652, applying the OATH approach (objective, associated climate impact, type of justice, harm) to assess the impacts of climate cases such as *Urgenda*, *Juliana*, and *Future Generations*.

[50] Andrew Bennett and Jeffery Checkel, *Process Tracing: From Metaphor to Analytic Tool (Strategies for Social Inquiry)* (CUP 2014).

recognizes that strategic litigation is a process that goes beyond the legal intervention itself and includes post-judgment implementation.[51] It is often based on case studies and interviews with involved or affected actors.[52] In some cases, funders of strategic litigation programmes require grantees to collect this impact information. For example, the Digital Freedom Fund (DFF)—a European strategic litigation funder—applies a framework of 'outcomes harvesting' with collection from a range of sources including those who were involved in the litigation. In this method, participants are asked to complete 'outcome statements' which must include: (1) a description of the change; (2) details of when and where the change happened; and (3) an explanation of who or what caused the change and how.[53]

While the conclusions of such qualitative studies about the contribution of litigation to particular outcomes can rarely be decisive,[54] they have the advantage of extending beyond a binary 'win/lose' understanding of a case's outcomes to recognize the 'multi-dimensional, multi-disciplinary, multi-stakeholder, iterative' processes by which strategic litigation produces impact.[55] However, a key constraint (which applies to both quantitative and qualitative studies) is often the time period selected for study,[56] which may be too short to pick up the often very long time frames of litigation and its effects.[57] Another question concerns *who* should bear the responsibility of conducting impact evaluation.[58] Although in practice this task is often undertaken post-hoc by scholars or evaluation consultancies, there is an argument that it should rather sit with the litigants, their legal team including funders, and the broader communities or movements of which they are a part.[59]

Bearing in mind these limitations and caveats, there are a number of consistent findings that emerge from assessments of the impact of litigation, and which seem to also characterize the climate litigation landscape. As described by the Open Society Justice Initiative's report on 'Strategic Litigation Impacts',[60] these include

[51] James Goldston and Erika Dailey, *Strategic Litigation Impacts: Insights from Global Experience* (Open Society Foundations 2018) 17. It may also encompass *ex ante* case planning as is the case for so-called 'systemic lawyering': Joana Setzer and Catherine Higham, *Global Trends in Climate Change Litigation: 2022 Snapshot* (Policy Report, Grantham Research Institute on Climate Change and the Environment 2022) 46 (hereafter Setzer and Higham, *2022 Snapshot*).

[52] Goldston and Dailey (n 51) 15.

[53] Digital Freedom Fund, 'Measuring the Impact of Strategic Litigation in Digital Rights Developing a Tool for the Field' (November 2019) <https://digitalfreedomfund.org/wp-content/uploads/2019/12/dff-litigation-impact-assessment-framework.pdf>.

[54] Langford, 'The Impact of Public Interest Litigation' (n 17) 515.

[55] Goldston and Dailey (n 51) 18.

[56] Langford, 'The Impact of Public Interest Litigation' (n 17) 515. See also Paul Gready, 'Reasons to Be Cautious About Evidence and Evaluation: Rights-Based Approaches to Development and the Emerging Culture of Evaluation' (2009) 1(3) Journal of Human Rights Practice 380.

[57] In this regard though see Duffy's Human Rights Litigation (HRL) impact framework (Helen Duffy, *Strategic Human Rights Litigation: Understanding and Maximising Impact* (Bloomsbury Publishing 2018) 48) which takes a 'long lens' approach in examining the role of courts in creating social change and may offer a good model for climate litigation impact assessment; see Nicola Silbert (n 8) 72.

[58] Setzer and Vanhala (n 6) 12.

[59] Michael McCann, *Rights at Work: Pay Equity Reform and the Politics of Legal Mobilization* (University of Chicago Press 1994).

[60] Goldston and Dailey (n 51) 29.

the insights that: strategic litigation is a process (including post-judgment implementation) not a single legal intervention; legal strategy needs to be coordinated with other actions; a strategy of filing mass or incremental cases is often more effective than seeking a single landmark judgment; some change, including through litigation, is the product of opportunistic and/or organic work whose value may become apparent only with the passage of time; and that strategic litigation is most effective when carried out for and in partnership with non-litigators involved in broader advocacy, campaign, or community work. Nonetheless, it remains the case that assessing the impact of litigation, including climate litigation

> is a complex undertaking in which multiple factors, including timeframe, causality, and individual perspective, can shape perceptions of strategic litigation's impact or lack thereof. This complexity is magnified when strategic litigation is examined in context, and considered as one possible tool of social change among many.[61]

## 2.3 Considering the Global South Context in Assessing Impact

As highlighted above, the literature on the assessment of impacts from litigation emphasizes the importance of context (social, political, legal) in evaluating the outcomes achieved (or able to be achieved) from climate and other strategic litigation. In assessing impact from cases brought in the Global South, these contextual factors are similarly likely to play a role in either fostering or retarding the potential for cases to achieve impact, as well as the time frames over which it is reasonable to expect impacts to manifest. In this section we offer some suggestions about what kinds of contextual factors will be important to consider in assessing the impact of Global South climate cases, as well as how assessment methods might need to be adjusted to yield a fuller picture of the ways in which a case or series of cases helps to generate change.

Given that good governance, civil society activism, and adherence to the rule of law are common foundations for environmental public interest litigation and climate litigation,[62] we would expect these measures to play a key role in shaping the impacts that can be achieved from climate cases in particular Global South countries. As Chapters 3 and 4 have identified, not all Global South countries are seeing climate cases emerge and among the countries where climate litigation is present there are clear 'frontrunner' jurisdictions in each region where there is a concentration of cases. What characterizes Global South countries that have seen climate litigation is the presence of adequate 'demand' factors that create tensions and disputes that will give rise to litigation (eg conflict between economic development priorities and

[61] ibid.
[62] ibid.

vast natural resources and inequities between different parts of the community in terms of who benefits or is most impacted by decisions around such issues) and sufficient 'supply' factors in the sense of an institutional structure that is well-functioning enough to encourage resort to the courts to adjudicate disputes. These demand and supply factors will also tend to influence the outcomes that can be achieved beyond the courtroom. For instance, active civil society coalitions who have knowledge of climate issues will be better able to develop the climate elements of broader environmental and natural resource use cases. Equally, institutional structures that have some capacity to ensure enforcement of rulings and implementation of policy changes will amplify the potential for litigation to achieve impact.

There may also be some frequently encountered contextual factors in Global South countries that are relevant in assessing the potential for climate cases to achieve impact. For instance, time frames for resolution of cases in court systems may be very lengthy in some Global South countries (although this can also be a feature of some Global North court systems and climate litigation, as in the United States). In the United Nations Environment Programme's (UNEP) First Report on the Environmental Rule of Law, the agency identified 'the slow pace of some court proceedings' as a significant challenge for implementation of effective environmental adjudication, a key element of the environmental rule of law.[63] Notably, the World Justice Project's Rule of Law Index for 2022 ranks many Global South countries as having weaker adherence to the rule of law in respect of civil justice (factor 7) compared with countries in the Global North.[64] This factor, inter alia, 'measures whether civil justice systems are accessible and affordable as well as free of discrimination, corruption, and improper influence by public officials . . . [and] whether court proceedings are conducted without unreasonable delays and whether decisions are enforced effectively'.[65] Lower rankings on the civil justice factor are particularly prevalent for countries in Southeast Asia (with the exception of Singapore and Malaysia) and South Asia.[66]

Looking at the 'significant constraints' for climate litigation in Global South countries, Joana Setzer and Lisa Benjamin identify, at a general level, the influence of factors such as limited human and financial resources, which mean countries 'lack the capacity to build and maintain effective environmental institutions, create strong scientific knowledge bases for environmental policymaking, effectively integrate environmental concerns into national economic development planning, and set up effective environmental monitoring and implementation schemes'.[67]

[63] United Nations Environment Program, *Environmental Rule of Law* (First Global Report, January 2019) 191.
[64] World Justice Project, 'Rule of Law Index' (2022) <https://worldjusticeproject.org/rule-of-law-index/global/2022/Civil%20Justice>.
[65] ibid.
[66] ibid.
[67] Joana Setzer and Lisa Benjamin, 'Climate Litigation in the Global South: Constraints and Innovations' (2020) 9(1) Transnational Environmental Law 77, 82.

As regards enforcement of environmental legislation, a lack of resourcing also frequently presents a barrier, and in some countries there are also considerable dangers faced by 'environmental defenders' in seeking to challenge government action.[68] However, Setzer and Benjamin also identify ways in which these constraints might be lessened, pointing to the role of '(i) access to justice in conjunction with the existence of progressive climate and/or environmental rights legislation, and (ii) judicial opportunism' (ie progressive and innovative judicial approaches). They find that '[w]hen these factors are combined, they have the potential to help actors in the Global South to overcome countervailing dynamics of significant capacity constraints in implementing environmental legislation and managing fragmented and under-resourced institutional structures, and can therefore contribute to progressive outcomes'.[69]

Studies such as Setzer and Benjamin's have underpinned more recent efforts by strategic climate litigation funders to direct funding towards capacity and 'field-building' activities to enhance the prospect for positive outcomes from climate cases brought in Global South countries.[70] Activities such as judicial training in environmental law and climate science, support provided to civil society organizations to build local capacity on climate litigation and climate science, and to develop sound case strategies can enhance the prospects for cases that are brought to be successful in and beyond the courtroom.

This points also to the importance of context in considering what types of impact from climate litigation might be sought and valued in Global South countries. For instance, public awareness raising and media attention might be less crucial (or even counterproductive) as an outcome from climate cases in some Global South countries than is often assumed to be the case in a Global North context. Conversely, measuring capacity and field-building impacts (although difficult to capture in existing assessment frameworks) may be more valuable as a demonstration of how litigation and associated activities are building the civil society and rule of law foundations for climate cases to be brought in the first place and then to have positive impacts. Equally, in countries where there are weak environmental institutions and poor implementation of laws, efforts that strengthen those institutions, for example to engage with climate science, or which support better implementation and enforcement of existing laws that will deliver climate co-benefits, might be particularly valued.

Assessments of impact from climate litigation in the Global South, and particularly attempts to consider 'alternative strategies' in evaluating the causal contribution of litigation to outcomes, also need to be cognizant of the different

[68] As Setzer and Benjamin note, ibid, this has not prevented strong outcomes from climate litigation in some of the Global South countries ranked amongst the most dangerous for environmental defenders, such as Brazil, Colombia, and the Philippines.

[69] Setzer and Benjamin (n 67) 80.

[70] See eg the website of re-grantor Foundation for International Law for the Environment (FILE) on the nature of its work supporting climate litigation: <www.filefoundation.org/work>.

socio-political context encountered in Global South countries compared with those of the Global North. In particular, 'alternatives' that may be present in the Global North may be far less viable in the Global South as effective strategies, meaning it may be more difficult to identify and assess counterfactuals in the Global South. This complexity is reflected in the Open Justice Initiative project on strategic litigation impacts, where Goldston and Dailey observe that:

> It is axiomatic that strategic litigation wields its greatest direct impacts when the rule of law is already largely respected. It may be equally true that such enabling environments offer alternatives to strategic litigation, from elections to independent media to public protest. By contrast, in societies with limited democratic space and fewer options to express dissent or subject official policy to independent scrutiny, litigation may play a singular role. In short, while strategic litigation may be more effective in democratic societies, it may be more significant in illiberal societies where it is often one of the few forms of advocacy permitted.[71]

## 3. IMPACTS FROM GLOBAL SOUTH CLIMATE LITIGATION

As the previous section highlighted, one of the difficulties in assessing the impact of litigation is the long time frames involved, both in bringing cases and securing judgment, and then in development of the litigation's 'beyond courtroom' effects. It is not unusual in Global North jurisdictions for a climate case to take several years to reach final judicial resolution (especially considering the possibility for appeals), with broader outcomes only beginning to be seen five or more years after the initial case filing.[72] These time frames can be considerably longer in many Global South jurisdictions because of chronic problems of delay in court systems and often poor implementation of judicial decisions.[73] Given that climate litigation itself is only a relatively recent phenomenon in the Global South, this complicates efforts to assess impact.

The attempt made to do so in this section of the chapter is fully cognizant of those limitations—its assessments of follow-on effects from particular Global South cases should be treated as provisional since an evaluation over a longer time period could materially affect the view of the overall impact of a case(s). Lacking the dedicated funding and time to undertake a longitudinal study of impact in this

[71] Goldston and Dailey (n 51) 16.

[72] See also Malcolm Langford, 'Introduction: Civil Society and Rights in Context' in Malcolm Langford and others (eds), *Symbols or Substance? The Role and Impact of Socio-Economic Rights Strategies in South Africa* (Cambridge University Press 2014) 1–32, noting that, '[i]n some socio-economic rights cases, key impacts occurred 5 to 10 to 20 years after the judgment requiring making any eventual assessment dependent on what can be viewed as a reasonable period'.

[73] Setzer and Benjamin (n 67).

book (completed as it was during the exigencies of the Covid-19 pandemic), we have also not sought to apply fully the assessment frameworks, such as process tracing, outlined in Section 2 above. Instead, we have focused on compiling evidence of early effects from a handful of high-profile Global South cases, drawing on relevant information available through: (1) documentary review of grey literature, such as newspaper articles, media releases, and other open access media commentary; (2) information from a small number of interviews with litigant participants and drawn from our own practice and scholarly networks;[74] and (3) insights identified in scholarship on these cases.[75]

The following sections detail those findings in respect of four case studies of the *Leghari* case (Pakistan), the *Earthlife Africa* case (South Africa), the *PSB (Climate Fund)* case (Brazil), and the *Future Generations* case (Colombia). Identified effects of these cases are separated out, as far as possible, according to the categories of direct and indirect effects, and material and symbolic effects discussed in Section 2. These effects include: (1) changes in law or policy following a decision; (2) behavioural changes exhibited, for instance, by private sector actors; (3) effects related to altering public perceptions or producing other social change effects; (4) the role of the case in shaping future litigation; and (5) implications of the litigation for implementation of the Paris Agreement or global climate governance more broadly. We have tried, where possible, to identify contextual factors relevant for assessing the impact of the cases and acknowledge that this case study approach focusing on individual cases may not be adequate to capture the cumulative or additive effects of a case when considered as part of a suite of litigation.

### 3.1 Case Study 1—*Leghari* (Pakistan)

*Leghari* is and remains one of the best-known examples of climate litigation in the Global South.[76] The content of the judgments issued by Judge Mansoor Ali Shah (then of the Lahore High Court) have been canvassed elsewhere in the book and will not be repeated here, other than as necessary to highlight the case's follow-on effects. Despite receiving significant scholarly attention following the first judgment in the case issued in 2015, there have been few articles on *Leghari* in recent years. Following the trend outlined in earlier sections of the chapter, there have

[74] In this regard we are very grateful for the feedback provided by participants at the Climate Litigation in the Global South workshop convened at National University of Singapore, Faculty of Law, 22–23 May 2023.

[75] As noted in the study by Peel, Palmer, and Markey-Towler, legal publications are variable in the extent to which they consider the direct and indirect effects stemming from climate change and even where there is analytical engagement with questions of impact this is often brief (eg confined to a section at the end of an article) and speculative (without extensive reference to empirical evidence or use of a rigorous methodological approach to data collection). See Peel, Palmer, and Markey-Towler (n 32) 23.

[76] *Ashgar Leghari v Federation of Pakistan* (2015) Lahore High Court Green Bench WP No 25501/201.

been even fewer articles that have attempted to trace the impact of the case following the court's closure of its supervisory programme in 2018.

### Direct, material effects for law or policy

As part of the remedy provided in its first judgment in 2015, the Lahore High Court made a series of orders aimed at ensuring the implementation of the Pakistani Climate Change Policy (the Policy) and its associated implementation framework (the Framework). This included, as highlighted in previous chapters, the establishment of a Climate Change Commission to monitor the progress of the Framework, comprised of representatives from relevant ministries, civil society, and scientific experts to monitor the government's progress. The court examined the Framework and Policy in detail and set out the specifics for how the Commission would work.

Unusually, compared to many Global North jurisdictions, but far less so in Pakistan and other South Asian countries, the court retained supervisory jurisdiction over the activities of the Commission.[77] During the time the Commission was operational, it deliberated for several months, identifying hundreds of initiatives the government of Punjab could implement that year. It divided the priorities into four categories, including timelines for implementation: (1) priority actions, within two years; (2) short term, within five years; (3) medium term, within 10 years; and (4) long term, within 20 years. The Commission also created sub-committees, including on: water resources management; agriculture; forestry, biodiversity, and wildlife; coastal and marine areas; disaster risk management; and energy.[78]

Writing in 2017, Esmeralda Colombo described the impact of this order as persuading Pakistani governments, both at the federal and provincial levels, to take steps toward implementing climate policy plans under the strict supervision of the court in 'a quite expeditious fashion'.[79] She commented:

> Not only did the court compel executive authorities to take certain actions required by the Framework for Implementation of Climate Change Policy but it also provided the Pakistani government with material knowledge on climate change threats and effects. Furthermore, it directed the executive to gather data and analyses on specific issues in a progressive mode, thus structuring the judicial process as an ongoing check on governmental action.[80]

[77] See Setzer and Benjamin (n 67) 92: 'Noting there is already a judicial precedent in Pakistan which lends itself to such interventionist judicial activity. In 2012, the High Court of Lahore ordered the establishment of the River Ravi Commission to manage the restoration of the river, held periodic hearings on the Commission's progress, and ordered full-scale implementation of the bioremediation project in 2015.'

[78] Emily Barritt and Boitumelo Sediti, 'The Symbolic Value of *Leghari v Federation of Pakistan*: Climate Change Adjudication in the Global South' (2019) 30(2) King's Law Journal 203.

[79] Esmeralda Colombo, 'Enforcing International Climate Change Law in Domestic Courts: A New Trend of Cases for Boosting Principle 10 of the Rio Declaration?' (2017) 35(1) UCLA Journal of Environmental Law and Policy 40.

[80] ibid 40.

The court subsequently released a supplementary decision in 2018 as part of this supervision, which found that from September 2015 to January 2017, 66% of priority actions from the Framework had been implemented.[81] As a result of this evidence, Chief Justice Shah, by now a Justice of the Supreme Court of Pakistan, dissolved the Commission in 2018 because it had 'achieved its functions', although the court left open the possibility that the case could be revived in the event of future breaches.[82] Justice Shah ordered the creation of a Standing Committee on Climate Change to act as a link between the court and the executive and to provide a check on implementation.

Joana Setzer and Lisa Benjamin note that a new Climate Change Act was passed by Pakistan in 2017 during the progress of the case, which they claim illustrates 'how climate litigation and regulatory progress can have a cooperative relationship'.[83] Writing in 2020, the authors remarked:

> While there is some scepticism as to whether the new Act will be fully implemented, the objects and reasons section of the Act notes the extreme vulnerability of the country to climate change, and the Act was passed as a result of renewed political will on the issue.[84]

Taking a somewhat longer-term view, however, it seems that this scepticism was well founded. The draft climate change policy for Punjab from 2016 remains a draft, and the initiatives contemplated by the Commission were only acted on while the Commission was still active (pre-2018) and under the scrutiny of the court.[85]

In a news article from 2022, Lahore-based environmental lawyer Ahmad Rafay Alam said in an interview that the case was 'demonstrably' not the start of a huge change in climate justice litigation. In the years since the judgment, he says, a significant amount of environmental litigation has taken place in Pakistan, but none has 'circled right back round to principles of climate justice that were talked about in the Asghar Leghari case'.[86] A local Pakistani lawyer in our networks also highlighted the gap between the high rhetoric of climate justice offered by the country's high courts in cases like *Leghari* and substantive action on the ground to improve the lives of local people and safeguard them from climate-related harms.[87] While seeing 'not enough' real-world impact from the litigation, this lawyer did

[81] Climate Litigation Accelerator, '*Leghari v. Pakistan*: Pioneering Climate Justice Litigation in the Global South' (2018) <https://clxtoolkit.com/casebook/leghari-v-pakistan/#impacts> accessed 10 May 2023.

[82] Silbert (n 8) 278.

[83] Setzer and Benjamin (n 67) 98.

[84] ibid.

[85] ibid.

[86] Rishika Pardikar, 'Courts or Polices: Which Matters More for Climate Action in South Asia?' (*The Third Pole*, 10 October 2022) <www.eco-business.com/news/courts-or-policies-which-matters-more-for-climate-action-in-south-asia/>.

[87] Participant in National University of Singapore workshop (n 74).

acknowledge the part that *Leghari* played in forcing the various departments of the federal government to have to be cognizant of climate change in their policy documents and inspiring the inclusion of 'new angles' in these policies.[88]

Some scholars drawing on frameworks for assessing the impacts of strategic human rights litigation—frameworks which have a long history of identifying 'negative impacts' arising from litigation—have also offered a more nuanced view of the overall direct, material effects for law and policy stemming from the *Leghari* litigation. Drawing upon Duffy's Human Rights Litigation (HRL) framework in her recent analysis of *Leghari*, Nicola Silbert considers the issue of a potential type of negative impact, namely, backlash or anti-regulatory effects.[89] She draws attention to 'recent investment in coal-fired power plants and a rollback of environmental regulations in Pakistan in the years following the dissolution of the Climate Commission', although noting that 'there has been no research into the connection between *Leghari* and these developments'.[90]

### Direct, symbolic effects for public perception

Beyond the practical policy outcomes of the case, *Leghari* has been lauded as a 'lodestar'[91] for climate litigation by Emily Barritt and Boitumelo Sediti, and as holding 'particular symbolic value in the context of the global South due to its recognition of climate justice in the context of human rights, [representing] a resourceful approach to judicial enforcement'.[92] This analysis suggests the case played a role in Pakistan, and perhaps more broadly in the Global South, in helping reframe climate change as an issue of human suffering and human rights, rather than as simply an environmental matter. At COP27 in Egypt, which followed soon after the devastating 2022 floods in Pakistan, the Pakistani Prime Minister emphasized this rights-based climate justice framing in advocating for establishment of a new funding mechanism to address loss and damage, a call that was successful at the summit.[93]

Birsha Ohdedar notes, however, that '[a]lthough the *Leghari* case has been noted for its "symbolic value" as a leading case at a global level, the more important question from a domestic perspective is how climate change litigation will go from symbolic to transformational'.[94] This echoes concerns from a local lawyer in our

[88] ibid.

[89] Silbert (n 8) 285, noting the tendency of climate litigation scholarship to focus on pro-regulatory impacts and hence the value of the HRL impact framework to help direct attention to this previously overlooked site of impacts.

[90] ibid.

[91] Barritt and Sediti (n 78).

[92] Silbert (n 8) 278.

[93] Nina Lakhani, 'We Couldn't Fail Them': How Pakistan's Floods Spurred Fight at Cop for Loss and Damage Fund' *The Guardian* (20 November 2022) <www.theguardian.com/environment/2022/nov/20/loss-and-damage-pakistan-flooding-climate-justice-cop27> accessed 7 May 2023.

[94] Birsha Ohdedar, 'Climate Change Litigation in India and Pakistan: Analyzing Opportunities and Challenges' in Ivano Alogna, Christine Bakker, and Jean-Pierre Gauci (eds), *Climate Change Litigation: Global Perspectives* (Brill Nijhoff 2021) 103, 122.

networks who sees no issue with the high-brow rhetoric of climate justice expressed in *Leghari* if it inspires greater access to justice but notes that little has been done by the government to implement actions that would mitigate the suffering of local communities when climate-related disasters occur.[95]

#### Indirect effects in shaping further litigation

*Leghari's* role in spurring similar rights-based litigation, particularly in the Global South, is an outcome often noted in the literature.

Barritt and Sediti, for example, assert that '[t]his precedential impact can already be identified' in subsequent cases such as *Ali v Federation of Pakistan*, a petition pending before the Supreme Court of Pakistan, brought by a seven-year-old petitioner challenging the approval of the development of Thar coalfield which could result in the production of 60 million metric tons of coal a year.[96] In particular, the authors note the similar framing of the claimant's legal strategy which relies on the government's Climate Change Framework and Policy to link GHG emissions from coal mining to violations of people's constitutionally protected rights.

Barritt and Sediti also see the potential for *Leghari* to have broader precedent-setting effects beyond Pakistan that 'are likely to be far reaching on both sides of the equator'.[97] Ohdedar identifies the *Pandey* case before the National Green Tribunal in India and the petition in Maria Khan in Pakistan challenging the lack of action and support for renewable energy as 'building on the success of *Leghari*'.[98]

Lawyer Ahmad Rafay Alam, speaking to Third Pole, echoes this optimism, saying that despite its lessened practical achievements over time, the *Leghari* litigation nonetheless 'throws open the great potential of the legal system . . . to protect the rights of people against fossil fuel polluters; it opens up the possibility that this system, which is essentially derived to protect private property rights, does have a function beyond that'.[99]

### 3.2 Case Study II—The *Earthlife Africa* Case (South Africa)

The *Earthlife Africa* case and the subsequent similar litigation it has inspired in South Africa have been discussed in previous chapters, particularly Chapter 4.[100] The *Earthlife Africa* case is generally described as South Africa's first climate change litigation, being the first time that South Africa's courts were required to

[95] ibid.
[96] Barritt and Sediti (n 78) 6.
[97] ibid 7.
[98] Ohdedar (n 94) 103.
[99] Rishika Pardikar, 'Courts or Polices: Which Matters More for Climate Action in South Asia?' (*The Third Pole*, 10 October 2022) <www.eco-business.com/news/courts-or-policies-which-matters-more-for-climate-action-in-south-asia/> accessed 10 May 2023.
[100] *Earthlife Africa Johannesburg v Minister of Environmental Affairs* 2017 2 All SA 519 (GP).

consider climate impacts and the place of climate change considerations in domestic law.[101] Not only did this case result in a court 'win' for the environmental group who brought the case, but *Earthlife Africa* is also an example of how a case can contribute directly to climate action by prohibiting the construction of carbon-intensive infrastructure, and, as such, supporting decarbonization.[102]

To recap the key facts of the case, *Earthlife Africa* concerned an application by Thabametsi Power Company for an environmental approval to construct a coal-fired power station in the Limpopo Province. Thabametsi's application for environmental authorization was granted early in 2015. Earthlife Africa Johannesburg (Earthlife), an environmental NGO, argued that the Department of Environmental Affairs (DEA) had failed to consider relevant information in granting the needed approvals—namely the climate impacts of the project—and that this was 'unlawful, irrational and unreasonable'.[103] The court found that 'on a plain reading of [the relevant legislation] . . . a climate change impact assessment is indeed relevant' and that the Minister ought to have considered the power station's climate change impacts before deciding whether to authorize it.[104]

This initial 'win' in court ended up being of short duration. In early 2018, the Environment Minister again upheld Thabametsi's environmental authorization, on the basis that the earlier approval plan had already assessed climate impacts. Environmental justice group groundWork then joined Earthlife in launching new court proceedings, arguing that the Minister had to exercise her own discretion and consider the project-specific climate impacts of the power plant instead of treating the country's electricity plan as determinative of her decision. Following further judicial review and private settlement, together with a public interest campaign pressuring banks to withdraw funding for the project, the High Court in Pretoria set aside the approval for a final time.[105]

### Direct, material effects for law or policy

Prior to the *Earthlife Africa* case being brought, from a statutory perspective the only probable mitigation action in relation to the reduction of GHGs in South Africa was via observation of national air quality standards.[106] In Marjoné Van Der Bank and Jaco Karsten's assessment, climate action in South Africa has been hindered by the lack of development of domestic law or regulations to oblige

[101] Helen Papacostantis, 'South Africa's Journey to Climate Change Regulation: Earthlife Africa Johannesburg v Minister of Environmental Affairs 2017 2 All SA 519 (GP)' (2021) 24 Potchefstroom Electronic Law Journal (PELJ) <http://dx.doi.org/10.17159/1727-3781/2021/v24i0a8007?.

[102] Kim Bouwer and Joana Setzer, *Climate Litigation as Climate Activism: What Works?* (The British Academy 4 November 2020) 13.

[103] Papacostantis (n 101).

[104] ibid.

[105] Marjoné Van Der Bank and Jaco Karsten, 'Climate Change and South Africa: A Critical Analysis of the *Earthlife Africa Johannesburg and Another v Minister of Energy and Others 65662/16 (2017)* Case and the Drive for Concrete Climate Practices' (2020) 13 Air, Soil and Water Research.

[106] ibid 6.

adherence through the adoption of measures to reduce GHGs.[107] As Papacostantis summarizes, 'while there is a policy aimed at regulating climate change, there is no specific piece of legislation that provides for climate change overall'.[108] Instead, legislation at the national level regulates the use of natural resources.[109]

A year after the *Earthlife Africa* case, however, the Climate Change Bill 2018 was first published for public comment on 8 June 2018. The Bill was then formally introduced to Parliament on 18 February 2022 by the Department of Forestry, Fisheries and the Environment (DFFA) after three years of review. The Bill's objectives included provision for an 'effective management of climate change impacts' and making a 'fair contribution to the global effort'.[110] Commentators Van Der Bank and Karsten argue that the objects and principles of the Climate Change Bill showed a clear intention 'that the legislator commenced to codify the current shortcomings as pronounced by the honourable court in *Earthlife*'.[111] Following the *Earthlife Africa* case, according to the same authors, 'it can be concluded that reasonable progress has been made in South Africa in addressing climate change concerns at national level [and that] [t]he key to unlocking this concern was the establishment and implementation of a climate change impact assessment'.[112]

In addition to the *Climate Change Bill*, two years after the *Earthlife Africa* case, the *Carbon Tax Act* was passed in 2019, which effectively imposes a tax on carbon dioxide GHG emissions.[113] It is the first piece of legislation in South Africa that speaks specifically to climate change, although, as Papacostantis notes, 'it is not a comprehensive law that covers all aspects of climate change in all sectors'.[114] Although this Act had its origins in a policy introduced in 2013, more progress towards finalizing the legislation appeared to occur after the *Earthlife Africa* case was handed down.[115] Van Der Bank and Karsten also see the case as having been instrumental in accelerating the release of more substantial legislative reforms.[116]

Other statements about the subsequent legal developments following the *Earthlife Africa* case are less categorical about outcomes and their links to the litigation. For instance, in an article published by the law firm Fasken, Joubert and Rushton, the authors stated: 'it is hoped that the introduction of the Bill and the implementation of the Carbon Tax Act 15 of 2019 will jumpstart South Africa's move towards a low-carbon, climate-resilient and ecologically sustainable economy and society and will help South Africa to achieve its commitments under

[107] ibid 5.
[108] Papacostantis (n 101) 8.
[109] ibid.
[110] ibid 10.
[111] Van Der Bank and Karsten (n 105) 7.
[112] ibid 9.
[113] Papacostantis (n 101) 8.
[114] ibid 10.
[115] ibid.
[116] Van Der Bank and Karsten (n 105) 9.

the Paris Agreement and the goals set during the United Nations Climate Change Conference of the Parties (COP26) in November 2021'.[117] The legal organization which represented Earthlife, the Centre for Environmental Rights, has also alluded to these impacts with statements such as: '[t]he *Thabametsi* judgment should also pave the way for better policy decisions, which include consideration of climate change impacts'.[118]

### Indirect effects for private sector behaviour and public opinion

Commentary in a media release from the Centre for Environmental Rights reveals how their campaigning contributed to shifting the public and private support for the Thabametsi mine, especially by encouraging the withdrawal of private sector financing for the project. The Centre launched new challenges on various aspects of the mine's other approvals after the Minister used her discretionary powers to override the technical barrier to the mine's approval. In the Centre's view:

> By November 2020, Thabametsi had reached the end of the road. The Development Bank of South Africa, the Public Investment Commission and the Industrial Development Corporation withdrew their financing for Thabametsi, and investors KEPCO and Marubeni announced their withdrawal too. Thabametsi agreed to settle the litigation with Earthlife and groundWork, and notified government that it was cancelling the project.[119]

Nicole Loser, head of the Centre for Environmental Rights' Pollution and Climate Change programme, noted the links between the litigation and the broader public campaign.[120] In her opinion the case not only created 'new jurisprudence and legal precedent to aid future campaigns' but also 'pushed for better decisions that align with Constitutional rights, backed up by science and economics'.[121] Likewise, Earthlife Africa spokesperson Makoma Lekalakala said the ruling sent 'a very strong message to government and to locals to take part in consultation before any decision takes place'.[122]

117 François Joubert and Julia Rushton, 'Key Aspects of South Africa's Much-Anticipated Climate Change Bill' (*Fasken*, 9 June 2022) <www.fasken.com/en/knowledge/2022/06/9-key-aspects-of-south-africas-much-anticipated-climate-change-bill> accessed 7 May 2023.

118 Centre for Environmental Rights, 'Celebrating a Major Climate Victory: Court Sets Aside Approval for Thabametsi Coal Power Plant' (1 December 2020, media release) <https://cer.org.za/news/celebrating-a-major-climate-victory-court-sets-aside-approval-for-thabametsi-coal-power-plant> accessed 7 May 2023.

119 ibid.

120 ibid.

121 ibid.

122 'Earthlife Africa Wins South Africa's First Climate Change Case' *Mail & Guardian* (8 March 2017) <https://mg.co.za/article/2017-03-08-earthlife-africa-wins-south-africas-first-climate-change-case/> accessed 7 May 2023.

However, despite these symbolic effects of the litigation, evidence of practical changes in government action remains limited. In September 2020, the Energy Minister's determination called for 1,500 MW of electricity to be generated from new coal plants, and the South African government still appears to be backing a new 3,300 MW coal plant.[123]

### Indirect effects in shaping further litigation

According to the Centre for Environmental Rights which brought the *Earthlife Africa* case, environmental justice organizations across South Africa have lost no time in instituting legal challenges to other planned, privately financed coal-fired power stations.[124] This is borne out by the record of pending cases, discussed in Chapter 4. For instance, the organization groundWork, also represented by the Centre for Environmental Rights, has instituted two new court applications in the Pretoria High Court against the Minister of Environmental Affairs and others, challenging the decisions of the Minister and the Department of Environmental Affairs to authorize proposed independent power producer KiPower and Khanyisa coal-fired power stations without a full assessment of the plants' climate change impacts.[125]

Interviews conducted by the authors also provide evidence of knowledge-sharing between South African groups and those in other Global South countries such as Indonesia, to shape similar challenges to coal-fired power plants in those jurisdictions.[126]

### Direct, symbolic effects for Paris Agreement implementation

Had the Thabametsi coal plant gone ahead, it would have significantly impacted South Africa's ability to meet its targets under the Paris Agreement, with an estimated 136.1 million tonnes of carbon dioxide equivalent emissions entering the atmosphere.[127] In this sense the case can be said to have had a direct effect in reducing GHG emissions—an important outcome for climate litigation, especially in an emerging economy emitter such as South Africa.

More generally, the case can be seen as embedding in domestic law South Africa's international climate obligations to 'mainstream climate change considerations into social, economic and environmental policy'.[128] This could be described as a direct, symbolic effect associated with reframing the country's legal approach

[123] Centre for Environmental Rights, 'Celebrating a Major Climate Victory' (n 118).

[124] ibid.

[125] ibid.

[126] Interview No 7 conducted via Zoom on 20 April 2022.

[127] Tracy-Lynn Humby, 'The Thabametsi Case: Case No 65662/16 Earthlife Africa Johannesburg v Minister of Environmental Affairs' (2018) 30(1) Journal of Environmental Law 145. See also Van Der Bank and Karsten (n 105) 9.

[128] Papacostantis (n 101) 9.

to the assessment and authorization of major infrastructure projects with the potential to contribute to climate change.

## 3.3 Case Study III—*PSB (Climate Fund)* Case (Brazil)

The *PSB* or '*Climate Fund*' case has been discussed earlier in other chapters, particularly in Chapter 4.[129] This decision was not only the first climate case to reach Brazil's Supreme Court, but it was also the first to be brought by a political party (four in total), the first to address insufficient governmental action to finance climate change mitigation and adaptation measures, and the first case at that level to include a public hearing to obtain evidence before handing down judgment.[130] While—as a more recent case—*PSB* has attracted less scholarly attention than other cases in the Global South, such as *Leghari*, its indirect and symbolic effects are arguably being felt more strongly.

### Indirect, symbolic effects for public opinion and social change

The context of environmental activism in Brazil prior to the *PSB* case is a significant consideration in assessing its impact. Joana Setzer and Lisa Benjamin, for example, point to 'budgetary constraints, inappropriately staffed agencies, as well as strong industrial and commercial lobbies' as factors hampering the enforcement of environmental legislation in Brazil.[131] They also document the significant threats facing environmental defenders in Brazil, including the killing of activists.[132] Against this backdrop, for the highest court in the land to then convene a public hearing akin to a town hall (which is broader than a hearing with interveners) to collect additional factual information before giving its ruling in the *PSB* case is extremely significant (and courageous). Joana Setzer stresses that:

> Having a Supreme Court convening such a hearing is unusual if not unprecedented. The Court noted that climate issues are not limited to the law; they are inherently interdisciplinary. Arguably, this approach helps to expand public participation in climate policy, a much-needed push towards climate democracy and access to information, including on points of science.[133]

[129] *Partido Socialista Brasileiro (PSB), Partido Socialismo e Liberdade (PSOL), Partido dos Trabalhadores (PT) e Rede Sustentabilidade v União Federal*, Arguição de Descumprimento de Preceito Fundamental ADPF 708, filed 6 May 2020.

[130] Joana Setzer, 'First Climate Case Reaches Brazil's Supreme Court' *News and Commentary* (*Grantham Research Institute on Climate Change and the Environment*, 30 September 2020) <www.lse.ac.uk/granthaminstitute/news/first-climate-case-reaches-brazils-supreme-court/> accessed 8 May 2023.

[131] Setzer and Benjamin (n 67) 82.

[132] ibid.

[133] Setzer, 'First Climate Case Reaches Brazil's Supreme Court' (n 130).

The *PSB* case can be seen as marking the start of a new phase of climate litigation in Brazil, which has evolved rapidly over a short time frame. While the initial phases of climate case law in Brazil placed climate change at the periphery and were directed at resisting the deregulatory push of the Bolsonaro government, the *PSB* case is a core climate case that may herald a more accountability-focused orientation in the jurisprudence.[134] Furthermore, the *PSB* case concerned a financial mechanism for giving effect to climate policy, which may signal the growing interest of litigants and NGOs in Brazil in using climate litigation to address questions of the relationship between finance (public and private) and effective climate action.

The most significant socio-political event that followed the case was clearly the defeat of Jair Bolsonaro and the re-election in October 2022 of President 'Lula' Luiz Inácio Lula da Silva, who campaigned largely on a pro-climate manifesto. In an article from Voice of America News, it was reported that 'Veteran leftist Lula has been hammering home the message that "Brazil is back" as a partner in the fight against climate change, after average annual deforestation in the Brazilian Amazon increased by more than 75% under Bolsonaro from the previous decade'.[135] Similarly, the BBC reported incumbent President Lula at the UN climate summit COP27 in Egypt as saying: 'We must stop this rush to the abyss. There is no climate security for the world without a protected Amazon' and that '[w]e will do whatever it takes to have zero deforestation and the degradation of our biomes.'[136]

Of course, linking a particular court decision—no matter how high-profile—to an election result is causally very perilous. However, the decision in the *PSB* case, including the inclusive public forum convened by the court as part of the hearing process, is consistent with changing public attitudes to climate and environmental issues in Brazil that were also reflected in the campaign platform of President Lula. Maria Tigre and Alex Goodman point, in addition, to the case's role in articulating the responsibility of the Supreme Court to intervene when other branches of government fail to uphold rights protections. They contend that the *PSB* case 'effectively expands the role of the judiciary in determining environmental and climate policies, following a global trend that has been observed worldwide'.[137]

[134] ibid.

[135] 'Britain Pledges $102 Million for Brazil's Amazon Fund' *Voice Of America* (Washington DC, 5 May 2023) <www.voanews.com/a/britain-pledges-102-million-for-brazil-s-amazon-fund-/7080964.html> accessed 8 May 2023.

[136] Georgina Rannard, 'COP27: Brazil is Back on the World Stage, Lula Tells Climate Summit' *BBC News* (Sharm el-Sheikh, 16 November 2022) <www.bbc.com/news/science-environ<ment-63625698> accessed 8 May 2023.

[137] Maria Antonia Tigre and Alex Goodman, 'ADPF708/Climate Fund. What to Expect from Brazil's First Public Hearing on Climate Policy?' ( *Global Network for Human Rights and the Environment*, 22 September 2020) <https://gnhre.org/2020/09/adpf708-climate-fund-what-to-expect-from-brazils-first-public-hearing-on-climate-policy/?fbclid=IwAR2ooNH9bbAYMgkQ-K534NX4pksW_VlqAuuffJXyqC-9NgZwtoFVMcndEKQ> accessed 8 May 2023.

## Direct, material effects for law or policy

The policy and legal implications of the *PSB* case are strongly bound up together with the change of President, which has marked an overhaul across environmental and climate change politics and policy in Brazil. Brazil has updated its Nationally Determined Contribution (NDC) target twice since the conclusion of the Paris Agreement and once since the *PSB* ruling. Its most recent (March 2022) update is stronger than its 2020 update, which Climate Action Tracker had ranked as 'Highly Insufficient'. However, Brazil's current commitments remain weaker than its original NDC, submitted in 2016. The March 2022 updated NDC still allows higher emissions than Brazil's first submission, in contrast to the Paris Agreement's principles of progression and no back-sliding.[138] Further updates are expected though as Lula's Presidency progresses,[139] which highlights the need for time to be allowed for the full effects of the decision to be assessed.

The key outcome sought by the parties in *PSB* was the disbursement of the Climate Fund, a fund established in 2009 as a financial instrument of the National Climate Policy Plan. As discussed in earlier chapters, the Climate Fund is mandated to direct its annually authorized budget to projects and studies working towards climate change mitigation and adaptation. However, in 2019 and 2020, the Bolsonaro administration failed to allocate the available resources.[140] The court found in *PSB* that the executive branch has a constitutional duty to execute and allocate the funds of the Climate Fund to mitigate climate change, based on both the separation of powers and the constitutional right to a healthy environment. The court further found that the judiciary, in turn, must act to avoid the regression of environmental protection.[141]

Maria Tigre argues that 'in practice, this means the government will have a duty to allocate funds to where they are most needed in climate mitigation, and the government's exercise of its discretion in allocating the funds will be potentially subject to judicial review'.[142] Whether this eventuates in practice remains to be seen and cannot be definitely evaluated at this early stage. While the new government of Luis Inácio Lula da Silva 'reopened the Amazon Fund with Norway and Germany on the first day in office', updates on the disbursement of the Climate Fund have been less clear,[143] and the website for the Fund does not appear to have

[138] Climate Action Tracker, 'Brazil' (5 September 2022) <https://climateactiontracker.org/countries/brazil/> accessed 8 May 2023.

[139] Miriam Mazza and others, 'Brazil Updates: Biden and Lula Advance Environmental and Climate Agendas, Brazil Makes Progress on Other Chemical Regulatory Initiatives' (*Lexology*, 7 March 2023) <www.lexology.com/library/detail.aspx?g=43f20afe-ae88-413f-933d-8935ff13f1df> accessed 8 May 2023.

[140] ibid.

[141] Maria Antonia Tigre, 'Advancements in Climate Rights in Courts Around the World' (*Climate Law: A Sabin Center Blog*, 1 July 2022) <https://blogs.law.columbia.edu/climatechange/2022/07/01/advancements-in-climate-rights-in-courts-around-the-world/> accessed 8 May 2023.

[142] ibid.

[143] Claudia Horn, 'Brazil's Amazon Fund: A "Green Fix" between Offset Pressures and Deforestation Crisis' (2023) 55(6) Antipode 1686.

been updated for several years.[144] In addition, despite the obligation to establish an annual plan on the use of the Climate Fund's resources and to hold periodic meetings of its governance bodies, as ordered by the court, evidence of this occurring is yet to appear publicly.

#### Direct, symbolic effects for Paris Agreement implementation

In its *PSB* ruling, the court adopted a highly innovative, rights-based approach where it held that: (1) the Paris Agreement is a human rights treaty, and for that reason enjoys '*supralegal* status'; and (2) based on Article 225, the separation of powers doctrine, and its international obligations, that the Brazilian government is constitutionally obliged to combat climate change and prevent regression of environmental protection.[139] The ruling also means that any laws made by the Brazilian government will be invalid if they contradict the Paris Agreement, and violating this ruling or the Paris Agreement is a violation of the country's Constitution and human rights protections.

Undoubtedly the *PSB* case demonstrates a novel approach to implementing international climate governance at the domestic level, which may have broader effects in framing how countries understand their obligations under the Paris Agreement.[145] Reflecting this sentiment, the Climate Litigation Accelerator programme at New York University states that the *PSB* case 'sends an important signal globally regarding the importance of the Paris Agreement in tackling the climate crisis and its binding nature'.[146]

### 3.4 Case Study IV—*Demanda Generaciones Futuras*/'*Future Generations*' (Colombia)

Like the *Leghari* case, the *Future Generations* case is another well-known Global South climate change case and has been discussed in previous chapters.[147] The Colombian Supreme Court's decision in the case, issued in April 2018, addressed the deforestation problem in the Colombian Amazon Rainforest, relying on the constitutional right to life, the right to public health, and the right to a healthy environment.[148] The ruling required the Colombian government to implement certain action plans to reduce deforestation to net zero by 2020, in line with Colombia's

[144] O Banco Nacional do Desenvolvimento, 'Fundo Clima' <www.bndes.gov.br/wps/portal/site/home/financiamento/produto/fundo-clima> accessed 8 May 2023.

[145] Horn (n 143).

[146] Climate Litigation Accelerator, '*PSB v. Brazil (on Climate Fund)*' (2020) <https://clxtoolkit.com/casebook/psb-v-brazil-on-climate-fund/#impacts> accessed 8 May 2023.

[147] *Demanda Generaciones v Minambiente* (Bogotá District Court 2018).

[148] Paola Andrea Acosta Alvarado and Daniel Rivas-Ramírez, 'A Milestone in Environmental and Future Generations' Rights Protection: Recent Legal Developments before the Colombian Supreme Court' (2018) 30(3) Journal of Environmental Law 519.

first NDC, which was submitted to the UN Framework Convention on Climate Change Secretariat in 2018.[149]

The *Future Generations* case, including the court's recognition of the rights of future generations, was hailed as 'one of the most robust environmental court rulings in the world' by Michael Gerrard, Director of the Sabin Center for Climate Change Law at Columbia University. This recognition of the case's potential indirect and symbolic effects stands in contrast to the fact that little if anything has been done to date to enforce the judgment, limiting its direct, material effects.

Again, contextual factors may be a highly relevant part of the story here, not least the fact that the litigation followed close on the heels of the end of the long-running civil war in Colombia, leading to greater access to and encroachment on the Amazon forested areas of the country. The decision is also still a relatively 'young' case, and it may take time for more significant impacts to materialize.

### Direct, material effects for law or policy

The landmark court ruling issued in *Future Generations* consisted of four main orders for the Colombian government to action. In particular, the Colombian government was ordered by the court: (1) to formulate short, medium, and long-term action plans to counteract deforestation in the Amazon; (2) to create an Intergenerational Pact for the Life of the Colombian Amazon; (3) to update and implement Municipal Plans for Territorial Planning across the region; and (4) to carry out an Action Plan to counter deforestation as reported by the Institute of Hydrology, Meteorology and Environmental Studies through policing, judicial, or administrative measures.[150] According to helpful analysis conducted in 2020 by Dejusticia, the organization who ran the case, none of these orders has been effectively implemented. Instead, 'reports indicate a low level of compliance, and the ruling's interpretation and implementation have generated many challenges and caused harm to residents'.[151]

As Joana Setzer and Lisa Benjamin remark, the implementation of ambitious judicial orders can be challenging, and this certainly seems to have been the experience with the *Future Generations* case. They also draw attention to the fact that, despite the decision, deforestation in the Amazon increased following the litigation, with about 75% of all deforestation in Colombia in the Amazonian region.[152] The government of Iván Duque took no action to curb this destruction, prompting the plaintiffs to seek a declaration that the government and other defendants failed to fulfil the orders of the Supreme Court.

[149] Charles Beauregard and others (n 49) 654.

[150] Acosta Alvarado and Rivas-Ramírez (n 148).

[151] Dejusticia, 'What Does the Government Need to Implement the Sentence Against Climate Change and Deforestation?' (2 December 2020) <www.dejusticia.org/que-le-hace-falta-al-gobierno-para-implementar-la-sentencia-contra-el-cambio-climatico-y-la-deforestacion/> accessed 6 May 2023.

[152] Setzer and Benjamin (n 67) 98.

In an email to the International Network for Economic, Social and Cultural Rights, Gabriela Eslava from Dejusticia outlined her assessment of the effects of the *Future Generations* case a year after the ruling, which suggested some positive outcomes and steps towards policy change. She noted:

> Due to the short deadlines ordered by the Supreme Court, by July 2018, the Ministry of Environment had carried out five workshops in five different municipalities of the Colombian Amazon, where it consulted with the young plaintiffs, communities, environmental authorities, and other key actors [on] the policies to fulfil the court orders. … In addition to the workshop participants, many young people and communities joined the cause through social networks, and the plaintiffs have become spokespeople for the case and the cause in their respective hometowns, being those most interested in ensuring full compliance with the Supreme Court's ruling. [153]

However, the workshops were conducted in the principal cities of the Amazon, meaning the affected communities in rural areas were not represented and overall community participation was low. Despite these deficiencies, the Ministry of the Environment submitted to the court a document entitled 'Action plan to reduce deforestation and address the effects of climate change in the Colombian Amazon—Sentence 4360 of 2018'. The document itself is only a draft, as the text affirms: 'it should be noted that, in the context of the new government and the current process of appointing officials who will be responsible for the implementation of the action plan, the present action plan must be validated to specify its alignment with the [new] government's priorities and the budgetary limitations for medium-term activities.'[154]

On 20 October 2020, Dejusticia released a comprehensive report which analysed compliance (or lack thereof) with the four mandates set out by the court in *Future Generations*. An update was provided on the organization's website in English on 2 December 2020 which summarized these findings. In essence its findings were that: (1) no action plan to counter deforestation in the Amazon had been formulated (with the existing draft still incomplete in important respects); (2) the Intergovernmental Pact for the Life of the Colombian Amazon did not exist and the required participation had not been generated; (3) the municipalities of the Amazon had not implemented their Land Management Plans and nor did they have action plans in place to counteract deforestation; and (4) there was little evidence of the national government carrying out actions aimed at reducing

[153] This is quoted in a case note in ESCR-Net, 'Colombian Supreme Court Rules to Protect Future Generations and Amazon Rainforest in Climate Change Case' <https://www.escr-net.org/caselaw/2019/stc-4360-2018#:~:text=Due%20to%20the%20short%20deadlines,actors%20the%20policies%20to%20fulfill> accessed 18 January 2024.

[154] Dejusticia (n 151).

deforestation, 'especially within the framework of Presidential Directive No. 10 of 2018, with which the Government intends to give compliance with judgment 4360 of 2018'.[155] It might be argued that, while this report clearly highlights shortcomings with respect to enforcement of the court's judgment, this is not dispositive of the impacts the decision may achieve, particularly over a longer period of time.

It appears that enforcement of the court's orders has been hindered by a lack of resources and adequate institutional structures, with the assignment of responsibilities for carrying out the decision posing particular difficulties. For instance, the Presidency of the Republic's spokesperson 'admitted that the issue of deforestation in the Amazon is not the direct responsibility of any of the high presidential councils', which has contributed to delaying compliance measures.[156]

Regarding the Territorial Arrangement Plans required of municipalities by the court, Alvarado and Rivas-Ramírez found this is likely to force all municipalities located in the Colombian Amazon to take actions to reduce deforestation, which they argue may lead to significant negative impacts.[157] The authors contend that the court ruling:

> means that all local authorities will be obliged to reformulate their local policies in order to address this judicial order, which overlooks local dynamics and the many ways in which deforestation is being or might be attended. More precisely, the judgment conflicts with territorial autonomies recognised both by the Constitution and the Organic Law of Territorial Management, which established that municipalities have vast discretionary powers to decide what to do on their local lands. This is only one of many orders received by the state, which will give rise to constitutional tensions within the Colombian legal order.[158]

This statement reflects a more general phenomenon complicating assessments of climate litigation's impact, with different views of the 'success' or otherwise of a case expressed depending on the positionality and interests of the person or groups consulted. Climate cases centred on disputes over exploitation of natural resources will inevitably have winners and losers who will perceive the follow-on effects of the litigation differently.

At the conclusion of the follow-up hearings on 22 October 2019, the judges supervising implementation of the *Future Generations* case remarked that they were not seeing concrete actions by the authorities. Indeed, they were concerned that '[t]he hearings will end and from the Government we have only heard that

[155] ibid.
[156] ibid.
[157] ibid.
[158] Acosta Alvarado and Rivas-Ramírez (n 148) 525.

they will begin to consider how to act. Reducing deforestation in the Amazon is a gigantic task . . . and we have done nothing but talk.'[159]

Others, though, take a more positive view of the outcomes of the *Future Generations* case, emphasizing its indirect and symbolic effects. For instance, in 2020, Luisa Gómez-Betancur argued that 'the fact that the Amazon ruling has not triggered the expected governmental actions does not signify that its outcomes are modest'.[160] Instead, Gómez-Betancur cites outcomes such as the adoption of the judicial reasoning in other cases, which demonstrates a promising 'ripple effect', while also affirming a continuing need for 'intricate, systemic, and multidimensional actions'.[161]

### Direct, symbolic effects for Paris Agreement implementation

The *Future Generations* case demonstrated a new approach in Colombia to implementing international law, which generated a positive reception from scholars (albeit that such assessments have not been repeated in literature that has come out after lack of implementation of the court's orders). In particular, scholarly assessments of the case remarked on the novelty of the court's relying on international law in its case law,[162] and praised the judge's approach of imposing 'an explicit obligation on the State to respect and fulfil international obligations [under the Paris Agreement]'.[163]

Similarly to the *PSB* case discussed above, *Future Generations* has been seen as lending a new human rights frame to interpretation of the Paris Agreement. As Acosta Alvarado and Rivas-Ramírez put it in a 2018 article, the case marked the Paris Agreement 'as an important legal basis for such action, and for the interpretation of constitutional rights, such as the right to life, health and a healthy environment, as well as rights of future generations'.[164]

## 3.5 Lessons from Case Studies for Assessment of Climate Litigation Impact

The four case studies considered above demonstrate the capacity of climate change cases in the Global South to yield direct and indirect, material and symbolic effects which (mostly) make a meaningful contribution to addressing the problem of climate change. Translating ambitious, innovative rulings into actions that change

[159] ibid.

[160] Luisa Gómez-Betancur, 'The Rights of Nature in the Colombian Amazon: Examining Challenges and Opportunities in a Transitional Justice Setting' (2020) 25(1) UCLA Journal of International Law and Foreign Affairs 41.

[161] ibid 74.

[162] Acosta Alvarado and Rivas-Ramírez (n 148) 524.

[163] ibid 525.

[164] ibid 526.

the situation on the ground in the countries concerned has proved a challenge in all cases, especially for those cases like *Leghari*, *PSB*, and *Future Generations* where plaintiffs are seeking more systemic law, policy, and social change.

It appears from this history so far that indirect, symbolic effects have been more readily achieved by the cases than material changes in the formulation and application of climate-related laws and direct effects that improve mitigation and adaptation outcomes on the ground. Common indirect, symbolic effects include reframing of climate issues and Paris Agreement obligations in human rights terms, inspiring a different public and governmental attitude towards the salience of climate issues in public policy, and providing a template for future cases to expand upon. However, it is worth stressing again that these are young trends, based upon the assessment of limited data, leaving open the potential for a fuller picture of impact to emerge over time.

## 4. FUTURE AREAS FOR DEVELOPMENT

Given the previous section's analysis of the early impacts of a selection of Global South climate cases, is it possible to make predictions about the future trajectories, pathways, and 'effectiveness' of Global South climate litigation? In many ways, this exercise remains much easier in the Global North as opposed to the Global South, with the longer history of climate litigation in the North allowing a better longitudinal assessment of what outcomes these cases are achieving or might plausibly achieve. For instance, the 'rights turn' in climate litigation[165] seems to be solidifying and gaining pace, alongside or in conjunction with efforts to bring 'framework' mitigation cases that challenge the level of ambition of governments' climate commitments or the adequacy of their actions taken to comply with these commitments.[166] While globally cases brought against governments remain the dominant type of climate litigation,[167] in the Global North there is a growing number of corporate climate cases on issues of climate risk disclosure by businesses, as well as greenwashing complaints.[168]

The impacts emerging from Global South climate litigation so far suggest that doubling down on existing, successful avenues for cases, coupled with a continuing focus on ensuring implementation of rulings, is likely to be the best strategy for the immediate future. The Global South jurisprudence has developed a distinctive focus on constitutional rights enforcement,[169] for example, that seems to provide a useful hook for many plaintiffs on which to hang calls for greater climate

[165] Peel and Osofsky, 'Climate Change Litigation' (n 3).
[166] Setzer and Higham, *2022 Snapshot* (n 51).
[167] ibid.
[168] ibid.
[169] Setzer and Benjamin (n 67) 98; Peel and Lin (n 1).

action—either on mitigation or adaptation, enforcement of existing environmental and climate policies, or the fulfilment of national Paris Agreement commitments.

Interest remains, particularly amongst funders of climate litigation, in the exploration of alternative avenues of litigation in the Global South, particularly corporate and commercial law claims that might target private rather than (just) public actor actors. There is also movement towards accessing regional courts and international law avenues, for example, with the advisory opinion request to the International Court of Justice (championed by Vanuatu),[170] the Chilean/Colombian request for a climate and human rights advisory opinion from the Inter-American Court of Human Rights,[171] and efforts by small island states in the Pacific and Caribbean to mobilize behind a request for an advisory opinion from the International Tribunal for the Law of the Sea.[172]

For the foreseeable future, however, the centre of gravity for climate litigation seems likely to remain in domestic jurisdictions. With this domestic focus, in many Global South countries where climate litigation is occurring or emerging, a major task continues to be creating and bolstering the environmental rule of law conditions such as access to justice, effective adjudication, civil society activism, and judicial knowledge of climate science and policy that ultimately provide the foundation for the achievement of progressive outcomes in climate change cases.

With these considerations in mind, the following sections outline four possible areas for future development of climate case law in the Global South. Here we engage in some crystal-ball gazing and this discussion should not be taken as definitive predictions of how the case law might evolve. Instead, we identify areas where there are currently gaps but the potential for additional development, as well as areas where there are already emergent trends that might be further built on in the future.

## 4.1 Cases on Adaptation and Damages

A feature of the Global South climate case law is its focus on mitigation issues, such as those associated with reducing emissions from carbon-intensive projects and from natural resource practices like deforestation that add carbon to the atmosphere.[173] Somewhat surprisingly, and despite the considerable adaptation needs of

[170] International Court of Justice 'Request for an Advisory Opinion of the International Court of Justice on the Obligations of States in Respect of Climate Change' (Press Release No 2023/21, 25 April 2023).

[171] Pablo Saavedra Alessandri, 'Request for an Advisory Opinion on Climate Emergency and Human Rights to the Inter-American Court of Human Rights form the Republic of Colombia and the Republic of Chile' (Petition, Inter-American Court of Human Rights, 9 January 2023).

[172] Commission of Small Island States on Climate Change and International Law, 'Request for Advisory Opinion' (12 December 2022).

[173] See further Chapter 3.

many countries in the Global South, we are yet to see the development of a significant adaptation jurisprudence, including cases seeking damages for climate harms suffered by communities.

The reasons for this gap are not clear and likely relate to a complex interplay of factors in different jurisdictions. One (speculative) hypothesis is that litigants may be reluctant to sue their own governments for adaptation failures which are contributed to by high emissions in the Global North and a lack of adequate international climate finance for adaptation and loss and damage.[174] However, as cases like *Leghari* show, there can often be major failures by governments in Global South countries to implement the provisions of their own adaptation policies. Resourcing constraints may understandably prevent full implementation, but climate impacts and harms may also flow from a failure to tackle known land use problems or because of a lack of political prioritization of the environmental rights of vulnerable and displaced communities. Adaptation lawsuits may thus be a way of bringing attention to these issues and increasing pressure on governments at all levels to safeguard the rights of their population to a clean and healthy environment to the best of their capacity and available resources.

Scholars and practitioners are also beginning to consider the possibility of climate cases seeking damages for climate harms.[175] These cases might be brought against the government in a Global South country, for instance, for failing to carry out measures to prevent foreseeable harms from climate-linked events such as major flooding. Or they might be brought as transnational cases against government or corporate actors in the Global North, either in international forums or in the courts of Global North countries if a clear link can be established between a particular government or company's actions or investments, and climate damages occurring in the Global South.[176]

## 4.2 Cases Seeking Accountability of Corporate Actors

As mentioned above, corporate climate litigation represents a definite trend line in the Global North,[177] but in the Global South we are not yet seeing many cases brought directly against companies. There are some such examples in the cohort of 'transnational cases' discussed in Chapter 4, but here there are generally coalitions

[174] United Nations Environment Programme, *Adaptation Gap Report 2022: Too Little, Too Slow—Climate Adaptation Failure Puts World at Risk* (Report, 2022).

[175] Joana Setzer and Catherine Higham, 'Global Trends in Climate Change Litigation: 2023 Snapshot' (Policy Report, Grantham Research Institute on Climate Change and the Environment 2023) 12 (hereafter Setzer and Higham, *2023 Snapshot*).

[176] *Luciano Lliuya v RWE AG* (2015) Case No 2 O 285/15, Essen Regional Court, is a test case in this regard.

[177] Setzer and Higham, *2023 Snapshot* (n 175).

of actors involved who are able to leverage the resources and courts located in wealthy countries to bring such cases.

Again, the reasons for limited corporate accountability cases in the Global South so far may be complex and are likely to vary depending on the particular socio-political and legal context of different countries. Business interests in developing counties may often be powerful and deeply connected to government, as has been the case for the agri-business corporate sector in Brazil or oil palm companies in Indonesia.[178] This may make the prospect of climate litigation against corporate actors unappealing or too difficult for litigants in Global South countries, as opposed to more 'tried and true' avenues for constitutional rights or environmental litigation against governments. In addition, boundaries between government and corporate actors may be blurred, for instance, in the case of state-owned enterprises (SOEs) which are a prominent part of the energy sector in many Southeast Asian developing countries.[179]

Nonetheless, cases like the *PSB* (*Climate Fund*) case in Brazil point the way to new potential pathways for climate cases seeking the accountability of corporate actors in the Global South. This case demonstrates a focus on issues of how climate action is financed which might lead to greater attention to possible corporate and commercial law avenues for bringing claims that scrutinize the disclosures and investment decisions of corporate or blended state/corporate actors.

## 4.3 Cases Involving Indigenous Peoples

As Chapter 5 discussed, Indigenous peoples have been involved in only a handful of Global South climate cases to date.[180] However, we would foresee the potential for this group of actors to become a more powerful force in Global South climate litigation as the acceleration of the clean energy transition, as well as climate impacts, put greater pressures on land use, including the territories of Indigenous peoples.[181]

Claims of Indigenous peoples in Global North countries brought before international bodies, such as the United Nations Human Rights Committee, illustrate the potential for Indigenous peoples to mount strong arguments for improved climate action based on the need to safeguard their right to self-determination and cultural rights, which are often closely tied to their relationship to the

[178] Daniel Chrisendo, Hermanto Siregar, and Matin Qaim, 'Loss and Damage and Climate Litigation: The Case for Greater Interlinkage' (2020) 30(1) Review of European, Comparative and International Environmental Law 16.

[179] Philippe Benoît and others, 'Decarbonization in State-Owned Power Companies: Lessons From a Comparative Analysis' (2022) 355(1) Journal of Cleaner Production 1.

[180] See Chapter 5.

[181] Maria Antonia Tigre, 'Climate Change and Indigenous Groups: The Rise of Indigenous Voices in Climate Litigation' (2022) 9(3) E-Publica 214.

land.[182] Some of the cases involving Indigenous peoples in Indonesia also illustrate the dual roles these groups can play in, on the one hand, opposing foreign investment projects that exploit and damage Indigenous held territories,[183] and also intervening on the side of the government to support efforts to enforce anti-deforestation laws.

An underexplored, but potentially fruitful, area for future climate case development in the Global South involving Indigenous peoples would be cases challenging development proposals of either governments or companies for a failure to ensure the prior informed consent of Indigenous land holders. The South African case of *Sustaining the Wild Coast* offers a potential template here, with the court in that litigation upholding a challenge to seismic surveys by Shell for the purpose of fossil fuel exploration, based on the failure to consult and involve adequately the Indigenous peoples whose cultural rights would be impacted by Shell's activities.[184]

## 4.4 Transnational cases

One of the questions raised by the current trends in Global South climate litigation—concentrated as the case law is in lower and upper middle-income countries—is whether we are likely to see an expansion of climate litigation to other countries in the Global South. The analysis in Chapter 4 suggests that any such expansion is closely tied to environmental rule-of-law factors, such as access to justice, a progressive judiciary, workable legal or rights avenues for cases, and a well-functioning civil society. For some countries in the Global South, realization of these 'preconditions' for climate litigation is not likely to be a near-term prospect.

We might also, though, flip this question on its head and ask whether it would in fact be desirable to see climate litigation grow much beyond the existing cohort of Global South countries where we see these cases emerging. From a strategic perspective, transnational cases challenging practices of investment that support the continuation of carbon-intensive forms of development in Global South countries might be a more fruitful approach and lead to more tangible direct and material effects from the litigation.

In Southeast Asia, for example, Japanese and South Korean public and private finance organizations continue to heavily support fossil fuel projects in surrounding

[182] *Daniel Billy v Australia (Torres Strait Islanders Petition)* (2019) CCPR/C/135/D/3624/2019.

[183] *Asmania v Holcim* (2022) <https://www.foei.org/pari-island-against-holcim/>.

[184] 'South Africa: CSOs & Indigenous Communities Welcome Court Decision to Revoke Shell's Exploration Licence' (Business and Human Rights Resource Centre 6 September 2022) <https://www.business-humanrights.org/en/latest-news/south-africa-csos-indigenous-communities-welcome-court-decision-to-revoke-shells-exploration-licence/> accessed 18 January 2024.

nations to help supply domestic energy needs.[185] China—through its Belt and Road initiative—has also emerged as a major investor in developing countries in Africa and the Pacific, often in projects that involve debt financing in exchange for natural and mineral resources, and that often are accompanied by carbon-intensive projects such as coal mines for energy supply.[186] Transnational cases in Global North courts, or regional and international tribunals, that seek to hold such investors or finance institutions accountable for the climate, environmental, and human rights impacts of their investments may be the most feasible pathway to challenge these practices, rather than attempts to mount cases in the often weaker courts of the Global South host countries.

## 5. CONCLUSION

This chapter represents an early attempt to engage holistically with the types of climate litigation impact and methods for their assessment in a Global South context. We would hope that the analysis offered in this chapter prompts further study and refinement of some of the concepts, frameworks, and assessments of impact that we have presented here. Key questions that would benefit from more detailed examination and thought include consideration of what types of impacts we might expect from Global South climate cases; whether assessment frameworks developed largely for Global North contexts are useful or can be adapted to aid an evaluation of the impacts from climate litigation in the Global South; and who is best placed, among the constellation of actors involved in seeding, bringing, and assessing Global South climate cases, to undertake the task of compiling information on their outcomes.

The importance of assessing the impacts of climate litigation, including in the Global South, is tied to the urgency of addressing the climate crisis and the concomitant necessity of investing effort and resources where they are likely to be most effective. In this regard, the IPCC has provided a baseline assessment that climate litigation can have positive outcomes for climate governance, which provides a rationale for continuing activity in this sphere, alongside other strategies of policy advocacy, social mobilization, community campaigns, and law reform efforts.

While there are significant constraints on the potential for climate litigation to achieve impact in many Global South countries—constraints which the case studies in Section 4 of this chapter illustrate—this renders even more remarkable the progressive, innovative case decisions that have emerged in a number of these

[185] Benoît and others (n 179).

[186] Jeremy Garlick, 'The Regional Impacts of China's Belt and Road Initiative' (2020) 49(1) Journal of Current Chinese Affairs 1.

countries. On the evidence available, climate litigation offers a promising strategy for achieving meaningful progress on climate change in many Global South countries, which should encourage courageous civil society groups, litigators, and campaigners to continue their efforts to further develop this body of jurisprudence and to ensure implementation of these decisions.

This book has sought to give a 'big picture' view of the landscape of Global South climate litigation, including where it is occurring, by which pathways, with which actors, and with what types of impacts. The fast development of the case law to date suggests that this is a dynamic and innovative field likely to continue to evolve rapidly in coming years. Potential directions in which the climate case law involving Global South countries might develop include through the exploration of more adaptation-focused and damages claims, moves to develop nascent approaches for securing greater corporate climate accountability, cases involving Indigenous people based on challenges to consultation processes for development occurring on their lands, and additional transnational cases that attack the root causes of unsustainable financing and develo.0pment practices that perpetuate carbon-intensive energy and development in countries of the Global South.

For this field studying litigation of climate change in the Global South, it seems there is a bright future along with the prospect of actions and outcomes that will make a real difference to addressing the causes and impacts of climate change.

# Index

Tables and figures are indicated by *t* and *f* following the page number